Securing Emerging Wireless Systems

Lower-Layer Approaches

Yingying Chen • Wenyuan Xu • Wade Trappe •
YanYong Zhang

Securing Emerging Wireless Systems

Lower-Layer Approaches

 Springer

Yingying Chen
Stevens Institute of Technology
Hoboken, NJ
USA

Wenyuan Xu
University of South Carolina
Columbia, SC
USA

Wade Trappe
Rutgers University
North Brunswick, NJ
USA

Yanyong Zhang
Rutgers University
North Brunswick, NJ
USA

ISBN   978-1-4419-4693-5   e-ISBN   978-0-387-88491-2

To Our Families

Preface

We live in an increasingly wireless world and, even though the benefits of tetherless communication are sure to attract a plethora of new applications and help bring communication to those who were never before connected to the broader world, these very same benefits can also serve as the means to cause damage upon individuals, enterprises and governments. Unlike traditional wired communication, where physically protecting the medium is, to a large part, possible by running cables underground and wires in walls, wireless communication is not able to physically protected in the conventional sense. Sure, wireless access points are placed in buildings that may be locked, or cellular basestations are protected by fences and security cameras, but the medium itself is trivially open to a broad array of threats and thus securing wireless communication necessitates a collection of tools that can suitable protect the wireless medium.

This book is about security for wireless networks. However, whereas most conventional approaches to security focus on *cryptographic* solutions that are applied in building families of interconnected network security protocols, this book instead focuses on complementary techniques that aim to invoke unique properties of wireless communications in order to add security to wireless systems.

In order to place this book in an appropriate context, it is perhaps best to think of a simple analogy. Consider a cocktail party where two people, Alice and Bob in the usual parlay of the security community, are trying to have a conversation. Their vocal chords create an audio waveform that travels through the air and which can be heard by others who are close enough to Alice and Bob. Their voices correspond to wireless communication signals, which propagate in all directions. If Alice and Bob want to make certain that there are no eavesdroppers, they must communicate in a way to make certain that others won't be able to hear or decipher their conversation. Another concern that they might have is that there may be an imposter in the room, seeking to imitate Alice, when starting a conversation with Bob. For such masquerading or spoofing attacks, Bob must use some information specific to Alice, such as her previous location or even some unique aspect of her voice, in order to discern between legitimate and illegitimate conversations. Or, yet another concern that Alice and Bob may have is that there may be unsociable people present at the party who constantly interrupt their conversation. In such cases, Alice and Bob must find a way to excuse themselves and resume their conversation at a different place.

In this book, we deal with these sort of challenges as they apply to wireless communication networks. When considering the standard protocol stack, the aspects that are unique to wireless communications exist at the lower layers of the network stack. Signals and their properties are representative of the physical layer, and take on special characteristics based on their location relative to other wireless devices and to the background environment. It is at the lowest layer of the protocol stack where transmitter location has its greatest impact. The link layer, or medium access control (MAC) layer, must cope with the fact that multiple communications (or conversations) might be carried on at the same time and, just as we have *social protocols* that govern our interactions with each other in a party, wireless networks must also employ suitable MAC protocols to allow for sharing of the wireless medium. Although all communications systems involve localized communications as the basic building block to communicating long distances, many different wireless systems, ranging from sensor networks to ad hoc networks, are characterized by their multi-hop routing protocols. It is at the routing layer where we may direct communication towards or away from certain areas in hopes of achieving improved security or privacy.

Throughout this book we will explore a variety of different *lower layer* strategies for securing wireless networks. Our solutions build largely upon non-cryptographic methods, though occasionally we will employ cryptography in our solutions to make them more robust and resilient to attacks. Applications of cryptography to securing wireless protocols is a necessary component to securing wireless systems, and we thus feel it is important to iterate up front that the methods presented in this book should not be considered a replacement to a well-designed network security protocol. The methods presented in this book will never replace the role of TLS or HTTPS. Instead, it is our viewpoint and belief that wireless systems can only be secured when the full spectrum of tools available to the wireless engineer are brought to bear on the problem. A toolbox that leaves out either cryptography or lower layer characteristics would correspond to an incomplete set of tools that might require more effort in order to achieve a comparable level of security when cryptography and lower layer security methods are combined.

The approaches to securing wireless systems that exploit lower layer phenomena is an emerging area of research in the wireless security community, and the material presented in this book is, in large part, a compilation of research that was conducted by the authors. However, there are many people who should be acknowledged for their efforts in conducting research that led to some of the material presented in the book. Notably, the authors would like to acknowledge their colleagues Qing Li, Pandurang Kamat, Shu Chen, and Zang Li, who conducted research on forge-resistant relationships, privacy-enhanced routing, and physical layer security as part of their thesis research. Additionally, the authors would like to acknowledge several other collaborators who have helped in different ways to make this research lively: Konstantinos Kleisouris, Eiman Elnahraway, John-Austin Francisco, Rob Miller, Ke Ma, Richard Martin, Rich Howard, and Ivan Seskar. Each, in their own way, has helped make the material presented in this book possible.

Contents

1
Introduction

Wireless technologies promise to change the manner by which we communicate and access information. Over the past decade, computing and networking has shifted from the static model of the wired Internet towards the new and exciting "anytime-anywhere" service model of the mobile Internet, where information will be gathered by wireless devices and made available to mobile users to consume or process on-the-go. At the heart of the technologies facilitating such pervasive computing are recent advancements in wireless technologies that will provide the ubiquitous communication coverage that is so necessary for mobile services. In fact, a quick survey of existing and emerging technologies reveals a vast heterogeneity of platforms. Radio-frequency identification (RFID) technology are being developed to support inventory monitoring and supply-chain management. Sensor networks are being deployed to monitor environmental conditions and national infrastructure (e.g. bridges and roads), as well as supporting defense-related applications involving battlefield monitoring. Wireless local area networks, as is best represented by "WiFi" (802.11) hotspots that are increasingly being deployed in public locations, have become popular for allowing users to access the Internet and surf the Web. Another class of wireless technologies that are gaining popularity are ad hoc networks, such as mobile ad hoc networks (MANETs) and mesh style networks, which can be built using affordable wireless hardware (e.g. 802.11) and promise to provide high-bandwidth at low cost and pervasive connectivity (unlike WLAN technologies). As a result, these ad hoc networks are becoming an attractive alternative to traditional cellular networks and have started to make their way out of the laboratory and into field tests, as has been evidenced by recent deployments where mesh networks are being deployed to support municipal and public communication needs. One further wireless technology that is on the horizon and that could potentially alter how wireless communication is carried out is cognitive radio technology, where the wireless device itself is highly-programmable and capable of adjusting its protocols and communication interfaces in order to adjust to opportunities that can improve the user's overall communication experience.

Overall, the combination of a broad array of wireless technologies and the fact that wireless devices can seamlessly blend into users lives, it is easy to predict that future wireless networks will gradually become the primary interface for consumer applications. Unfortunately, in spite of the vast array of emerging wireless technologies, there is a major hurdle facing the successful deployment of these wireless technologies– the issue of ensuring secure and trustworthy communication across these networks. Even for existing wireless networks, security is often cited as a major technical barrier that must be overcome before widespread adoption of mobile services can occur. The popular WiFi or 802.11 wireless local-area network technology was initially based on a standard with relatively weak wireless security called WEP, resulting in major security concerns as the equipment was deployed in offices and homes. Further, 3G cellular data services have also been shown to have limited security capabilities. Moreover, it has become clear that end-to-end security solutions (such as SSL and IPSec), which were originally designed for the wired Internet, have limited applicability to the unique problems associated with wireless networks. Lastly, as wireless devices become increasingly pervasive and essential, they will become both the target for attack and the very weapon with which such an attack can be carried out.

Securing wireless networks is distinctly different from securing traditional "wired" networks. In part, this is due to the fact that wireless networks are open networks: open because the radio medium is a broadcast medium, which means adversaries can easily witness anything that is transmitted; and open because wireless devices are commodity items, allowing adversaries to easily purchase low-cost wireless devices and, with little effort, use these platforms to launch a variety of attacks. In spite of these differences, the approach that the security community has typically taken to secure wireless networks has been to translate traditional network security services to the wireless domain. Although the application of conventional cryptographic protocols to wireless networks is essential, such an approach is incomplete- it leaves out a broad array of characteristics that are unique to wireless communications and that can be used to enhance the security of wireless systems.

In short, although traditional approaches to network security are important to securing wireless systems, they cannot protect against the full range of threats facing wireless networks, nor provide a complete toolbox to protect wireless networks. In fact, there are a broad variety of properties related to wireless systems that can serve as powerful sources of domain-specific information that can augment traditional security mechanisms.

The objective of this book is to highlight the importance of new paradigms for securing wireless systems that take advantage of wireless-specific properties to thwart security threats. Notably, this book will focus on security problems that involve properties unique to wireless systems, such as the characteristics of radio propagation, or the location of communicating entities, or the properties of the medium access control layer. We will also examine how one may use the properties of wireless systems to overcome these threats. For example, we will examine detection mechanisms and highlight defense strategies that cope with threats against wireless localization infrastructure. We will examine attacks on wireless networks that exploit entity identity (i.e. spoofing attacks) and discuss methods for overcoming these threats by using relationships inherent in sequences of packet transmissions or by using location information associated with a transmitter. We will also examine jamming and radio interference attacks that can undermine the availability of wireless communications, and provide techniques that can restore network connectivity in spite of radio interference. Privacy threats, where an adversary seeks to infer spatial and temporal contextual information surrounding wireless communications, will be examined and defense mechanisms that incorporate routing obfuscation and in-network buffering will be examined. Additionally, we will explore new paradigms of physical layer security for wireless systems, which can support authentication and confidentiality services by exploiting fading properties unique to wireless communications.

1.1   Book Overview

This book focuses on securing wireless communication. At the most basic level, this deals with assuring that communications between entities are secure and authentic, that services are accessed by those who are intended to access them, and that the availability of the communication infrastructure should be maintained in spite of a broad array of threats. In order to accomplish these objectives, a theme throughout this book is that such secure wireless communication should take advantage of specific properties of wireless communications that are generally not used in conventional cryptographic approaches to network security. Looking forward, the book covers a range of topics in security for wireless communications, ranging from the trustworthiness of location information to issues of identity in wireless networks to guaranteeing the availability of wireless services. A break down of the chapters is:

- Chapter 2: Location is a critical property surrounding wireless communications and, as such, can serve as the basis for building new applications, as well as the basis for constructing unique defenses for wireless networks. In this chapter, we will provide an overview of wireless localization schemes. We will start by surveying localization strategies based on different localization infrastructures, ranging methodologies, mapping/inference functions, and aggregate or singular domains. We will then introduce the concepts of point-based and area-based localization schemes and their pros and cons. Finally, we will wrap up this chapter by summarizing a list of adversarial attacks that are unique to wireless localization and which could threaten the validity and trustworthiness of localization results.

- Chapter 3: On the Robustness of Localization Algorithms to Signal Strength Attacks: It has been noted that localization algorithms based on signal strength are susceptible to non-cryptographic attacks, which consequently threatens their viability for many wireless and sensor network applications. In this chapter, we examine a representative set of localization algorithms and evaluate their robustness to attacks in which an adversary attenuates or amplifies the signal strength at one or more landmarks. We study both point-based and area-based methods that employ received signal strength for localization, and propose several performance metrics that quantify the estimator's precision and error, including Hlder metrics, which quantify the variability in position space for a given variability in signal strength space. Further, we provide a trace-driven evaluation of these algorithms, where we measured their performance as we applied attacks on real data from two different buildings.

- Chapter 4: In Chapter 2, we show that the performance of localization algorithms degrades significantly under physical attacks. Compromised localization results are a serious threat due to their adverse impact on applications. In this chapter we provide several detection schemes for wireless localization attacks that can provide a means for a location-oriented service to decide whether the localization system is being attacked. We first formulate a theoretical foundation for the attack detection problem using statistical significance testing. Next, we define test metrics for two broad localization approaches: multilateration and signal strength. We then derive both mathematical models and analytic solutions for attack detection for any system that utilizes those approaches. We also provide additional test statistics that are specific to a diverse set of algorithms. Moreover, the effectiveness and the suitability of the proposed attack detection schemes for a variety of wireless networks is supported through evaluations across both an 802.11 (WiFi) network as well as an 802.15.4 (ZigBee) network.

- Chapter 5: Attacks to localization infrastructure are diverse in nature, and there may be many unforeseen attacks that can bypass traditional security countermeasures. This chapter will be dedicated to discussing robust statistical techniques that make localization attack-tolerant. For multilateration-based localization, we provide an adaptive least squares and least median squares position estimator that has the computational advantages of least squares in the absence of attacks and is capable of switching to a robust method when being attacked. Moreover, we introduce robustness to fingerprinting-based localization through the use of a median-based distance metric.

- Chapter 6: In this chapter we explore the challenges of supporting new forms of location-based access control. Contrary to conventional access control approaches where access to objects/services is solely based on the user's identity, spatio-temporal access control allows for objects/services to be accessed only if the user is in the right place at the right time. By inverting the role of a wireless sensor network, where we allow sensor nodes to actuate the environment by emitting location-specific information, we can support location-based access control. Additionally, we outline a challenge-response mechanism for verifying user location in a centralized spatio-temporal access control mechanism. Finally, we show that spatio-temporal access control policies can be specified using finite automata.

- Chapter 7: Wireless networks are vulnerable to spoofing attacks, which further allows for many other forms of attacks on the networks. Although the identity of a node can be verified through cryptographic authentication, authentication is not always possible because it requires key management and additional infrastructural overhead. In this chapter, we propose that it is desirable to have a scheme complementary to traditional authentication that can detect device spoofing with no dependency on cryptographic material. We present the strategy of using forge-resistant relationships associated with transmitted packets to detect anomalous (i.e., spoofing) activity. The proposed strategy is generic and can be realized through several mechanisms, such as the use of monotonic relationships in the sequence number fields or the enforcement of statistical characteristics of legitimate traffic. We illustrate how these forge-resistant consistency checks can be augmented with a measurement of the threat severity in order to facilitate multi-level classification. The proposed methods do not require the explicit use or establishment of cryptographic material, and thus are suitable for scenarios where the maintenance of keying material is not practical. Finally, we present the validation of these methods for anomalous traffic scenarios involving multiple sources sharing the same MAC address through experiments conducted on the WINLAB ORBIT wireless testbed.

- Chapter 8: In this chapter we propose a method for detecting spoofing attacks, as well as locating the positions of adversaries performing the attacks. None of the current methods in the literature provide the ability to localize the positions of the spoofing attackers after detection of anomalous activities. We describe an approach that can achieve both attack detection and attack localization by exploiting location information surrounding the attacker. First, we present an attack detector for wireless spoofing that utilizes K-means cluster analysis. Next, we describe how we integrate our attack detector into a real-time indoor localization system, which is also capable of localizing the positions of the attackers. The proposed method in this chapter can be built into existing communication methods, and thus does not add any overhead to the wireless devices and sensor nodes.

- Chapter 9: Wireless networks are built upon a shared medium that makes it easy for adversaries to launch jamming-style attacks. In this chapter, we will first provide a brief history of jamming, and the related literature. Then, we will briefly summarize the problem of radio interference for various types of wireless networks. In particular, we provide examples of how easily RF-interference can disrupt wireless communications, and give a high-level overview of a generic strategy to defend against interference, which consists of a multi-phased approach involving the detection of interference and then the use of an evasion strategy.

- Chapter 10: Jamming disrupts link connectivity and adversely affects communication between wireless devices. An important starting point for coping with the threat of radio interference is to understand the types of jamming threats that are possible, as well as the severity of their impact on communication performance. This chapter examines the feasibility of launching jamming attacks/radio interference in wireless sensor or ad hoc networks. In particular, we will first examine the effect of interference from a communication capacity perspective, and then present four jamming strategies that might be used against wireless networks and measure their effectiveness after implementing all the strategies on a testbed of MICA2 motes.

- Chapter 11: In order to cope with jamming, it is necessary to detect malicious disruption of link connectivity. In this chapter we will examine methods that can be employed by the wireless network in order to detect the presence of jamming. Through extensive experimentation with various detection methods, we find that employing a consistency check between at least two properties of the network is necessary to obtain reasonable detection accuracy.

- Chapter 12: After detecting jamming attacks/radio interference, the next step is to repair the network communication in the presence of radio interference. Thus, in this chapter, we will examine such a repair strategy, which we call channel surfing, requiring wireless devices in the network adapting their channel assignments to restore network connectivity. In particular, we explore two different approaches to channel surfing: coordinated channel switching, where the entire sensor network adjusts its channel; and spectral multiplexing, where nodes in a jammed region switch channels and nodes on the boundary of a jammed region act as radio relays between multiple spectral zones.

- Chapter 13: One of the most notable challenges threatening the successful deployment of sensor systems is privacy. Although many privacy-related issues can be addressed by security mechanisms, one sensor network privacy issue that cannot be adequately addressed by network security is the privacy of the context information surrounding the "events of interest" in sensor networks. One such privacy concern is source-location privacy, where adversaries may use RF localization techniques to perform hop-by-hop trace back to the source sensor's location. This chapter will provide a formal model for the source-location privacy problem in sensor networks. We will show that most of the current routing protocols cannot provide efficient source-location privacy even at a high communication cost. We will then propose a new routing technique, i.e., phantom routing, which is capable of protecting the source's location, while not incurring a noticeable increase in communication and energy overhead.

- Chapter 14: The context surrounding 'events of interest' (e.g. sensor measurements or, more generally, any wireless communication) may be sensitive and therefore should be protected from eavesdroppers. Beyond source-location privacy, another important such privacy is temporal privacy. An adversary armed with knowledge of the network deployment, routing algorithms, and the base-station (data sink) location can infer the temporal patterns of interesting events by merely monitoring the arrival of packets at the recipient, thereby allowing the adversary to remotely track the spatio-temporal evolution of a sensed event. In this chapter, we introduce the problem of temporal privacy for delay tolerant sensor networks and propose adaptive buffering at intermediate nodes on the source-sink routing path to obfuscate temporal information from an adversary.

- Chapter 15: Although conventional cryptographic security mechanisms are essential to the overall problem of securing wireless networks, these techniques do not directly leverage the unique properties of the wireless domain to address security threats. The properties of the wireless medium are a powerful source of domain specific information that can complement and enhance traditional security mechanisms. In this chapter, we present methods which utilize the fact that the radio channel decorrelates rapidly in space, time and frequency to establish new forms of authentication and confidentiality. Such methods operate at the physical layer and can be used to facilitate cross-layer security paradigms.

- Chapter 16: This chapter will conclude the book, providing a summary of the methods mapped out in the book, as well as provide a look forward to future security methods that are important to be developed by researchers for securing emerging wireless systems.

Part I

Secure Localization

2

Overview of Wireless Localization

2.1 Introduction

Wireless networks are changing the way we work, study, and interact with each other. As wireless networks become increasingly prevalent, they make integrating new information types into applications possible. Location information is one such information source that is very important for many applications. Localization refers to determining the physical position of a wireless device or a sensor node which can be either static or mobile. The location information can be one-dimensional (e.g., location on a long airport corridor), two-dimensional (e.g., location on one floor in a hospital), or three-dimensional (e.g., location within a multi-level shopping mall). For example, in the public arena, doctors want to use location information to track and monitor patients in medical facilities; for wild life observation, biologists can put tags on animals and perform habitat tracking; first responders can track victims and each other during an emergency. In the enterprise domain, location-based access control is needed for accessing the proprietary corporate materials in restricted areas or rooms. For example, during meetings, certain documents may need to be sent only to laptops within the involved conference rooms, which requires location-aware content delivery. In addition, asset tracking also relies on location information. These examples show that accurately positioning nodes in wireless and sensor networks is important as the location of sensors is a critical input to many high-level networking tasks and applications.

With recent great advances in wireless technology, there are three wireless communication standards that have conjoined with our everyday life and have further promised to realize location-based services: First, Wireless Local Area Networks (WLANs) usually refer to networks based on *WiFi* technology functioning according to IEEE 802.11 standards [1]. The normal infrastructure for a WiFi network consists of one or more Access Points (APs), which has the ability to do wireless transmission and also serves as a gateway to a wired network. WiFi devices can thus connect to the Internet and talk to each other through APs. The most popular WiFi devices are laptops and Personal Digital Assistants (PDAs). Second, *Bluetooth* technology uses IEEE 802.15.1 standards [2] and is designed for lower power consumption, and thus has a relatively shorter range around 10 meters. It is mostly used for communication between devices closely located to each other. Currently many devices support Bluetooth including laptops, cell phones, headsets, mouses, and digital cameras. Finally, *ZigBee* implements IEEE 802.15.4 standards [3] and targets for sensor networks with embedded applications such as environmental monitoring, data collection, and intruder detection. Because of the nature of embedded applications, the corresponding devices utilizing ZigBee protocol are required to be small. The current available ones are about the size of a quarter [4,5].

In wireless and sensor networks, there are various physical modalities can be employed to perform localization such as Received Signal Strength (RSS), Time of Arrival (TOA), Angle of Arrival (AOA), Hop Counts, and etc.. Among the localization techniques, utilizing RSS is especially attractive since it can reuse the existing deployment of wireless communication networks, rather than require a specialized localization infrastructure such as ultrasound or infrared methods. This provides tremendous cost savings. Also, all current standard commodity radio technologies, such as 802.11, 802.15.4, and Bluetooth provide it, and thus the same algorithms can be applied across different platforms. Further, based on the information obtained from physical modalities, different principles can be used to determine the positions of sensors. There has been active research in developing localization algorithms using lateration [6–10], angulation [11], probabilistic approaches [12, 13], and statistical supervised learning techniques [14–16]. We detail these efforts in Section 2.2.

However, in spite of the utility of the location information, it is only useful if the location information is accurate and trustworthy. As more location-dependent services are deployed, they will increasingly become tempting targets for malicious attacks. Unlike traditional systems, the localization infrastructure is sensitive to a variety of attacks, ranging from conventional to non-cryptographic , that can subvert the utility of location information. Conventional attacks, where an adversary injects false messages, can be isolated and protected against using traditional cryptographic methods such as authentication. However, there is a completely orthogonal set of attacks that are non-cryptographic, where the measurement process itself can be corrupted by adversaries. For instance, an adversary could introduce an absorbing barrier between the transmitter and the target, changing the underlying propagation physics. As the signal propagates through the barrier, it is attenuated, and hence the target would observe a significantly lower received signal strength. Consequently, the receiver would conclude that it is further from the transmitter than it actually is. On the other hand, wormhole attacks tunnel through a faster channel to shorten the observed distance between two nodes. Unfortunately, these non-cryptographic attacks can not be addressed by traditional security services. Thus, it is desirable to study the impact of these attacks on localization algorithms and explore methods to detect and further to eliminate these attacks from the network. This is the focus of this book. We are motivated to develop solutions that can be integrated into early generations of localization systems, so that we will not have to apply patchwork solutions to solve security threats that arise after localization systems are deployed.

In this chapter, we begin in providing an overview of wireless localization approaches in Section 2.2. Then we show severe impacts of non-cryptographic attacks to localization results and discuss methods to verify localization estimates in Section 2.3. Next, in Section 2.4 we study existing work on access control models and present the importance of location-based access control. Further, we point out identity-based spoofing attacks are a serious threat in the network and review conventional methods and a few new approaches to address spoofing attacks in Section 2.5.

2.2 Wireless Localization

There has been much activity toward developing localization technique for wireless and sensor networks. We cannot cover the entire body of works in this section. Rather, we give a short overview of the different localization strategies in this section.

Localization approaches can be categorized using various taxonomies. Range-based localization algorithms involve measuring physical properties that can be used to calculate the distance between a sensor node and an anchor point whose location is known. Time of Arrival (TOA) is an important property that can be used to measure range, and arises in GPS [6]. The Time Difference of Arrival (TdOA) [17] is also widely used, and has been used in MIT's Cricket [17], and appeared in [18,19]. In addition, APS [20] pointed out that the Angle of Arrival (AOA) can be used to calculate the relative angle between two nodes, which can be further used to calculate the distance. The RSSI value of the received signal, together with the signal propagation model, is also a good indicator of the distance between two nodes [16,21]. Other properties of arriving signals can also be exploited. One interesting example is to use visual cuing [22], which tries to determine the position and orientation of a mobile robot from visual cues obtained from color cylinders strategically placed in the field of the view.

Range-free localization algorithms do not require the measurement of physical distance-related properties. For example, one can count the number of hops between a sensor node and an anchor point, and further convert the hop counts to physical distances, such as in [7,23–25]. As another example, a sensor node can estimate its location using the centroid of those anchor nodes that are within its radio range, such as in Centroid [26]. Similarly, APIT [27] employs an area-based estimation scheme to determine a node's location. Compared to range-based localization algorithms, these schemes do not require special hardware, and their accuracies are thus lower as well.

Another classification method relates how a node is mapped to a location. Lateration approaches [6–10], try to solve a set of equations involving distances to landmarks (i.e., nodes with known locations) ; angulation uses the angles from landmarks [11]; while probabilistic approaches [12,13] use statistical inferences, and statistical supervised learning techniques [14–16] utilize training data to help inference the location estimation. Among them, scene matching strategies [12,13,15,16,28] are originated from machine learning techniques. Usually a radio map of the environment is constructed, either by measuring actual samples, using signal propagation models, or some combination of the two. A node then measures a set of radio properties (often just the RSS of a set of landmarks), the *fingerprint*, and attempts to match these to known location(s) on the radio map. These approaches are almost always used in indoor environments because signal propagation is extensively affected by reflection, diffraction and scattering, and thus ranging or simple distance bounds cannot be effectively employed. Matching fingerprints to locations can be cast in statistical terms [12,13], as a machine-learning classifier problem [28], or as a clustering problem [16].

Finally, a third dimension of classification extends to aggregate or singular algorithms. Aggregate approaches [29,30] use collections of many nodes in the network in order to localize (often by flooding), while localization of a node in singular methods only requires it to communicate to a few landmarks with known locations.

In addition, some research have experimented with using ultrasound , infrared , or a combination of infrared and RSS for localization [17,19,31–33]. The goal is to reach centimeter accuracy. These work use specialty hardware or have limited range as in the infrared technology. As such, they can only be deployed in highly engineered and controlled areas, and hence have not become very popular.

Also, this is different from our goals: we conjecture that sacrificing little accuracy for scalability would create more practical positioning systems that are easier to bootstrap. We focus our work on two broad localization mechanisms: multilateration and signal strength. Multilateration clearly applies to both single and multi-hop range-based approaches, while signal strength can be applied to a wider variety of both range-based and scene matching algorithms.

2.3 Secure Localization

There has been considerably less work on the problem of ensuring the trustworthiness of wireless localization. In this section, we first describe attacks that are unique to localization and then review research methods that developed for secure localization.

Different localization methods are built upon the measurement of some basic properties. In Table 2.1, we enumerate several properties that are used by localization algorithms, along with different threats that may be employed against these properties. The threats we identify are specific to localization, and are primarily *non-cryptographic* attacks that are directed against the measurement process. Consequently, these attacks bypass conventional security services.

We note, however, that there are many classical security threats that may be launched against a wireless or sensor network, which can have an adverse affect on the localization process. For example, a Sybil attack can disrupt localization services by allowing a device to claim multiple identities. In order to address the Sybil attack, one may employ entity verification techniques, such as radio resource testing or the verification of key sets, which were presented in [34]. In general, for attacks that are cryptographic in nature, there are extensive efforts to migrate traditional security services, such as authentication, to the sensor platform in order to handle these threats.

Even so, though, it should be realized that it is unlikely that any single technique will remove all possible threat models and, in spite of the security countermeasures that are employed, many adversarial attacks will be able to bypass security layers. To address threats that are *non-cryptographic*, or threats that bypass conventional security countermeasures, we take the viewpoint that attack detection and statistical robustness need to be introduced into the wireless localization process.

We now explore several of these threats. We start by looking at methods that employ time of flight. The basic concept behind time of flight methods is that there is a direct relationship between the distance between two points, the propagation speed, and the duration needed for a signal to propagate between these two points. For time of flight methods, an attacker may try to bias the estimation of distance to a larger value by forcing the observed signal to come from a multipath. This may be accomplished by placing a barrier sufficiently close to the transmitter and effectively removing the line-of-sight signal. Another technique that may be used to falsely increase the distance estimate occurs in techniques employing round-trip time of flight. Here, an adversarial target that does not wish to be located by the network receives a transmission and holds it for a short time before retransmitting. An attack that skews the distances to smaller values can be accomplished by exploiting the propagation speed of different media. For example, in CRICKET [17], the combination of an RF signal and an ultrasound signal allows for the estimation of distance since the acoustic

TABLE 2.1. Properties employed by different localization algorithms and attacks that may be launched against these properties.

Property	Example Algorithms	Attack Threats
Time of Flight	Cricket	Remove direct path and force radio transmission to employ a multipath; Delay transmission of a response message; Exploit difference in propagation speeds (speedup attack, transmission through a different medium).
Signal Strength	RADAR SpotON Nibble	Remove direct path and force radio transmission to employ a multipath; Introduce different microwave or acoustic propagation loss model; Transmit at a different power than specified by protocol; Locally elevate ambient channel noise.
Angle of Arrival	APS	Remove direct path and force radio transmission to employ a multipath; Change the signal arrival angel by using reflective objects, e.g., mirrors; Alter clockwise/counter-clockwise orientation of receiver (up-down attack).
Region Inclusion	APIT SerLoc	Enlarge neighborhood by wormholes; Manipulate the one-hop distance measurements; Alter neighborhood by jamming along certain directions.
Hop Count	DV-Hop	Shorten the routing path between two nodes through wormholes; Lengthen the routing path between two nodes by jamming; Alter the hop count by manipulating the radio range; Vary per-hop distance by physically removing/displacing nodes.
Neighbor Location	Centroid SerLoc	Shrink radio region (jamming); Enlarge radio region (transmit at higher power, wormhole); Replay; Modify the message; Physically move locators; Change antenna receive pattern.

signal travels at a slower propagation velocity. An adversary located near the target may therefore hear the RF signal and then transmit an ultrasound signal that would arrive before the original ultrasound signal can reach the receiver [35].

As another example, consider a location system that uses signal strength as the basis for location. Such a system is very closely tied to the underlying physical-layer path loss model that is employed (such as a free space model where signal strength decays in inverse proportion to the square of distance). In order to attack such a system, an adversary could introduce an absorbing barrier between the transmitter and the target, changing the underlying propagation physics. As the signal propagates through the barrier, it is attenuated, and hence the target would observe a significantly lower received signal strength. Consequently, the receiver would conclude that it is further from the transmitter than it actually is.

Hop count based localization schemes [24] usually consist of two phases. In the first phase, per-hop distance is measured. In the second phase, anchor points flood beacons to individual sensor nodes, which count the number of hops between them, and these hop counts are translated into physical distances. As a result, adversaries can initiate attacks as follows: (1) manipulate the hop count measurement, and (2) manipulate the translation from

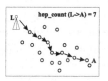 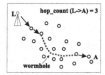 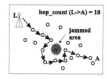

FIGURE 2.1. (Left) Operation of localization using hop count, (Middle) Wormhole attack on hop count methods, and (Right) Jamming attack on hop count methods.

hop count to physical distance. A number of tricks can be played to tweak hop count measurements, ranging from PHY-layer attacks, such as increasing/decreasing transmission power, to network layer attacks that tamper with the routing path. Since PHY-layer attacks have been discussed earlier, we now focus on some possible network layer attacks, namely jamming [36] and wormholes [37]. By jamming a certain area between two nodes, beacons may take a longer route to reach the other end (as shown in Figure 2.1), which increases the measured hop count. While jamming may not always increase the hop count, for it may not block the shortest path between the two nodes, the other type of attacks, which involve wormhole links, are more harmful because they can often significantly shorten the shortest path and result in a much smaller hop count. Figure 2.1 illustrates such a scenario: the shortest path between anchor L and node A has 7 hops, while the illustrated wormhole brings the hop count down to 3. Consequently, these attacks can also affect the translation from hop count to physical distance. In addition, if adversaries can manage to physically remove or displace some sensor nodes, even correct hop counts are not useful for obtaining accurate location calculations.

Localization methods that use neighbor locations are built upon the implicit assumption that neighbors are uniformly distributed in space around the wireless device. These localization methods, such as the Centroid method, can be attacked by altering the shape of the received radio region. For example, an attacker can shrink the effective radio region through blocking some neighbors by introducing a strong absorbing barrier around several neighbors. Another approach to shrinking the radio region is for an adversary to employ a set of strategically located jammers. Since these neighbors are not heard by the wireless device, the location estimate will be biased toward the unblocked side.

Cryptographic threats on localization can be addressed through traditional security services [38–44], e.g. authentication. However, the above described attacks are non-cryptographic, where the measurement process itself are corrupted by adversaries. And these non-cryptographic attacks can not be addressed by traditional security services. In order to address the non-cryptographic attacks, different strategies are required. [45, 46] proposed distance bounding protocols for verification of node positions. [47] proposed the Verifiable Multilateration mechanism which is based on the distance bounding protocols for secure position computation and verification. [48] uses hidden and mobile base stations to localize and verify location estimates. [49] uses both directional antennas and distance bounding to achieve security. Compared to all these methods, which employ location verification and discard location estimate that indicates under attack, [9, 50, 51] try to eliminate attack effects and still provide accurate localization. [9] makes use of the data redundancy and robust statistical methods to achieve reliable localization in the presence of attacks. [50] proposes to detect attacks based on data inconsistency from received beacons and to use a greedy search or voting algorithm to eliminate the malicious beacon information.

2.4   Location-based Access Control

The conventional literature on access control models can be broken into several main categories: identity-based access control , role-based access control , and context-based access control . Location-based access control may be considered as a specialized form of context-based access control [52, 53].

Research into supporting location-based access control has primarily focused on the issue of providing secure and robust position information. In [35], the authors listed a few attacks that might affect the correctness of localization algorithms along with a few countermeasures. SecRLoc [54] employs a sectored antenna, and presented an algorithm that makes use of the property that two sensor nodes that can hear from each other must be within the distance $2r$ assuming r is fixed in order to defend against attacks. [48] uses hidden and mobile base stations to localize and verify location estimate. Since such base station locations are hard for attackers to infer, it is hard to launch an attack, thereby providing extra security. [49] uses both directional antenna and distance bounding to achieve security. Compared to all these methods, which employ location verification and discard location estimate that indicates under attack, [50] try to eliminate the effect of attack and still provide good localization.

There has been less work devoted to developing the remaining components needed for spatio-temporal access control , and there has only been a few efforts that have tried to develop location-based access control systems. In [55], the objective is to provide location-based access control to a resource(in this case, the wireless links). The authors propose a Key-independent Wireless Infrastructure (KIWI) where, during the handshake period, KIWI challenges a client who is intended to access the resource with a set of nonces. Only when the client send back the proof that it correctly received all the nonces, can it complete the authentication handshake. In [56, 57], the PAC architecture is proposed in order to provide location-aware access control in pervasive computing environments. Although PAC preserves user anonymity and does not expose a user's exact location, one drawback is that it uses only coarse geographical location areas.

In the area of developing formal access control models, most of the efforts in the literature have focused on general context-sensitive access control models, such as presented in [58]. Here, a formal model is presented which consists of context, policy, request, and algorithm sets. Context information is represented as a six dimensional vector, which includes *time* and *location* as two of the dimensions. This work provides some high-level outlines of authorization and access control protocols that can be based upon their model. Similarly, in [59], the Context Sensitive Access Control framework was presented, which provided a comparison of different access control mechanisms and context verification mechanisms. Two other related efforts were presented in [60] and [61]. In [60], a comprehensive RBAC model is presented that employs location information in its formal model, while [61] employs temporal information in its model.

One important issue related to STAC is user privacy, as techniques that acquire the user's exact location also can expose this information to unwanted parties [62]. Location privacy has recently been studied in the context of location-based services [63, 64]. In [63, 64], a distributed anonymity algorithm was introduced that serves to remove fine levels of detail that could compromise the privacy associated with user locations in location-oriented services. For example, a location-based service might choose to reveal that a group of users is at a specific location (such as an office), or an individual is located in a vague location (such as in a building), but would not reveal that a specific individual is located at a specific location. Duri examined the protection of telematics data by applying a set of privacy and security techniques [65].

2.5 Coping with Identity Fraud

In wireless networks, attackers can gather useful identity information during passive monitoring and utilize the identity information to launch identity-based spoofing attacks. For instance, it is easy for a wireless device to acquire a valid MAC address and masquerade as another device. The 802.11 protocol suite provides insufficient identity verification during message exchange, including most control and management frames. Therefore, the adversary can utilize this weakness and request various services as if it were another user. Identity-based spoofing attacks are a serious threat in the network since they represent a form of identity compromise and can facilitate a series of traffic injection attacks. There has been active research addressing spoofing attacks as well as those facilitated by adversaries masquerading as another wireless device. In this section, we give a short overview of the attacks that is based on identity-spoofing , and the traditional methods and several new approaches to address spoofing attacks.

An adversary can launch a deauthetication attack. After a client chooses an access point for future communication, it must authenticate itself to the access point before the communication session starts. Both the client and the access point are allowed to explicitly request for deauthentication to void the existing authentication relationship with each other. Unfortunately, this deauthentication message is not authenticated. Therefore, an attacker can spoof this deauthentication message, either on the clients behalf, or on the access points behalf [66]. The adversary can persistently repeat this attack and completely prevent the client from transmitting or receiving. Further, An attacker can utilize identity spoofing and launch the Rogue Access Point (AP) attack against the wireless network. In the Rogue AP attack, the adversary first sets up a rogue access point with the same MAC address and SSID as the legitimate access point, but with a stronger signal. When a station enters the coverage of the rogue AP, the default network configuration will make the station automatically associate with the rogue access point, which has a stronger signal. Then the adversary can take actions to influence the communication. For example, it can direct fake traffic to the associated station or drop the requests made by the station. Besides the basic packet flooding attacks, the adversary can make use of identity-spoofing to perform more sophisticated flooding attacks on access points, such as probe request, authentication request, and association request flooding attacks [67].

The traditional security approach to cope with identity fraud is to use cryptographic authentication. An authentication framework for hierarchical, ad hoc sensor networks is proposed in [40] and a hop-by-hop authentication protocol is presented in [38]. Additional infrastructural overhead and computational power are needed to distribute, maintain, and refresh the key management functions needed for authentication. [41] has introduced a secure and efficient key management framework (SEKM). SEKM builds a Public Key Infrastructure (PKI) by applying a secret sharing scheme and an underlying multicast server group. [44] implemented a key management mechanism with periodic key refresh and host revocation to prevent the compromise of authentication keys. In addition, binding approaches are employed by Cryptographically Generated Addresses (CGA) to defend against the network identity spoofing [68, 69].

2.6 Conclusion

In this chapter, we have given an overview on wireless localization. We have discussed different localization approaches based on various taxonomies. We then provided a survey of the potential non-cryptographic attacks that are unique to localization and described the work on the problem of ensuring the trust-worthiness of wireless localization.

Further, we pointed out that location-based access control is a new computing paradigm. Different from the conventional identity-based approaches, the wireless infrastructure can facilitate location-aware computing paradigms, where services are only accessible if the user is in the right place at the right time.

In the next chapters, we provide a thorough study on the robustness of a broad array of localization algorithms to attacks that corrupt signal strength readings in Chapter 3. The characterization of the response of algorithms provides important insights to be taken into consideration by system designers when choosing localization systems for deployment. From the robustness study, we observed that attackers can cause large localization errors using simple techniques. Hence, we must detect the presence of attacks in the network. We examine attack detection problems in Chapter 4 and provide robust localization approaches in Chapter 5. The closest works to the attack detection methods presented in Chapter 4 are [50, 70]. A general location anomaly detection scheme is described in [70] that relied on the neighbor information to detect inconsistencies. However, it assumes a highly dense network where the positions of the nodes follow a Gaussian distribution, which is contrary to the structure of many deployed systems where much lower densities are typical. The LLS approach described in Chapter 4 is more general than the ARMMSE approach in [50]. Further, the LLS approach provides a broader choices of detectors than that work.

In Chapter 6 we describe location-based access control by inverting the role of a wireless sensor network, where we allow sensor nodes to actuate the environment by emitting location-specific information.

Finally, we described the traditional security approach to cope with identity fraud as well as new approaches that use non-cryptographic schemes to detect indentity fraud. Due to the limited resources in wireless and sensor nodes, and the infrastructural overhead needed to maintain the authentication mechanisms, it is not always desirable to use authentication. In this book, we examine new approaches to detect spoofing attacks in wireless networks. Chapter 7 have introduced a security layer that is separate from conventional network authentication methods. It developed forge-resistant relationships based on packet traffic by using packet sequence numbers, traffic interarrival, one-way chain of temporary identifiers, and signal strength consistency checks to detect spoofing attacks. Further, in Chapter 8 we describe a novel approach that can not only detect spoofing attacks, but also localize the positions of the spoofing attackers after detection. The spoofing detector is integrated into a real-time localization system which can both detect the spoofing attacks, as well as localize the adversaries in wireless and sensor networks.

3

On the Robustness of Localization Algorithms to Signal Strength Attacks

3.1 Introduction

Accurately localizing sensor nodes is a critical function for many higher level applications such as health care monitoring, wildlife animal habitat tracking, emergency rescue and recovery, location-based access control, and location-aware content delivery. Over the past few years, many localization algorithms have been proposed to localize wireless devices and sensors and provide location information to new classes of location-oriented applications. Out of the myriad of localization methods, algorithms that use Received Signal Strength (RSS) as the basis of localization are very attractive options as using RSS allows the localization system to reuse the existing communication infrastructure, rather than requiring the additional cost needed to deploy specialized localization infrastructure, such as ceiling-based ultrasound, GPS, or infrared methods [17, 19, 31]. In particular, all commodity radio technologies, such as 802.11, 802.15.4, and Bluetooth provide RSS values associated with packet reception, and thus localization services can easily be built for such systems. Further, RSS-based localization is attractive as the techniques are technology-independent: an algorithm can be developed and applied across different platforms, whether 802.11 or Bluetooth. In addition, it provides reasonable accuracy with median errors of 1 to 5 meters [15]. However, as more location-dependent services are deployed, they will increasingly become tempting targets for malicious attacks. Adversaries may alter signal strength measurements for the purpose of accessing services that are based on location information (e.g. WLAN access may only be granted to devices inside of a building.)

Unlike traditional systems, the localization infrastructure is sensitive to a variety of attacks, ranging from conventional to non-cryptographic, that can subvert the utility of location information. Conventional attacks, where an adversary injects false messages, can be isolated and protected against using traditional cryptographic methods, such as authentication. However, there is a completely orthogonal set of attacks that are non-cryptographic, where the measurement process itself can be corrupted by adversaries. Unfortunately, these non-cryptographic attacks cannot be addressed by traditional security services. Thus, it is desirable to study the impact of these attacks on localization algorithms and explore methods to detect and further to eliminate these attacks from the network. Although there has been recent research on securing localization [9, 45–48, 50], to date there has been no study on the robustness of the existing generation of RSS-based localization algorithms to physical attacks.

Rather than jumping to the immediate conclusion that all RSS-based localization systems are vulnerable, we believe that a thorough performance evaluation of existing RSS-based localization schemes is warranted. Such an evaluation would represent a valuable contribution to a wireless sensor network designer as it would help drive protocol decisions, and allow the engineer to decide whether more complicated secure localization algorithms are truly necessary. In this chapter, we detail an investigation into the susceptibility of a wide range of signal strength localization algorithms to attacks on the Received Signal Strength (RSS). Specifically, we examine the response of several localization algorithms to unanticipated power losses and gains, i.e. attenuation and amplification attacks. In these attacks, the attacker modifies the RSS of a sensor node or landmark, for example, by placing an absorbing or reflecting material around the node or landmark. Notably, we expand the set of attack scenarios to include amplification or attenuation attacks on combinations of landmarks, as well as analyze the results of simultaneous amplification and attenuation on multiple landmarks. We investigate both point-based and area-based algorithms that utilize RSS to perform localization. In order to evaluate the robustness of these algorithms, we provide a generalized characterization of the localization problem, and then present several performance metrics suitable for quantifying performance, including estimator angle bias, estimator distance error, and estimator precision. Additionally, an essential contribution of our work is the introduction of a new family of localization performance metrics, which we call Hölder metrics. These metrics quantify the susceptibility of localization algorithms to perturbations in signal strength readings. We use worst-case and average-case versions of the Hölder metric, which describe the maximum and average variability as a function of changes in the RSS. We then experimentally evaluate the performance of a wide variety of localization algorithms after applying attenuation and amplification attacks to real data measured from two different office buildings.

Using experimentally observed localization performance, we found that the error for a wide variety of algorithms scaled with surprising similarity under attack. The single exception was the Bayesian Networks algorithm, which degraded slower than the others in response to attacks against a single landmark and was attack resistant when simultaneously localizing multiple devices without using training data under all-landmark attacks. In addition to our experimental observations, we found a similar average-case response of the algorithms using our Hölder metrics. However, we observed that methods which returned an average of likely positions had less variability and are thus less susceptible than other methods.

We also observed that all algorithms, except Bayesian Networks without using training data, degraded gracefully, experiencing linear scaling in localization error as a function of the amount of loss or gain (in dB) an attack introduced. This observation applied to various statistical descriptions of the error, led us to conclude that no algorithm "collapses" in response to an attack. This is important because it means that, for all the algorithms we examined, there is no tipping point at which an attacker can cause gross errors. In particular, we found the mean error of most of the algorithms for both buildings scaled between 1.3-1.8 ft/dB when all the landmarks were attacked simultaneously, and 0.5-0.8 ft/dB when a single landmark was attacked. Additionally, the performance of the mean response of algorithms with multiple landmarks under attack is between the all-landmark attack and the single landmark attack, which scaled at 0.4-1.4 ft/dB. Further we observed that mixed attacks with simultaneous attenuation and amplification cause the mean response of algorithms to move faster, ranging from 0.2-2.3 ft/dB. More powerful effects were witnessed when the mixed attack was applied to landmarks that were further apart from each other. We also showed experimentally that RSS can be easily attenuated by 15 dB, and that, as a general rule of thumb, very simple signal strength attacks can lead to localization errors of 20-30 ft.

Finally, we conducted a detailed evaluation of area-based algorithms as this family of algorithms return a set of potential locations for the transmitter. Thus, it is possible that these algorithms might return a set with a larger area in response to an attack and could have less precision (or more uncertainty) under attack. However, we found all three of our area-based algorithms shifted the returned areas rather than increased returned area. Further, one of the algorithms, the Area Based Probability (ABP) scheme, significantly shrank the size of the returned area in response to very large changes in signal strength.

The rest of this chapter is organized as follows. We begin, in Section 3.2, by giving an overview of the algorithms used in our performance study and discuss how signal strength attacks can be performed in Section 3.3. In Section 3.4, we provide a formal model of the localization problem as well as introduce the metrics that we use in this chapter. We then examine the performance of the algorithms through an experimental study in Section

3.5, and discuss the Hölder metrics for these algorithms in Section 3.6. Finally, we conclude in Section 3.7.

3.2   Localization Algorithms

Signal strength is a common physical property used by a widely diverse set of algorithms. For example, most fingerpriting approaches utilize the RSS, e.g. [16, 28], and many multilateration approaches [14] use it as well. Although these algorithms provide several-meter level accuracy, using the RSS is an attractive approach, because the existing wireless infrastructure can be reused — this feature presents a tremendous cost savings over deploying localization-specific hardware. In this chapter we thus focus on localization algorithms that employ signal strength measurements. In this section, we provide an overview of a representative set of algorithms selected for conducting performance analysis under attack. These algorithms use either deterministic or probabilistic methods for location estimation.

There are several ways to classify localization schemes that use signal strength: range-based schemes , which explicitly involve the calculation of distances to landmarks; and RF fingerprinting schemes whereby a radio map is constructed using prior measurements, and a device is localized by referencing this radio map. In this work, we focus on indoor signal strength based localization algorithms utilizing these approaches. We can further break down the algorithms into two main categories: point-based methods, and area-based methods.

3.2.1   Point-based Algorithms

Point-based methods return an estimated point as a localization result. Here we describe a few representative point-based schemes for our study.

RADAR (R1): A primary example of a point-based method is the RADAR scheme [16]. In R1, multiple base stations are deployed to provide overlapping coverage of an area, such as an office building. During set up, a mobile host with known position broadcasts beacons periodically, and the signal strength readings are measured at a set of fixed landmarks. Collecting together the averaged signal strength readings from each of the landmarks for different transmitter locations provides a radio map. After training, localization is performed by measuring a wireless device's RSS at each landmark, and the vector of RSS values is compared to the radio map. The record in the radio map whose signal strength vector is closest in the Euclidean sense to the observed signal strength vector is declared to correspond to the location of the transmitter. Variations of RADAR, such as *Averaged RADAR* (R2) which returns the average of the closest 2 fingerprints and *Gridded RADAR* (GR) that uses the Interpolated Map Grid

(IMG) as a set of additional fingerprints over the basic RADAR have been proposed in [15].

Highest Probability (P1): The P1 method uses a probabilistic approach by applying the statistical Bayes' rule to return the point with the highest probability in the pre-constructed radio map as the location estimation result [13]. There are variations of Highest Probability. *Averaged Highest Probability* (P2) returns the mid-point of the top 2 training fingerprints. And like GR, *Gridded Probability* (GP) uses fingerprints based on an IMG [15].

3.2.2 Area-based Algorithms

On the other hand, area-based algorithms return a *most likely* area in which the true location resides. One of the major advantages of area-based methods compared to point-based methods is that they return a region, which has an increased chance of capturing the transmitter's true location. We study 3 area-based algorithms [14, 15], two of them, Simple Point Matching (SPM) and Area Based Probability (ABP), use an Interpolated Map Grid (IMG) and perform scene matching (fingerprint matching) for localization; and the other, Bayesian Networks (BN), is a multilateration algorithm.

Simple Point Matching (SPM): In SPM, the floor is divided into small tiles. The strategy behind SPM is to find a set of tiles that fall within a threshold of the RSS for each landmark independently, then return the tiles that form the intersection of each landmark's set. We define the threshold as

$$s_i \pm q, \tag{3.1}$$

where s_i is the expected value of the RSS reading from Landmark i and q is an expected noise level. One way to choose q is to use the maximum of the standard deviation σ with

$$\sigma = max\{\sigma_{ij}; i \in \{1..numberoflandmarks\}, j \in \{1..numberofpoints\}\}. \tag{3.2}$$

SPM [15] is an approximation of the Maximum Likelihood Estimation (MLE) method.

Area Based Probability (ABP): ABP returns a set of tiles bounded by a probability that the transmitter is within the returned tile set. The probability is called the confidence α and it is adjustable by the user. ABP assumes the distribution of RSS for each landmark follows a Gaussian distribution with mean as the expected value of RSS reading vector **s**. The Gaussian random variable from each landmark is independent. ABP then computes the probability of the transmitter being at each tile L_i, with $i = 0...L$, on the floor using Bayes' rule:

$$P(L_i|\mathbf{s}) = \frac{P(\mathbf{s}|L_i) \times P(L_i)}{P(\mathbf{s})}. \tag{3.3}$$

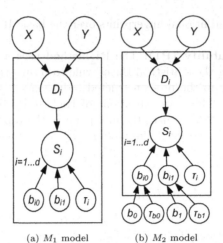

(a) M_1 model   (b) M_2 model

FIGURE 3.1. Bayesian Networks localization algorithm: Bayesian Graphical Models using plate notation.

Given that the transmitter must be at exactly one tile satisfying $\sum_{i=1}^{L} P(L_i|$ s$) = 1$, ABP normalizes the probability and returns the most likely tiles up to its confidence α [15].

Bayesian Networks (BN): BN localization is a multilateration algorithm that encodes the signal-to-distance propagation model into the Bayesian Graphical Model for localization [14]. In BN, the overall joint density of $x \in X$, where x is a random variable, only depends on the parents of x, denoted $pa(x)$:

$$p(X) = \prod_{x \in X} p(x|\mathrm{pa}(x)). \qquad (3.4)$$

Once $p(X)$ is computed, the marginal distribution of any subset of the variables of the network can be obtained as it is proportional to the overall joint distribution.

Figure 3.1 presents two Bayesian Network algorithms, M_1 and M_2, that we used for our analysis. Each rectangle is a "plate", and shows a part of the network that is replicated; in our case, the nodes on each plate are repeated for each of the n landmarks whose locations are known. The vertices X and Y represent location; the vertex S_i is the RSS reading from the ith landmark; and the vertex D_i represents the Euclidean distance between the location specified by X and Y and the ith landmark. The value of S_i follows a signal propagation model $S_i = b_{i0} + b_{i1} \log D_i$, where b_{i0}, b_{i1} are the parameters specific to the ith landmark. The distance $D_i = \sqrt{(X - x_i)^2 + (Y - y_i)^2}$ in turn depends on the location (X, Y) of the measured signal and the coordinates $(x_i,\ y_i)$ of the ith landmark. The networks model noise and outliers by modeling the S_i as a Gaussian distribution around the above propagation model, with variance τ_i:

$$S_i \sim N(b_{i0} + b_{i1} \log D_i, \tau_i). \tag{3.5}$$

The initial parameters (b_{i0}, b_{i1}, τ_i) of the model are unknown, and usually the training data is used to adjust the specific parameters of the model according to the relationships encoded in the network. Through Markov Chain Monte Carlo (MCMC) simulation, BN returns the sampling distribution of the possible location of X and Y as the localization result.

M_1 utilizes a simple Bayesian Network model as depicted in Figure 3.1 (a), and requires location information in the training set in order to give good localization results. The M_2 model is hierarchical shown in Figure 3.1 (b) by making the coefficients of the signal propagation model have common parents. The BN M_2 algorithm can localize multiple devices simultaneously with no training set, leading to a zero-profiling technique for location estimation.

The algorithms we have described in this section are summarized in Table 3.1. Although there are a variety of other signal strength based localization algorithms that may be studied, we believe that our results are general and can be applied to other point-based and area-based methods.

3.3 Conducting Signal Strength Attacks

In this section, we study the feasibility of conducting signal strength attacks. We first discuss the possible attacks on signal strength. We then provide experimental results for signal strength going through various materials. Finally, we derive an attack model for our performance analysis of the robustness of localization algorithms.

3.3.1 Signal Strength Attacks

The first step to tackle a security problem is to put oneself in the role of the adversary and attempt to understand the attacks. To attack signal-strength based localization systems, an adversary must attenuate or amplify the RSS readings. This can be done by applying the attack at the transmitting device, e.g. simply placing foil around the 802.11 card; or by directing the attack at the landmarks. For example, we may steer the lobes and nulls of an antenna to target selected landmarks. A broad variety of attenuation attacks can be performed by introducing materials between the landmarks and sensors as presented in Chapter 2.

In order to support the claim that physical attacks on received signal strength are feasible and capable of significantly affecting the results of a localization algorithm, we first examined the possibility of signal strength attacks. Next, we report results of actual experiments to quantify the effectiveness of various ways of attenuating/amplifying signal strength.

TABLE 3.1. Algorithms under study

Algorithm	Abbreviation	Description
Area-Based		
Simple Point Matching	SPM	Maximum likelihood matching of the RSS to an area using thresholds.
Area Based Probability	ABP-α	Bayes rule matching of the RSS to an area probabilistically bounded by the confidence level α%.
Bayesian Network (M_1, M_2)	BN	Returns the most likely area using a Bayesian network approach.
Point-Based		
RADAR	R1	Returns the closest record in the Euclidean distance of signal space.
Averaged RADAR	R2	Returns the average of the top 2 closest records in the signal map.
Gridded RADAR	GR	Applies RADAR using an interpolated grid signal map.
Highest Probability	P1	Applies maximum likelihood estimation to the received signal.
Averaged Highest Probability	P2	Returns the average of the top 2 likelihoods.
Gridded Highest Probability	GP	Applies likelihoods to an interpolated grid signal map.

3.3.2   Experimental Results of Attacks

Our experiments were performed in our laboratory in the 3rd floor of the CoRE building at Rutgers University, as shown in Figure 3.4 (a). There are 4 landmarks deployed in the 3rd floor of CoRE. We measured the RSS of beacon signals coming from each of the landmark. The RSS readings were collected using a laptop with an Orinoco Silver wireless card, using `iwlist` to sample the signal strength. In order to mitigate the effect of fluctuations, we collected samples once every second for 10 minutes, and averaged the signal strength over 600 samples.

As noted earlier, an adversary may attack the signal strength by attenuating or amplifying the RSS readings. This can be done either at the receiver or at the transmitter. Our aim is to find the results of power loss in dB by simple attacks. Therefore, in the experiments, we placed various obstruction materials close to the laptop's wireless card and measured the RSS values from each landmark at the laptop. The following obstructions

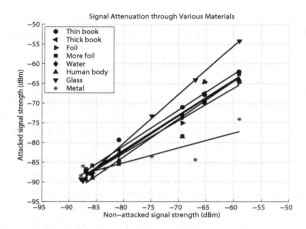

FIGURE 3.2. Signal strength when going through a barrier.

were used: a thin book, a thick book, a layer of metal foil, three layers of foil (referred to as more foil), a mug filled with water (referred to as water), a glass mug (referred to as glass), a metal cabinet (referred to as metal), and a human body. These materials are easy to access and attacks utilizing these kind of materials can be simply performed with low cost. The original signal strength values, together with the signal strength measurements in the presence of these objects, are provided in Figure 3.2. The points represent the measured data from experimental results for various materials, while the lines are the linear least-squares fitting. The results are intended to show the feasibility of using such materials for attacks. As we would expect, highly attenuating materials such as metal box or foil have a large impact on signal strength, whereas other materials do not affect the signal much. A more comprehensive study of propagation loss through common materials can be found in [71], and we note that more powerful attenuation loss is possible by using more advanced materials (such as RF-absorptive carbon fabric). Finally, we note that these results also imply that amplification is possible by removing a barrier (e.g. a door) of the corresponding material or through antenna-based methods.

3.3.3 Attack Model

Based upon the results in Figure 3.2, we further see that there is a linear relationship between the unattacked signal strength and the attacked signal strength in dB for various materials. The linear relationship implies that there is an easy way for an adversary to perform and control the effect of an attack on the observed signal strength by appropriately selecting different materials. Specifically, we envision that an adversary may suitably introduce and/or remove barriers of appropriate materials so as to attenuate and amplify the signal strength readings at one or more landmarks. Due to the observed linear relationship illustrated in Figure 3.2, we refer to this as the "linear attack model".

In the rest of this chapter, we will use the linear attack model to describe the effect of an attack on the RSS readings at one or more landmarks. The resulting attacked readings are then used to study the consequent effects on localization for the algorithms surveyed above. In particular, in this study, we apply our attacks to individual landmarks, which might correspond to placing a barrier directly in front of a landmark, as well as to the entire set of landmarks, which corresponds to placing a barrier around the transmitting device. Similar arguments can be made for amplification attacks, whereby usually barriers are removed between the source and receivers. Moreover, we apply attenuation, amplification, or a mixture of simultaneous attenuation and amplification attacks to multiple landmarks and study the performance of localization algorithms. The broad collection of our attack scenarios has covered the set of possibilities that an adversary could attempt to accomplish. Although there are many different and more complex signal strength attack methods that can be used, we believe their effects will not vary much from the linear signal strength attack model we use in this chapter, and note that such sophisticated attacks could involve much higher cost to perform.

3.4   Measuring Attack Susceptibility

The aim of a localization attack is to perturb a set of signal strength readings in order to have an effect on the localization output. When selecting a localization algorithm, it is desirable to have a set of metrics by which we can quantify how susceptible a localization algorithm is to varying levels of attack by an adversary. In this section, we shall provide a formal specification for an attack, and present several measurement tools for quantifying the effectiveness of an attack.

3.4.1   A Generalized Localization Model

In order to begin, we need to specify a model that captures a variety of RF-fingerprinting localization algorithms. Let us suppose that we have a domain D in two-dimensions, such as an office building, over which we wish to localize transmitters. Within D, a set of n landmarks have been deployed to assist in localization. A wireless device that transmits with a fixed power in an isotropic manner will cause a vector of n signal strength readings to be measured by the n landmarks. In practice, these n signal strength readings are averaged over a sufficiently large time window to remove statistical variability. Therefore, corresponding to each location in D, there is an n-dimensional vector of signal readings $\mathbf{s} = (s_1, s_2, \cdots, s_n)$ that resides in a range R.

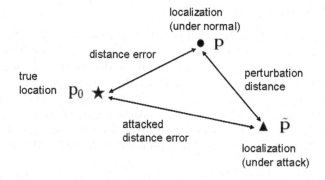

FIGURE 3.3. Interpretation of distances in location estimation.

This relationship between positions in D and signal strength vectors defines a fingerprint function $F : D \rightarrow R$ that takes our real world position (x, y) and maps it to a signal strength reading $\mathbf{s}$. F has some important properties. First, in practice, F is not completely specified, but rather a finite set of positions (x_j, y_j) is used for measuring a corresponding set of signal strength vectors $\mathbf{s}_j$. Additionally, the function F is generally one-to-one, but is not onto. This means that the inverse of F is a function G that is not well-defined: There are holes in the n-dimensional space in which R resides for which there is no well-defined inverse.

It is precisely the inverse function G, though, that allows us to perform localization. In general, we will have a signal strength reading $\mathbf{s}$ for which there is no explicit inverse (e.g. perhaps due to noise variability). Instead of using G, which has a domain restricted to R, we consider various pseudo-inverses G_{alg} of F for which the domain of G_{alg} is the complete n-dimensional space. Here, the notation G_{alg} indicates that there may be different *algorithmic* choices for the pseudo-inverse. For example, we shall denote G_R to be the RADAR localization algorithm. In general, the function G_{alg} maps an n-dimensional signal strength vector to a region in D. For point-based localization algorithms, the image of G_{alg} is a single point corresponding to the localization result. On the other hand, for area-based methods, the localization algorithm G_{alg} produces a set of likely positions.

An attack on the localization algorithm is a perturbation to the correct n-dimensional signal strength vector $\mathbf{s}$ to produce a corrupted n-dimensional vector $\tilde{\mathbf{s}}$. Corresponding to the uncorrupted signal strength vector $\mathbf{s}$ is a correct localization result $\mathbf{p} = G_{alg}(\mathbf{s})$, while the corrupted signal strength vector produces an attacked localization result $\tilde{\mathbf{p}} = G_{alg}(\tilde{\mathbf{s}})$. Here, $\mathbf{p}$ and $\tilde{\mathbf{p}}$ are set-valued and may either be a single point or a region in D.

3.4.2 Attack Susceptibility Metrics

We wish to quantify the effect that an attack has on localization by relating the effect of a change in a signal strength reading s to the resulting change in the localization result $\mathbf{p}$. We shall use $\mathbf{p}_0$ to denote the correct location of a transmitter, $\mathbf{p}$ to denote the estimated location (set) when there is no attack being performed, and $\tilde{\mathbf{p}}$ to denote the position (set) returned by the estimator after an attack has affected the signal strength. Figure 3.3 illustrates the relationship between the true location and the estimated locations. There are several performance metrics that we will use:

- **Estimator Angle Bias:** The perturbation on the signal strength vector caused by an attack will result in the variability of location estimation in the physical space. We want to investigate the bias along the angular dimension. That is, if we plot the relative error position in polar coordinates, for an unbiased estimator the error would have an equal probability of falling along any angle. However, when attacking a single landmark, we may expect an angular bias to be introduced. The estimation angle bias is studied by calculating the estimated position for different experimental trials, and comparing these results, in a spatial sense, to the true position. An angularly-unbiased algorithm should uniformly cover the 360 degrees around the true location. For area-based methods, we replace $\tilde{\mathbf{p}}$, which is a set, with its median (along the x and y dimensions separately) to get a point. The angular bias is an important metric as it can serve as an indication as to whether an attacker can skew the localization result in a specific direction - algorithms with more angular bias are more skewable and hence worse choices for deployment since an adversary can use this knowledge to its advantage.

- **Estimator Distance Error:** An attack will cause the magnitude of $\mathbf{p}_0 - \tilde{\mathbf{p}}$ to increase. For a particular localization algorithm G_{alg} we are interested in the statistical characterization of $\|\mathbf{p}_0 - \tilde{\mathbf{p}}\|$ over all possible locations in the building. The characterization of $\|\mathbf{p}_0 - \tilde{\mathbf{p}}\|$ depends on whether a point-based method or an area-based method is used, and can be described via its mean and distributional behavior. For a point-based method, we may measure the cumulative distribution (cdf) of the error $\|\mathbf{p}_0 - \tilde{\mathbf{p}}\|$ over the entire building. For area-based metrics, we calculate the CDF of the distance between the median of the estimated locations $\tilde{\mathbf{p}}_{med}$ and the true location, i.e. $\|\mathbf{p}_0 - \tilde{\mathbf{p}}_{med}\|$.

The CDF provides a complete statistical specification of the distance errors. It is often more desirable to look at the average behavior of the error. For point-based methods, the average distance error is simply $E[\|\mathbf{p}_0 - \tilde{\mathbf{p}}\|]$, which is just the average of $\|\mathbf{p}_0 - \tilde{\mathbf{p}}\|$ over all locations. Area-based methods allow for more options in defining the average distance error. First, for a particular value of $\mathbf{p}_0$, $\tilde{\mathbf{p}}$ is a set of points.

For each $\mathbf{p}_0$, we get a collection of error values $\|\mathbf{p}_0 - \mathbf{q}\|$, as $\mathbf{q}$ varies over points in $\tilde{\mathbf{p}}$. For each $\mathbf{p}_0$, we may extract the minimum, 25th percentile, median, 75th percentile, and maximum. These quartile values of $\|\mathbf{p}_0 - \mathbf{q}\|$ are then averaged over the different positions $\mathbf{p}_0$.

- **Estimator Precision:** An area-based localization algorithm returns a set $\mathbf{p}$. For localization, precision refers to the size of the returned estimated area. This metric quantifies the average value of the area of the localized set $\mathbf{p}$ over different signal strength readings $\mathbf{s}$. Generally speaking, the smaller the size of the returned area, the more precise the estimation is. When an attack is conducted, it is possible that the precision of the answer $\tilde{\mathbf{p}}$ is affected.

- **Precision vs. Perturbation Distance:** The perturbation distance is the quantity $\|\mathbf{p}_{med} - \tilde{\mathbf{p}}_{med}\|$. The precision vs. perturbation distance metric depicts the functional dependency between precision and increased perturbation distance.

- **Hölder Metrics:** In addition to error performance, we are interested in how dramatically the returned results can be perturbed by an attack. Thus, we wish to relate the magnitude of the perturbation $\|\mathbf{s} - \tilde{\mathbf{s}}\|$ to its effect on the localization result, which is measured by $\|G_{alg}(\mathbf{s}) - G_{alg}(\tilde{\mathbf{s}})\|$. In order to quantify the effect that a change in the signal strength space has on the position space, we borrow a measure from functional analysis [72], called the Hölder parameter (also known as the Lipschitz parameter) for G_{alg}. The Hölder parameter H_{alg} is defined via

$$H_{alg} = \max_{\mathbf{s},\mathbf{v}} \frac{\|G_{alg}(\mathbf{s}) - G_{alg}(\mathbf{v})\|}{\|\mathbf{s} - \mathbf{v}\|} \qquad (3.6)$$

where $\mathbf{s}$ and $\mathbf{v}$ are all the possible combinations of signal strength vectors in signal space. For continuous G_{alg}, the Hölder parameter measures the maximum (or worst-case) ratio of variability in position space for a given variability in signal strength space. Since the traditional Hölder parameter describes the worst-case effect an attack might have, it is natural to also provide an average-case measurement of an attack, and therefore we introduce the average-case Hölder parameter

$$\overline{H}_{alg} = \mathrm{avg}_{\mathbf{s},\mathbf{v}} \frac{\|G_{alg}(\mathbf{s}) - G_{alg}(\mathbf{v})\|}{\|\mathbf{s} - \mathbf{v}\|}. \qquad (3.7)$$

These parameters are only defined for continuous functions G_{alg}, and many localization algorithms are not continuous. For example, if we look at G_R for RADAR, the result of varying a signal strength reading is that it will yield a *stair-step* behavior in position space, i.e. small changes will map to the same output and then suddenly, as we continue changing the signal strength vector, there will be a change to a new position estimate (we have switched over to a new Voronoi cell in signal space). In reality, this behavior does not concern us too much, as we are merely concerned with whether adjacent Voronoi cells map to close positions. We will revisit this issue in Section 3.6. Finally, we emphasize that Hölder metrics measure the perturbability of the returned results, and do not directly measure error.

3.5   Experimental Results

In this section we present our experimental results. We first describe our experimental method. Next, we examine the impact of attacks on the RSS to localization bias and localization error under different attacking scenarios. We then quantify the algorithms' linear responses to RSS changes. Finally, we present a precision study that investigates the impact of attacks on the returned areas for area-based algorithms.

3.5.1   Experimental Setup

Figure 3.4 shows our experimental set up. The floor map on the left, (a) is the 3rd floor of the CoRE building at Rutgers, which houses the computer science department and has an area of 200x80ft (16000 ft^2). The other floor shown in (b) is an industrial research laboratory (we call it the Industrial Lab), which has an area of 225x144ft (32400 ft^2). The stars are the training points, the small dots are testing points, and the larger squares are the landmarks, which are 802.11 access points. Notice that the 4 CoRE landmarks are more co-linear than the 5 landmarks in the Industrial Lab. Next, we perform a trace-driven simulation study to apply our linear attack model to the experimental data collected from these two buildings.

For both attenuation and amplification attacks, we ran the algorithms but modified the measured RSS of the testing points collected from these two office buildings. Specifically, we altered the RSS by +/-5 dB to +/-25 dB, in increments of 5 dB. We experimented with different ways to handle signals that would fall below the detectable threshold of -92 dBm for our wireless cards. We found that substituting the minimal signal (-92 dBm) produced about the same localization results and did not require changing the algorithms to special case missing data.

We experimented with different training set sizes, including 20, 35, 60, 85, 115, 145, 185, 215, 245, 253, and 286 points. Experimental data was collected at a total of 286 locations in the CoRE building and at a total of 253 locations in the Industrial Lab. Although there are some small differences, we found that the behavior of the algorithms matches previous results [15] and varied little after using 115 training points. We therefore chose to use a training set size of 115 for this study.

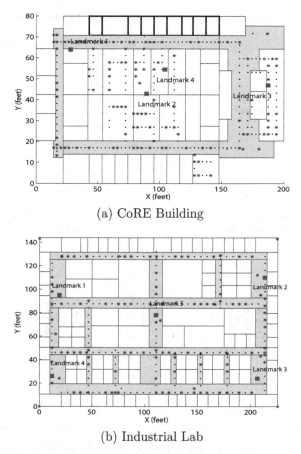

(a) CoRE Building

(b) Industrial Lab

FIGURE 3.4. Deployment of landmarks and training locations on the experimental floors

3.5.2 Localization Angle Bias

In this section, we study the angular bias of the localization schemes introduced by signal strength attacks. For the Industrial Lab, Figure 3.5(a) shows the localization result of ABP under no attack for the relative estimation positions to the true locations, setting as the origin, over all the localization attempts. The normal performance of the algorithms are unbiased with the localization results uniformly distributed around the true locations.

Figure 3.5(b) is the relative position estimation results under 25dB attenuation attack on all landmarks, while Figure 3.5(c) and Figure 3.5(d) show the attacked results on single landmarks, landmark 1 and landmark 3, respectively. Figure 3.4(b) shows that landmark 1 and landmark 3 are placed in diagonal positions across the Industrial Lab. We have observed

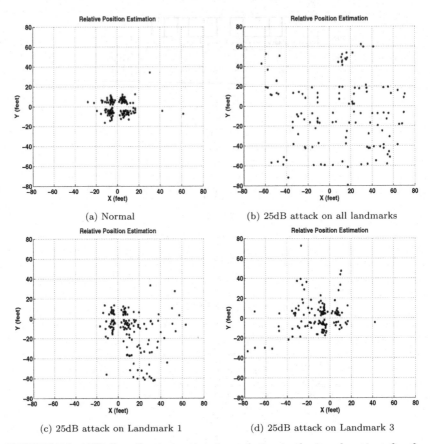

FIGURE 3.5. ABP: Localization estimation relative to the true locations for the Industrial Lab.

that signal strength attacks have affected the localization schemes by introducing angular bias on the results with the location estimation more likely to be in the fourth quadrant relative to the true location when landmark 1 is attacked, as shown in Figure 3.5(c). Because landmark 1 is placed in the upper left corner in the building floor map shown in Figure 3.4, signal attenuation on landmark 1 made the localization system think the sensor node is farther away from landmark 1, and thus the resulting localization results under attack have been pushed into the fourth quadrant. This effect has been proved by examining the localization results when landmark 3 is under attack. As presented in Figure 3.5(d), the relative localization results are mostly in the second quadrant since landmark 3 is placed in the lower right corner of the building floor map. Further, as expected, for simultaneous landmark attacks, the localization results are distributed around the true locations randomly, but with much larger estimation errors as presented in

Figure 3.5(b). We have observed similar effects for the other algorithms in the Industrial Lab and the CoRE building.

3.5.3   Localization Error Analysis

In this section, we analyze the estimator distance error through the statistical characterization of $\|\mathbf{p}_0 - \tilde{\mathbf{p}}\|$ by presenting the error CDFs of all the algorithms as a function of attenuation and amplification attacks. The CDF provides a complete statistical specification of the distance errors. Specifically, we study the localization error under four attack scenarios: an all-landmark attack; a single landmark attack; attacks involving multiple landmarks; and attacks involving simultaneous amplification and attenuation on multiple landmarks.

As a baseline, Figure 3.6(a) shows the normal performance of the algorithms for the CoRE building and (e) shows the results for the Industrial Lab. For the area-based algorithms, the median tile error is presented, as well as the minimum and maximum tile errors for ABP-75. For BN, we present the results using the simple Bayesian Network M_1 algorithm, denoted as BNmed in the plot. Note that the results from Bayesian Network M_2 are, in fact, better than M_1, and are comparable to the results for the RADAR scheme R1. However, for the sake of clarity of the plot, we have chosen to only present the results of M_1. As in previous work [15], the algorithms all obtain similar performance, with the exception of BN, which slightly under-performs the other algorithms.

First, we look at the performance of localization algorithms under an all-landmark attack. Figures 3.6(b) and 3.6(c) show the error CDFs under simultaneous landmark attenuation attacks of 10 and 25 dB for CoRE, respectively, while Figure 3.6(f) and 3.6(g) show the similar results in the industrial lab. First, the bulk of the curves shift to the right by roughly equal amounts: no algorithm is qualitatively more robust than the others. Comparing the two buildings, the results show that the industrial lab errors are slightly higher for attacks at equal dB, but again, qualitatively the impact of the building environment is not very significant.

Figures 3.6(d) and 3.6(h) show the error CDFs for the CoRE and Industrial Lab under a 10 dB amplification attack. The results are qualitatively symmetric with respect to the outcome of the 10dB attenuation attack. We found that, in general, comparing amplifications to attenuations of equal dB, the errors were qualitatively the same.

An interesting feature is that in CoRE the minimum error for ABP-75 also shifts to the right by roughly the same amount as the other curves. Figures 3.6(a) and 3.6(e) show that, in the non-attacked case, the minimum tile error for ABP-75 is quite small, meaning that the localized node is almost always within or very close to the returned area. However, under attacks, the closest part of the returned area moves away from the true location at the same rate as the median tile. We observed similar effects for

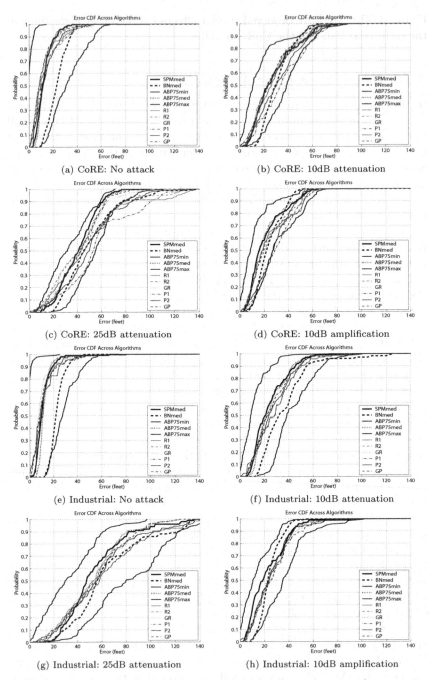

FIGURE 3.6. Error CDF across localization algorithms when attacks are performed on all the landmarks.

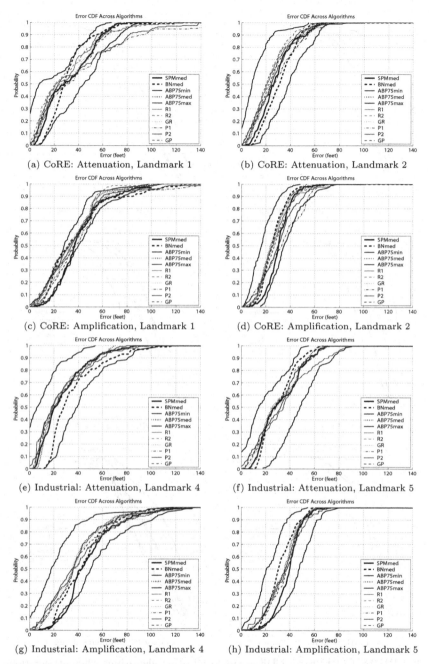

FIGURE 3.7. Error CDF across localization algorithms when attacks are performed on an individual landmark. The attack is 25dB of signal attenuation and signal amplification respectively.

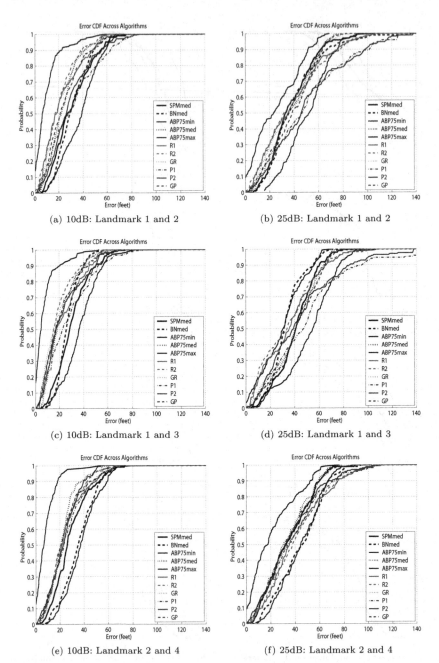

FIGURE 3.8. CoRE: Error CDF across localization algorithms when attenuation attacks are performed on multiple landmarks.

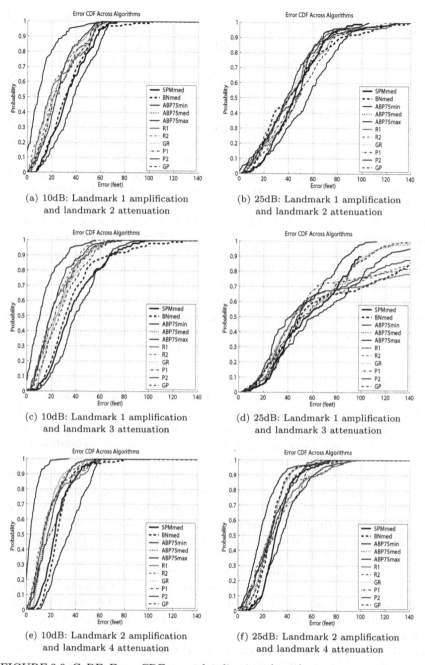

(a) 10dB: Landmark 1 amplification
and landmark 2 attenuation

(b) 25dB: Landmark 1 amplification
and landmark 2 attenuation

(c) 10dB: Landmark 1 amplification
and landmark 3 attenuation

(d) 25dB: Landmark 1 amplification
and landmark 3 attenuation

(e) 10dB: Landmark 2 amplification
and landmark 4 attenuation

(f) 25dB: Landmark 2 amplification
and landmark 4 attenuation

FIGURE 3.9. CoRE: Error CDF across localization algorithms when amplification
and attenuation attacks are simultaneously performed on multiple landmarks.

the SPM and BN algorithms. We noticed that under large attacks around 25dB, the median error CDF curves in the Industrial Lab have similar performance to those from the CoRE building, but there are two curves that seem to be outliers, namely ABP75min and ABP75max. These two curves represent the best and the worst cases from the ABP algorithm and we see that they are not moving at the same speed as the median errors, when compared with the results of the CoRE building. This tells us that the variance/spread of the performance of area-based algorithms in the Industrial Lab has increased under an all-landmark attack, but that the average behavior is consistent across the two buildings.

We then examine attacks against a single landmark. We found attacks against certain landmarks had a much higher impact than against others in the CoRE building. Figure 3.7(a) and 3.7(b) show the difference in the error CDF by comparing attacks of landmarks 1 and 2. Figure 3.4(a) shows that landmark 1 is at the left end of the building, while landmark 2 is in the center and is close to landmark 4. The tail of the curves in Figure 3.7(a) are much worse than for 3.7(b), showing that when landmark 1 is attacked, significantly more high errors are returned. Figures 3.7(c) and 3.7(d) show a similar effect for amplification attacks. This is because landmark 1 is at one end of the building alone. The contribution of the signal strength reading from landmark 1 plays an important role in localization, while the contribution of landmark 2 can be reduced by the contribution from the nearby landmark 4 when under attack.

The Industrial Lab results in Figures 3.7(e)-(h) show much less sensitivity to landmark placement compared to the CoRE building. Figure 3.4(b) shows that landmark 5 is centrally located and we initially suspected this would result in increased attack sensitivity. However, the error CDFs show that the remaining 4 landmarks provide sufficient coverage: as landmark 5 is attacked, the error CDFs are not much different from attacking landmark 4. The landmark placement in the CoRE building is colinear (to maximize the signal coverage in the floor), while the landmark placement in the Industrial Lab is more close to the optimal landmark placement for location accuracy. We believe that the better landmark placement for localization [73] in the Industrial Lab can account for the localization performance being less sensitive to landmark placement under attack.

Next, we study attacks on more than one landmark, but not on all landmarks. Figure 3.8 presents the localization results in the CoRE building when attenuation attacks are performed on multiple landmarks, specifically on landmark pairs, *1 and 2*, *1 and 3*, and *2 and 4*. We found that attacks on landmark pair *1 and 3* shown in Figure 3.8(d) cause larger errors compared to results in Figures 3.8(b) and 3.8(f) when attacking landmark pairs *1 and 2*, and *2 and 4*. Since landmarks 1 and 3 are placed at two ends of the building alone, the contribution of the RSS reading from these two landmarks is significant compared to the readings from landmark 2 and 4, which are closely placed and can cover each other. In general, the impact of the multiple landmark attacks on the localization performance is between the performance of a single landmark attack and an all-landmark attack.

Fourthly, we look at the attack scenario that the adversary simultaneously performs both amplification and attenuation attacks on multiple landmarks. The localization results are presented in Figure 3.9 for the CoRE building. For a direct comparison, we present results when mixed attacks are applied on landmark pairs, *1 and 2*, *1 and 3*, and *2 and 4*. We should expect that such an attack would be more effective in falsifying the location results, and this is what we observe. But, beyond this, we observe that the performance depends heavily upon which landmarks are attacked. We found that if the attacked landmarks are close to each other such as landmark 2 and 4, which are located in the center of the building, the effects of amplification and attenuation attacks are canceled out. Thus the impact of mixed attacks does not lead to significant perturbation in the localization results, as shown in Figure 3.9(f), which is about the same as under single landmark attacks displayed in Figure 3.7. However, if the attacked landmarks are farther away from each other, such as landmark 1 and 3, which are located at opposite ends of the building, the simultaneous amplification and attenuation attacks can be very harmful and cause larger localization errors for all the algorithms presented in Figure 3.9(d). The behavior of the error CDFs in Figure 3.9(d) is qualitatively different than others with very long tails. The effect of the amplification attack on landmark 1 and the attenuation attack on landmark 3 pushed the localization results further in one direction, and thus introduced large localization bias.

The four attack scenarios we studied have covered a broad collection of possible combinations of signal strength attacks. We found that simultaneously attacking all landmarks has more impact on localization performance than attacking an individual landmark. Further, simultaneous amplification and attenuation attacks on certain landmarks can cause qualitatively larger errors than other kinds of attacks. Most importantly, we observed that none of the localization algorithms outperforms the others for the attacks we examined.

3.5.4 Linear Response

In this section, we show that the average distance error, $E[\|\mathbf{p}_0 - \tilde{\mathbf{p}}\|]$, of all the algorithms scales in a linear way to attacks. That is, the mean localization error changes linearly with respect to the size of the signal strength change introduced in dB (recall dB is a log-scaled change in power).

Figure 3.10 plots the median error vs. RSS attenuation for an all-landmark attack in Figure 3.10(a) and 3.10(e), and for individual landmarks in the other figures. Figure 3.11 plots the median localization error under simultaneous signal strength attenuation and amplification attacks on multiple attacks. Points are data derived from experimental results, and the lines are linear least-squares fits. For BN, in this section we only present the

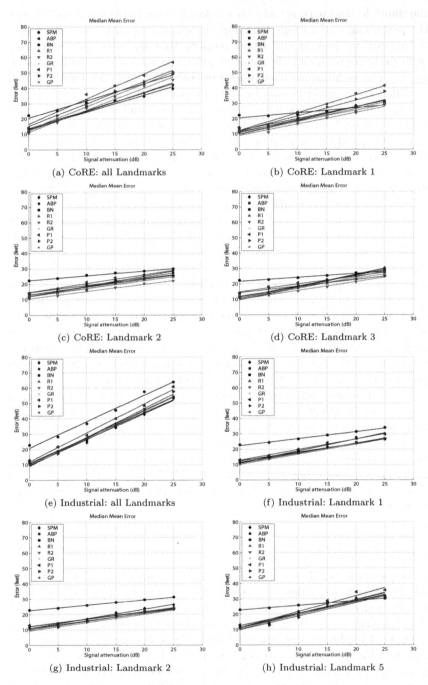

FIGURE 3.10. Average location estimation error across localization algorithms under signal strength attenuation attack.

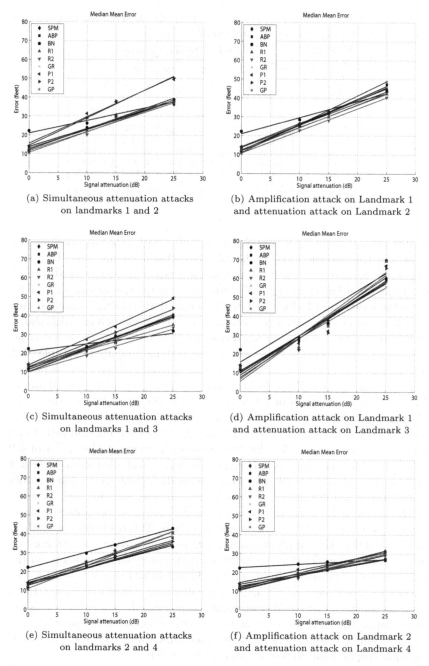

FIGURE 3.11. CoRE: Average location estimation error across localization algorithms under simultaneous signal strength attenuation and amplification attacks on multiple landmarks.

TABLE 3.2. CoRE: Slopes of Average Error from Linear Regression for attenuation attacks on all landmarks and individual landmark

Buildings	CoRE: attenuation attack				
Landmarks	All	1	2	3	4
Area-Based					
SPM	1.1048	0.8331	0.662	0.7816	0.6244
ABP-75	1.1656	0.7783	0.5049	0.7052	0.384
BN	1.1157	0.3287	0.3065	0.2544	0.493
Point-Based					
R1	1.4922	0.7006	0.5151	0.5702	0.7941
R2	1.4327	0.7534	0.4687	0.5732	0.7425
GR	1.1896	0.8440	0.5033	0.7357	0.7124
P1	1.6306	1.1597	0.5728	0.5026	0.3644
P2	1.4505	1.0123	0.464	0.4251	0.3063
GP	1.2359	0.8915	0.6028	0.8103	0.4595
Average	1.3131	0.8113	0.5111	0.5954	0.5423

results from Bayesian Network M_1 to compare with other algorithms. But note that the performance for the M_2 algorithm is comparable. The most important feature is that, in all cases, the median responses of all the algorithms fits a line extremely well, with an average R^2-statistic of 0.97 for both the CoRE and Industrial Lab. The mixed attacks with amplification attack on landmark 1 and attenuation attack on landmark 3 in CoRE shown in Figure 3.11(d) is an exceptional case with R^2 of 0.86 as the worst case.

Comparing the slopes across all the algorithms presented in Tables 3.2, 3.3, and 3.4, we found a mean change in positioning error vs. signal attenuation of 1.55 ft/dB under an all-landmark attack with a minimum of 1.3 ft/dB and maximum of 1.8 ft/dB. For the single landmark attack, the slope was substantially less, 0.64 ft/dB, although BN degrades consistently less than the other algorithms at 0.44 ft/dB. Under attenuation attacks on multiple landmarks, the localization algorithms move at the speed of 0.9 ft/dB to 1.4 ft/dB, which is between the results of a single landmark attack and an all-landmark attack. However, the median error moves faster under simultaneous amplification and attenuation attacks on landmark 1 and 3, at the speed of 1.8 - 2.2 ft/dB as shown in Table 3.4. We note the mean error tops out when the attack strength is 25dB. This confirms our analysis in Figure 3.9(d) that applying simultaneous amplification and attenuation attacks on landmarks that are farther apart causes larger impacts on the performance of localization schemes, although in practice it is hard for an adversary to conduct simultaneous amplification and attenuation attacks without using sophisticated equipment. In general, the linear fit results are quite important as it means that no algorithm has a cliff where the average positioning error suffers a catastrophic failure under attack. Instead, it remains proportional to the severity of the attack.

While the median error characterizes the overall response to attacks, it does not address whether an attacker can cause a few, large errors. We examined the response of the maximum error as a function of the strength of the attack on an all-landmark attack, i.e. how the 100th percentile error

TABLE 3.3. Industrial: Slopes of Average Error from Linear Regression for attenuation attacks on all landmarks and individual landmark

Buildings	Industrial Lab: attenuation attack					
Landmarks	All	1	2	3	4	5
Area-Based						
SPM	1.6901	0.7753	0.6283	0.5485	0.6455	0.9103
ABP-75	1.6479	0.5615	0.4852	0.4146	0.5469	0.8072
BN	1.7249	0.4528	0.3487	0.5215	0.5615	0.3094
Point-Based						
R1	1.8823	0.6827	0.4837	0.4286	0.5867	1.0356
R2	1.8816	0.6524	0.5394	0.4000	0.5861	0.8800
GR	1.7860	0.6514	0.5410	0.4668	0.6331	0.9358
P1	1.8854	0.6856	0.4710	0.4532	0.5881	1.0390
P2	1.8802	0.6448	0.5431	0.4023	0.5875	0.8861
GP	1.7666	0.6148	0.4976	0.4800	0.6213	0.8553
Average	1.7939	0.6357	0.504	0.4573	0.5952	0.8510

TABLE 3.4. CoRE: Slopes of Average Error from Linear Regression for mixed attacks of signal attenuation and amplification on multiple landmarks

Buildings	attenuation attacks			amplification and attenuation attacks		
Landmarks	1 and 2	1 and 3	2 and 4	1 and 2	1 and 3	2 and 4
Area-Based						
SPM	1.0054	1.1328	0.8836	1.3358	1.9556	0.8018
ABP-75	0.9740	1.1050	0.8125	1.3670	1.8628	0.5778
BN	0.6716	0.3965	0.8401	0.8665	1.8868	0.1812
Point-Based						
R1	1.0392	0.9069	1.1326	1.1895	2.2731	0.7522
R2	1.1013	0.9222	1.2148	1.1841	2.2552	0.7633
GR	1.0276	1.1559	0.9196	1.2337	1.8046	0.7642
P1	1.4142	1.4104	0.9683	1.2414	2.0808	0.6492
P2	1.4735	1.2330	0.9054	1.1921	2.0606	0.5472
GP	1.1003	1.2246	0.9271	1.5197	1.9138	0.7387
Average	1.0897	1.0541	0.9560	1.2367	2.0104	0.6417

scales as a function of the change in dB under an all-landmark attack. The all-landmark attack corresponds to a common attack scenario. It is thus desirable to study the worst-case situation under an all-landmark attack. We note that this characterization is not the same as, nor is directly related to, the Hölder metrics. Those metrics define the rates of change between physical and signal space within the localization function itself, while here we characterize the change in the estimator error to the change in signal, i.e. $\|\mathbf{p}_0 - \tilde{\mathbf{p}}\|/\|\mathbf{s} - \mathbf{v}\|$.

Figure 3.12 plots the worst-case error for each algorithm as a function of signal dB for the CoRE building under an all-landmark attack. The figure shows that almost all the responses are again linear, with least-squares fits of R^2 values of 0.84 or higher, though SPM does not have a linear response. The second important point is the algorithms' responses vary, falling into three groups. BN, R1 and R2 are quite poor, with the worst case error scaling at about 4 ft/dB. P1 and P2, are in a second class, scaling at close to 3 ft/dB. The gridded algorithms, GP and GR, as well as ABP-75 fair

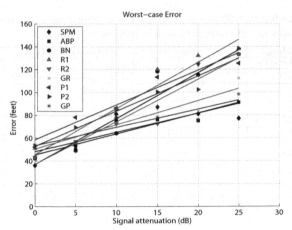

FIGURE 3.12. CoRE: Maximum error as a function of attack strength from an all-landmark attack.

better, scaling at 2 ft/dB or less. Finally, SPM is in a class by itself, with a poor linear fit (R^2 of 0.61) and the maximum error topping out at about 85 ft after 15 dB of attack.

Examining the error CDFs and the maximum errors, we can see that most of the localizations move fairly slowly in response to an attack, at about 1.5 ft/dB. However, for some of the algorithms, particularly BN, R1 and R2, the top part of the error CDF moves faster, at about 4 ft/dB. What this means is that, for a few selected points, an attacker can cause more substantial errors of over 100 ft. However, at most places in the building, an attack can only cause errors with much less magnitude.

Figure 3.10 shows that BN is more robust compared to other algorithms for individual landmark attacks. Recall BN uses a Monte-Carlo sampling technique (Gibbs sampling) to compute the full joint-probability distribution for not just the position coordinates, but also for every node in the Bayesian network. Under a single landmark attack we found the network reduces the contribution of network nodes directly affected by the attacked landmark to the full joint-probability distribution while increasing other landmarks' contributions. In effect, the network "discounts" the attacked landmark's contribution to the overall joint-density because the attacked data from that landmark is highly unlikely given the training data.

To show this effect we developed our own Gibbs sampler so that we could observe the relative contributions of each node in the Bayesian network to the final answer. Figure 3.13 shows the percentage contribution for each landmark to the overall joint-density. For instance, in CoRE, the contribution of each landmark starts almost uniformly. When Landmark 1 is under attack, the contribution of Landmark 1 goes from 0.25 down to 0.15.

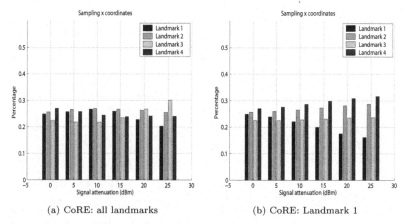

(a) CoRE: all landmarks   (b) CoRE: Landmark 1

FIGURE 3.13. Contribution of each Landmark during sampling in the BN algorithm under attenuation attacks.

3.5.5   Precision Study

In this section, we examine the area-based algorithms' precision in response to attacks. Figure 3.14 shows a localization example of the area-based algorithms in the CoRE building. The actual point is shown as a big dot and the convex hulls of the returned areas are outlined. Normally, the SPM and ABP algorithms perform similarly, while the BN algorithm has a much different profile by returning the sampling distribution of the possible estimation. Under signal strength attacks, we observed that the returned areas are reduced and shifted from the true location.

Figure 3.15 shows the CDF of the precision (i.e. size of the returned area) for different area-based algorithms under attack for all the landmarks in CoRE and Industrial Lab. We found that overall the algorithms did not become less precise in response to attacks, but rather, the algorithms tended to shift and shrink the returned areas. Figure 3.15(a) shows a small average shrinkage for SPM in the CoRE building, and likewise, 3.15(b) shows a similar effect for BN.

ABP-75 had the most dramatic effect. Figures 3.15(c) and 3.15(d) show the precision versus the attack strength for both buildings. The shrinkages are quite substantial. We found that, under attack, the probability densities of the tiles shrank to small values that were located on a few tiles– reflecting the fact that an attack causes there not to be a likely position to localize a node. We also found that this effect held for amplification attacks, as is shown in Figure 3.15(d). The shrinking precision behavior may be useful for attack detection, although a full characterization of how this effect occurs remains for future work.

Examining this effect further, Figure 3.16 presents the precision vs. the perturbation distance $\|\mathbf{p}_{med} - \tilde{\mathbf{p}}_{med}\|$, with a least squares line fit. Fig-

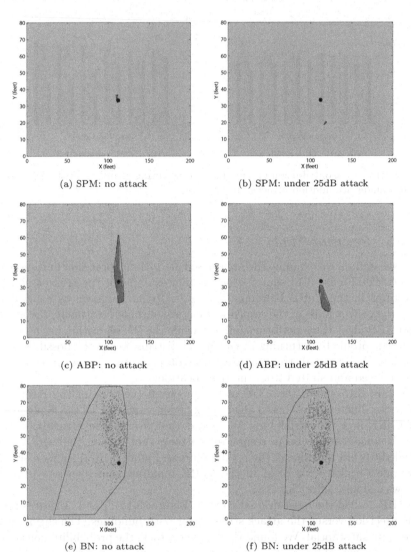

(a) SPM: no attack

(b) SPM: under 25dB attack

(c) ABP: no attack

(d) ABP: under 25dB attack

(e) BN: no attack

(f) BN: under 25dB attack

FIGURE 3.14. CoRE: Comparison of localization results from the area-based algorithms for a testing point.

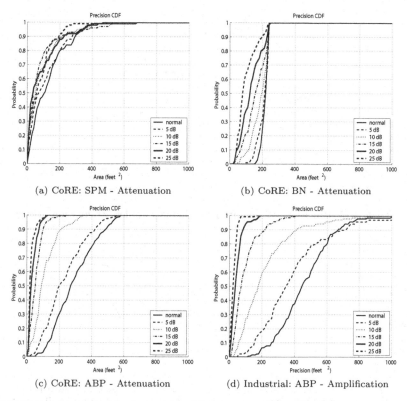

FIGURE 3.15. Analysis of precision CDF across area-based algorithms. The attack is performed on all the landmarks.

ure 3.16(a) shows the effect when attacking all landmarks on the CoRE building. Figure 3.16(b) shows a downward trend, but much weaker, when one landmark is under attack. We observed similar results for the Industrial Lab. We see mostly linear changes in precision in response to attacks, although with great differences between the algorithms. The figures show that the decrease in precision as a function of dB is particularly strong for ABP-75.

3.5.6   Robust Multi-device Localization

As presented in Section 3.2, the Bayesian Network M_2 algorithm can simultaneously localize multiple devices with no training data. Figure 3.17 presents the error CDFs of M_2 when simultaneously positioning 171 devices under a normal operational situation and with a 25dB attack applied to all landmarks (i.e., for all signals coming into each landmark) respectively. As shown in Figure 3.17 (a) the performance when simultaneously

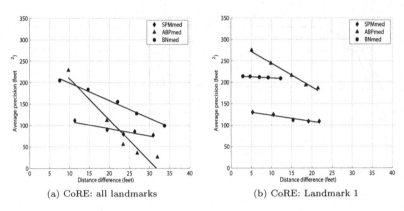

(a) CoRE: all landmarks

(b) CoRE: Landmark 1

FIGURE 3.16. Precision vs. perturbation distance under attenuation attack.

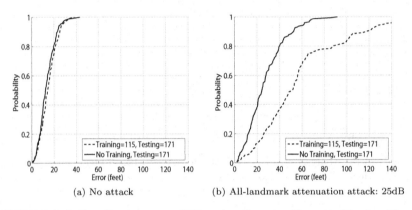

(a) No attack

(b) All-landmark attenuation attack: 25dB

FIGURE 3.17. Localization error CDFs using Bayesian Network M_2 algorithm.

localizing multiple devices (No Training, Testing=171) is very similar to that of positioning only one device at a time (Training=115, Testing=171). However, we found that under an all-landmark attack with 25dB severity, the CDF curve of localizing a single device with training data shifts to the right largely, with the same trend as presented in Figure 3.6, but the curve of localizing multiple devices without training data moves at a much slower speed. This indicates that BN M_2 is more robust than all the other algorithms under attacks, especially under realistic operation conditions involving multiple, simultaneous localization tasks.

Usually Bayesian Networks utilize the training data to predict their model parameters. When no training data is used for localizing multiple transmitting devices, M_2 relies on all the RSS readings from multiple devices to adjust the parameters specified in the model. Under the case of all-landmark attacks, The RSS readings of multiple devices are all corrupted

TABLE 3.5. Analysis of (worst-case) H and (average-case) $\overline{H}$

Algorithms	CoRE: H	LAB: H	CoRE: $\overline{H}$	LAB: $\overline{H}$
Area-Based				
SPM	23.7646	11.0659	1.8856	2.3548
ABP-75	20.0347	23.0652	1.8548	2.3424
BN	31.7324	14.9168	2.0595	2.5873
Point-Based				
R1	36.2400	20.7846	1.9750	2.3677
R2	19.8586	8.7313	1.9138	2.3058
GR	35.9880	20.6886	1.9691	2.3628
P1	20.8832	20.7846	1.9793	2.3683
P2	19.8586	8.7313	1.9178	2.3058
GP	21.8303	20.6886	1.9649	2.2882

and shifted by a constant. Thus the model parameters predicted by M_2 should be qualitatively similar to those predicted under the normal situation without attack. Therefore, the BN M_2 algorithm is attack-resistant if there is a massive attack that all signals coming into each landmark are being attacked by adversaries.

3.6 Discussion about Hölder Metrics

In the previous section we examined the experimental results, and looked at the performance of a set of representative localization algorithms in terms of error and precision. We now focus on the performance of these localization algorithms in terms of the Hölder metrics. The Hölder metrics measure the variability of the *returned* answer in response to changes in the signal strength vectors.

We first discuss the practical aspects of measuring H and $\overline{H}$ for different algorithms. In Section 3.4, the Hölder parameters are defined by calculating the maximum and average over the entire n-dimensional signal strength space. In practice, it is necessary to perform a sampling technique to measure H and $\overline{H}$. Additionally, as noted earlier, the definition of H and $\overline{H}$ are only suitable for (Hölder) continuous functions G_{alg}. In reality, several localization algorithms, such as RADAR, are not continuous and involve the tessellation of the signal strength space into Voronoi cells V_j, and thus only a discrete set of localization results are produced (image of V_j under G_{alg}). Hence, for any $\mathbf{s} \in V_j$ we have $G_R(\mathbf{s}) = (x_j, y_j)$. Unfortunately, for neighboring Voronoi cells, we may take $\mathbf{s} \in V_j$ and $\mathbf{v} \in V_i$ such that they are arbitrarily close (i.e. $\|\mathbf{s} - \mathbf{v}\| \to 0$), while $\|G_R(\mathbf{s}) - G_R(\mathbf{v})\| \neq 0$. In such a case, the formal calculation of H and $\overline{H}$ is not possible. However, for our purposes, we are only interested in measuring the notion of adjacency of Voronoi cells in signal space yielding *close* localization results. Thus, our calculation of H and $\overline{H}$ is only performed over the centroids of the various

Voronoi cells for localization algorithms that tessellate of signal strength space.

The Hölder parameters for the different localization algorithms are presented in Table 3.5. Examining these results, there are several important observations that can be made. First, if we examine the results for $\overline{H}$ we see that, for each building, all of the algorithms have very similar $\overline{H}$ values. Hence, we may conclude that the average variability of the returned localization result to a change in the signal strength vector is roughly the same for all algorithms. This is an important result as it means, regardless of which RF fingerprinting localization system we deploy, the average susceptibility of the returned results to an attack is essentially identical.

However, if we examine the results for H, which reflects the worst-case susceptibility, then we see that there are some differences across the algorithms. First, comparing H and $\overline{H}$ for both point-based and area-based algorithms, we see that the worst-case variability can be much larger than the average variability. Additionally, the point-based methods appear to cluster. Notably, RADAR (R1) and Gridded Radar (GR) have similar performance across both CoRE and the Industrial Lab, while averaged RADAR (R2) and averaged Highest Probability (P2) have similar performance across both buildings. A very interesting phenomena is observed by looking at the algorithms that returned an average of likely locations (R2 and P2). Across both buildings these algorithms exhibited less variability compared to other algorithms. This is to be expected as averaging is a smoothing operation, which reduces variations in a function. This observation suggests that R2 and P2 are more robust from a worst-case point-of-view than other point-based algorithms.

3.7   Conclusion

In this paper, we provided a performance analysis of the robustness of RF-based localization algorithms to attacks that target signal strength measurements. We first examined the feasibility of conducting signal amplification and attenuation attacks, and observed a linear dependency between non-attacked signal strength and attacked signal strength readings for different barriers placed between the transmitter and a landmark receiver. We then provided a set of performance metrics for quantifying the effectiveness of an attenuation/amplification attack and their impacts on the localization. Our metrics included localization angular bias, localization error, the precision of area-based algorithms, and a new family of metrics, called Hölder metrics, that quantify the variability of the returned location results versus change in signal strength vectors.

We conducted a trace-driven evaluation of a representative set of point-based and area-based localization algorithms where the linear attack model was applied to data measured in two different office buildings. We investigated the impact of both signal attenuation as well as signal amplification attacks on a sensor node or landmark by applying signal perturbations to individual landmarks, multiple landmarks, and all landmarks. We found that the localization error scaled similarly for all algorithms under attack, except for the Bayesian Networks algorithm. Large localization errors are introduced under severe attacks, resulting in 20-30 feet location perturbation under an attack strength of 15dB. Further, we found that, when attacked, area-based algorithms did not experience a degradation in precision although they experienced degradation in accuracy and more uncertainty in location estimation. One important observation is that Bayesian Networks are more robust under both an individual landmark attack when positioning a single device, as well as an all-landmark attack when localizing multiple devices simultaneously.

We then examined the variability of the localization results under attack by measuring the Hölder metrics. We found that most algorithms had similar average variability, but those methods that returned the average of a set of most likely positions exhibited less variability. This result suggests that the average susceptibility of the returned results to an attack is essentially identical across point-based and area-based algorithms, though it might be desirable to employ either area-based methods or point-based methods that perform averaging in order to lessen the worst-case effect of a potential attack. Additionally this investigation indicates that the performance of most of the RSS-based localization algorithms degrades significantly under signal strength attacks, and consequently that network designers need to resort to more complicated secure localization algorithms for dealing with potential attacks in an uncontrolled environment.

4

Attack Detection in Wireless Localization

4.1 Introduction

In this chapter, we examine the problem of detecting attacks on wireless localization. We present a general formulation for attack detection using statistical significance testing and then build tests that are applicable to broad classes of multilateration and signal strength-based methods, as well as several other test statistics that are unique to a variety of different localization algorithms.

Multilateration is a popular localization approach that uses Least Squares (LS) techniques to perform localization [6–8, 10], and has the desirable property of supporting mathematical analysis, in part because LS-based regression has well-known statistical descriptions when operating near ideal conditions. By examining Linear Least Squares (LLS) , we build a mathematical model and derive an analytic solution for attack detection using the residuals of an LLS regression. We show that attack detection using LLS is easy to conduct and is suitable for both single-hop and multi-hop ranging methods because it is independent of the ranging modality used by the localization system.

On the other hand, many signal strength based algorithms [15, 16] rely on either statistical inference or machine-learning in the context of scene matching to perform localization, and consequently do not yield closed-form solutions. However, for algorithms based on signal strength, we found that the minimum distance between an observation and the database of signal strength vectors is a good test statistic to perform attack detection. One key advantage of our approach for signal strength based methods is that the detection phase can be performed before localization.

To evaluate the effectiveness of our attack detection mechanisms we first present experimental results illustrating the feasibility of physical attacks on localization. We then conducted a trace driven evaluation using both an 802.11 (WiFi) network as well as an 802.15.4 (ZigBee) network in two real office buildings. In particular, we applied signal strength attenuation and amplification, using a linear attack model obtained from our experiments, to the Received Signal Strength (RSS) readings collected from these two office buildings. We evaluated the performance of our attack detection schemes using detection rates and receiver operating characteristic curves. Our experimental results provide strong evidence of the effectiveness of our attack detection schemes with high detection rates, over 95%, and low false positive rates, often under 5%. Surprisingly, we found that most of the attack detection schemes provide qualitatively similar performance. This shows that the different localization systems have similar attack detection capabilities.

The rest of the chapter is organized as follows. We study the feasibility of attacks and present our experimental methodologies in Section 4.2. We present our generalized theoretical formulation for the attack detection problem in Section 4.3. We next derive an analytic solution for attack detection using Least Squares in Section 4.4. Using common features for attack detection in signal strength based algorithms is presented in Section 4.5. We study the test statistics that are specific to a variety of different algorithms in Section 4.6. Then, we provide a discussion in Section 4.7. Finally, we conclude in Section 4.8.

4.2   Feasibility of Attacks

In this section we provide background on how attackers can impact the localization system. We next discuss the feasibility of conducting these attacks on signal strength, and provide the experimental methodology that we use to evaluate our attack detection mechanisms later in this chapter.

4.2.1   Localization Attacks

Localization mechanisms are built upon different ranging modalities, such as RSS, TOA, AOA, and hop count. These all rely on the measurement of the physical properties of the wireless system. Adversaries can apply non-cryptographic attacks against the measurement processes, bypassing conventional security services, and as a result can affect the localization performance. For example, wormhole attacks tunnel through a faster channel to shorten the observed distance between two nodes [74]. An attenuation attack would decrease the radio range, and thus potentially lengthen the

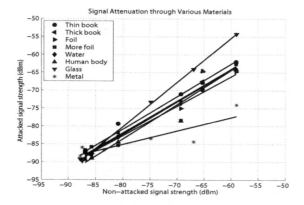

FIGURE 4.1. Linear attack model on received signal strength for various media.

hop-count. Compromised nodes may delay response messages to disrupt distance estimation. RSS readings can be altered due to attenuation or amplification of the signal strength by an adversary as presented in Chapter 3. For a survey of the potential non-cryptographic attacks that are unique to localization, please refer to Chapter 2.

4.2.2 Signal Strength Attacks

We choose to use RSS as the ranging modality for localization algorithms. An adversary can attack the wireless node directly or compromise the landmarks involved in localization by attenuating or amplifying the signal strength readings. Based on our experimental attacks using real materials, we will use the linear attack model presented in Chapter 3 (i.e. a material causes a constant percentage power loss independent of distance) as shown in Figure 4.1 to describe the effect of an attack on the RSS readings at the wireless device or at the landmarks. As presented in the figure, these attacks are easy to conduct with low cost materials. The linear relationship implies that it is easy for an adversary to control the effect of an attack on the observed signal strength by appropriately selecting different materials.

4.2.3 Experimental Methodology

In order to study the generality of our attack detection approaches, we have conducted experiments in two office buildings, one is the 3rd floor of the Computer Science building at Rutgers University (CoRE) as shown in Figure 4.2 (a) and the other is in a floor of an industrial research lab (Industrial Lab) as presented in Figure 4.2 (b). In Figure 4.2 (a), the experiments are performed for both an 802.11 (WiFi) network as well as an 802.15.4 (Zig-Bee) network. For the 802.11 (WiFi) network, there are 4 landmarks shown in red squares deployed in a collinear manner to maximize signal strength

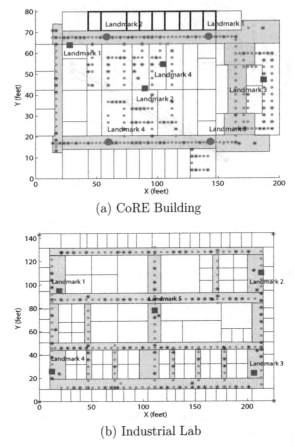

(a) CoRE Building

(b) Industrial Lab

FIGURE 4.2. Layout of the experimental floor

coverage. While for the 802.15.4 (ZigBee) network, there are 4 landmarks shown in magenta circles placed in a square set to maximize localization accuracy [73]. For experiments conducted in the industrial lab, as depicted in Figure 4.2 (b), we only used an 802.11 (WiFi) network with 5 landmarks. The small green dots are the localization testing points and the small blue stars are the training points. We will present the results of our experiments for each of the proposed attack detection cases in its associated section in this chapter. Across all experiments, we have performed a trace-driven evaluation by either attenuating or amplifying RSS readings collected from these two buildings.

4.3   Generalized Attack Detection Model

In this section we first propose a general formulation for the localization attack detection problem. We then introduce metrics for evaluating the effectiveness of our approaches.

4.3.1   Localization Attack Detection

In general, the error of a localization algorithm is defined as the distance between the true location $\mathbf{x} = [x, y]^T$ and the estimated location $\hat{\mathbf{x}}$, $D_{err} = \|\mathbf{x} - \hat{\mathbf{x}}\|$. We found in Chapter 3 that under physical attacks, the localization error D_{err} increases significantly. However, D_{err} is not directly available during run-time, and the challenge in attack detection is to devise strategies for detecting localization attacks that do not use localization errors.

We propose to formulate location attack detection as a statistical significance testing problem, where the null hypothesis is

$$\mathcal{H}_0 : \text{normal (no attack)}.$$

In significance testing, a test statistic $\mathbf{T}$ is used to evaluate whether observed data belongs to the null-hypothesis or not. For a particular significance level, α (defined as the probability of rejecting the hypothesis if it is true), there is a corresponding *acceptance region* Ω such that we declare the null hypothesis valid if an observed value of the test statistic $\mathbf{T}^{\mathbf{obs}} \in \Omega$, and reject the null hypothesis if $\mathbf{T}^{\mathbf{obs}} \notin \Omega$ (i.e. declare an attack is present if $\mathbf{T}^{\mathbf{obs}} \in \Omega^c$, where Ω^c is the *critical region* of the test). In our attack detection problem, the region Ω and decision rule is specified according to the form of the detection statistic $\mathbf{T}$ (for example, when using distance in signal strength space for $\mathbf{T}$, the decision rule becomes comparison against a threshold), and rejection of the null hypothesis corresponds to declaring the presence of an attack.

4.3.2   Effectiveness

In order to evaluate the effectiveness of our attack detection methods, we will utilize the following performance metrics:

Cumulative Distribution Function (CDF): The CDF of the test statistic $\mathbf{T}$ provides the sensitivity of $\mathbf{T}$ under attack. Based on the CDF, we can study the feasibility of using $\mathbf{T}$ for attack detection.

Detection Rate (DR): An attack may cause the significance test to reject $\mathcal{H}_0$. We are thus interested in the statistical characterization of the attack detection attempts over all the localization attempts. The Detection Rate is defined as the percentage of localization attempts that are determined to be under attack, i.e.:

$$DR = \frac{N_{attack}}{N_{total}} \tag{4.1}$$

where N_{total} is the total number of localization attempts and N_{attack} is the number concluded under attack by detection. Note that when the signal is attacked, the detection rate corresponds to the probability of detection P_d, while under normal (non-attack) conditions it corresponds to the probability of declaring a false positive P_{fa} . We will examine DR as a function of the attack strength.

Receiving Operating Characteristic (ROC) curve: To evaluate an attack detection scheme we want to study the false positive rate P_{fa} and probability of detection P_d together. The ROC curve is usually used to measure the tradeoff between false-positives and correct detections. The ROC curve is a plot of attack detection accuracy against the false positive rate. It can be obtained by varying the detection thresholds.

4.4   Using Least Squares

In this section we provide mathematical analysis for attack detection in multilateration algorithms. We first provide background in using LS to perform localization. Next, based on the properties of the LLS estimator, we define an attack detection scheme that utilizes regression residuals, and give an analytic formulation to specify the acceptance region Ω. Finally, the experimental results are presented to evaluate the effectiveness of the detection scheme.

4.4.1   Localization

To perform localization with LS requires 2 steps: ranging and lateration.

Ranging Step: Recent research has seen a host of variants on the ranging step such as RSS, TOA, TDOA, and hop count. Our attack detection approach works with any ranging modality.

Lateration Step: From the estimated distances d_i and known positions (x_i, y_i) of the landmarks, the position (x, y) of the localizing node can be found by finding $(\hat{x}, \hat{y})$ satisfying:

$$(\hat{x}, \hat{y}) = arg \min_{x,y} \sum_{i=1}^{n} [\sqrt{(x_i - x)^2 + (y_i - y)^2} - d_i]^2 \qquad (4.2)$$

where n is the total number of landmarks. We call solving the above problem *Nonlinear Least Squares*, or NLS. Solving the NLS problem requires significant complexity and is difficult to analyze. We may approximate the NLS solution and linearize the problem [73] into the system $\mathbf{Ax} = \mathbf{b}$, where:

$$\mathbf{A} = \begin{pmatrix} x_1 - \frac{1}{n}\sum_{i=1}^{n} x_i & y_1 - \frac{1}{n}\sum_{i=1}^{n} y_i \\ \vdots & \vdots \\ x_n - \frac{1}{n}\sum_{i=1}^{n} x_i & y_n - \frac{1}{n}\sum_{i=1}^{n} y_i \end{pmatrix} \qquad (4.3)$$

and

$$b = \frac{1}{2} \begin{pmatrix} (x_1^2 - \frac{1}{n}\sum_{i=1}^{n} x_i^2) + (y_1^2 - \frac{1}{n}\sum_{i=1}^{n} y_i^2) \\ -(d_1^2 - \frac{1}{n}\sum_{i=1}^{n} d_i^2) \\ \vdots \\ (x_n^2 - \frac{1}{n}\sum_{i=1}^{n} x_i^2) + (y_n^2 - \frac{1}{n}\sum_{i=1}^{n} y_i^2) \\ -(d_n^2 - \frac{1}{n}\sum_{i=1}^{n} d_i^2) \end{pmatrix}. \tag{4.4}$$

Note that $\mathbf{A}$ is described by the coordinates of landmarks only, while $\mathbf{b}$ is represented by the distances to the landmarks together with the coordinates of landmarks. We call the above formulation of the problem *Linear Least Squares*, or LLS. The estimate of $\mathbf{x} = [x, y]^T$ is done via

$$\mathbf{x} = (\mathbf{A}^T\mathbf{A})^{-1}\mathbf{A}^T\mathbf{b} \tag{4.5}$$

In addition to its computational advantages, the LLS formulation allows for tractable statistical analysis, as we shall now see.

4.4.2 The Residuals

In practice, there are estimation errors from the ranging step. The LLS formulation can be refined as a linear regression , $\mathbf{b} = \mathbf{Ax} + \mathbf{e}$, where $\mathbf{e}$ corresponds to model errors. The localization result is then $\hat{\mathbf{x}} = (\mathbf{A}^T\mathbf{A})^{-1}\mathbf{A}^T\mathbf{b}$, and the fitted values $\hat{\mathbf{b}}$ corresponding to the observed values $\mathbf{b}$ are given by

$$\hat{\mathbf{b}} = \mathbf{A}\hat{\mathbf{x}} = \mathbf{A}[(\mathbf{A}^T\mathbf{A})^{-1}\mathbf{A}^T\mathbf{b}] = \mathbf{A}(\mathbf{A}^T\mathbf{A})^{-1}\mathbf{A}^T\mathbf{b}. \tag{4.6}$$

Further, we define the vector of residuals $\hat{\mathbf{e}}$ as

$$\hat{\mathbf{e}} = \mathbf{b} - \hat{\mathbf{b}} = [1 - \mathbf{A}(\mathbf{A}^T\mathbf{A})^{-1}\mathbf{A}^T]\mathbf{b}. \tag{4.7}$$

When the regression model is performing well we may assume that the model errors are Gaussian [75, 76]. Under this assumption, the residuals also follow a Gaussian distribution, $\mathbf{N}(\mu, \Sigma)$, since the residuals are a linear combination of the elements of $\mathbf{b}$ and $\mathbf{e}$. Here, μ is the mean vector and Σ is the covariance matrix. We choose the residuals $\hat{\mathbf{e}}$ as the test statistic $\mathbf{T}$, and will build our attack detection scheme by using the statistical properties of $\hat{\mathbf{e}}$ when LLS is operating in a desirable performance regime.

4.4.3 The Detection Scheme

The LLS attack detection is performed after localization. The residuals are correlated Gaussian random variables and the multivariate Gaussian distribution of $\hat{\mathbf{e}}$ can be expressed as:

$$f(\hat{\mathbf{e}}) = \frac{1}{(\sqrt{2\pi})^n |\Sigma|^{\frac{1}{2}}} e^{-\frac{1}{2}(\hat{\mathbf{e}} - \mu)^T \Sigma^{-1}(\hat{\mathbf{e}} - \mu)}. \tag{4.8}$$

In order to determine whether the location result is compromised by adversaries, we perform attack detection through significance testing. We can define an acceptance region in $\hat{\mathbf{e}}$ space by

$$\Omega = \{\hat{\mathbf{e}} : Pr(\{\mathbf{T} : (\mathbf{T} - \mu)^T \mathbf{\Sigma}^{-1}(\mathbf{T} - \mu) >$$

$$(\hat{\mathbf{e}} - \mu)^T \mathbf{\Sigma}^{-1}(\hat{\mathbf{e}} - \mu)\}) > \alpha\}.$$

In practice, after performing localization using LLS, we have an observed value of residuals $\hat{\mathbf{e}}^{\mathbf{obs}}$. Testing the null hypothesis, we can decide that the localization is under attack if the probability $P = 1 - M < \alpha$, where

$$M = \frac{1}{(\sqrt{2\pi})^n |\mathbf{\Sigma}|^{\frac{1}{2}}} \int \cdots \int_E e^{-\frac{1}{2}(\hat{\mathbf{e}}-\mu)^T \mathbf{\Sigma}^{-1}(\hat{\mathbf{e}}-\mu)} d\hat{e}_1...d\hat{e}_n \qquad (4.9)$$

and E is the integration region defined by $(\hat{\mathbf{e}} - \mu)^{\mathbf{T}} \mathbf{\Sigma}^{-1}(\hat{\mathbf{e}} - \mu) \leq X^2$ with

$$X^2 = (\hat{\mathbf{e}}^{\mathbf{obs}} - \mu)^{\mathbf{T}} \mathbf{\Sigma}^{-1}(\hat{\mathbf{e}}^{\mathbf{obs}} - \mu).$$

We can express the term

$$\begin{aligned}
(\hat{\mathbf{e}} - \mu)^{\mathbf{T}} \mathbf{\Sigma}^{-1}(\hat{\mathbf{e}} - \mu) &= (\hat{\mathbf{e}} - \mu)^{\mathbf{T}} \mathbf{D}^{\mathbf{T}} \mathbf{D}(\hat{\mathbf{e}} - \mu) \\
&= (\mathbf{D}(\hat{\mathbf{e}} - \mu))^{\mathbf{T}}(\mathbf{D}(\hat{\mathbf{e}} - \mu)) \\
&= \mathbf{y}^{\mathbf{T}}\mathbf{y}. \qquad (4.10)
\end{aligned}$$

Substituting $\mathbf{y} = \mathbf{D}(\hat{\mathbf{e}} - \mu)$ into Equation (4.9), we get

$$\begin{aligned}
M &= \frac{1}{(\sqrt{2\pi})^n} \int \cdots \int_{E'} e^{-\frac{1}{2}\mathbf{y}^{\mathbf{T}}\mathbf{y}} dy_1...dy_n \\
&= \frac{1}{(\sqrt{2\pi})^n} \int \cdots \int_{E'} e^{-\frac{1}{2}\sum_{i=1}^n y_i^2} dy_1...dy_n \qquad (4.11)
\end{aligned}$$

with E' defined by $\mathbf{y}^{\mathbf{T}}\mathbf{y} \leq X^2$. Changing to polar coordinates, we get

$$\begin{aligned}
M &= \frac{1}{(\sqrt{2\pi})^n} \int_0^X \int_0^{2\pi} \int_0^{\pi} \cdots \int_0^{\pi} [e^{-\frac{r^2}{2}} r^{n-1} dr d\phi_1 \\
&\quad sin\phi_2 d\phi_2...sin^{n-2}\phi_{n-1} d\phi_{n-1}] \\
&= \frac{1}{(\sqrt{2\pi})^n} \int_0^X e^{-\frac{r^2}{2}} r^{n-1} dr \times \int_0^{2\pi} d\phi_1 \\
&\quad \times \prod_{i=2}^{n-1} \int_0^{\pi} sin^{i-1}\phi_i d\phi_i \\
&= \frac{2}{(\sqrt{\pi})^{n-2}} \times A_{r,n} \times \prod_{i=2}^{n-1} B_i \qquad (4.12)
\end{aligned}$$

with

$$A_{r,n} = \frac{1}{(\sqrt{2})^n} \int_0^X e^{-\frac{r^2}{2}} r^{n-1} dr$$

and

$$B_i = \int_0^\pi sin^{i-1} \phi_i d\phi_i.$$

Using $v = r^2/2$, we have

$$A_{r,n} = \frac{1}{2} \int_0^{\frac{X^2}{2}} e^{-v} v^{\frac{n-2}{2}} dv = \frac{1}{2} \times \Gamma(\frac{n}{2}, \frac{X^2}{2}) \qquad (4.13)$$

where Γ is the incomplete gamma function. Since

$$B_i = \beta(\frac{i}{2}, \frac{1}{2}) = \frac{\Gamma(\frac{i}{2})}{\Gamma(\frac{i+1}{2})} \times \sqrt{\pi}. \qquad (4.14)$$

Through further simplification, we can get

$$\prod_{i=2}^{n-1} B_i = (\sqrt{\pi})^{n-2} \times \frac{1}{\Gamma(\frac{n}{2})}. \qquad (4.15)$$

Hence, substituting Equations (4.13) and (4.15) into (4.12), we obtain the probability mass

$$M = \frac{\Gamma(n/2, X^2/2)}{\Gamma(n/2)}.$$

We then further obtain the probability by $P = 1 - M$. Based on the definition in Section 4.3, if the probability is sufficiently low, i.e. $P < \alpha$, then $\hat{e}^{obs}$ belongs to the critical region Ω^c and we can conclude that the location result is under attack.

4.4.4 Experimental Evaluation

In this section we present the evaluation of the effectiveness of the attack detection scheme. We chose RSS as the ranging modality and performed signal strength attacks according to the experimental methodologies described in Section 4.2.

The average ranging error as a function of the severity of signal strength attacks is shown in Figure 4.3(a). We know that the relationship between the RSS error and the ranging error is multiplicative with distance [73]. Even small random perturbation in RSS readings can cause large ranging errors due to this multiplicative factor. We observed this effect in Figure 4.3(a); the ranging error increases superlinearly to attack severity. Figure 4.4 presents DR vs. the ranging errors when tested against significance level $\alpha = 0.01$ and $\alpha = 0.05$. We found that under a normal situation,

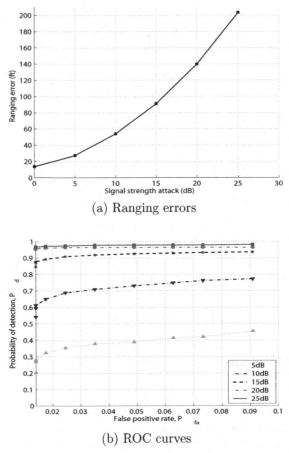

(a) Ranging errors

(b) ROC curves

FIGURE 4.3. CoRE 802.11: (a) Ranging errors under the signal strength attacks (b) LLS residuals: Receiver Operating Characteristic (ROC) curves

where the ranging errors are less than 15 feet, the false alarm probability, P_{fa}, is less than 1.5% and 2.5% for $\alpha = 0.01$ and $\alpha = 0.05$ respectively. Large signal strength attacks, greater than 15dB, can cause ranging errors larger than 90 feet, and then the detection rates are more than 90%. These results strongly indicate that using residuals in LS as a test statistic for attack detection is effective.

Further, the ROC curves in Figure 4.3(b) show that for false positive rates less than 10%, the detection rates are above 90% and close to 99% when the attack strength increases to 20dB and 25dB. This shows that if the adversary wants to cause a large localization error, it is almost certain that our attack detection mechanism will detect it. For small attacks of less than 5dB, the detection rates are about 40%. In this case, it is difficult to distinguish whether the anomaly in the test statistic is caused by attacks or by measurement errors since the RSS readings can fluctuate around 5dB

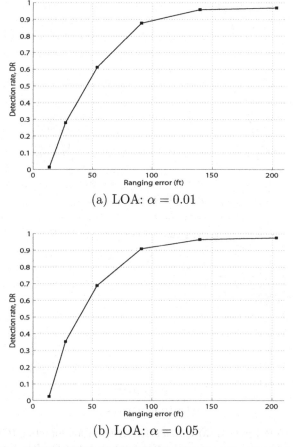

(a) LOA: $\alpha = 0.01$

(b) LOA: $\alpha = 0.05$

FIGURE 4.4. CoRE 802.11, LLS residuals: effectiveness of attack detection

due to environmental effects. However, for such small attacks, because the resulting impact on the final localization result was shown to be small in Chapter 3, the consequences of failing to detect such attacks would likely be small as well.

4.5   Distance In Signal Space

RSS is a common physical property used by a widely diverse set of algorithms. For example, most scene matching approaches utilize the RSS, e.g. [16, 28], and many multilateration approaches [14] use it as well. In spite of its several meter-level accuracy, using the RSS is an attractive approach because it can re-use the existing wireless infrastructure — this feature presents a tremendous cost savings over deploying localization-specific

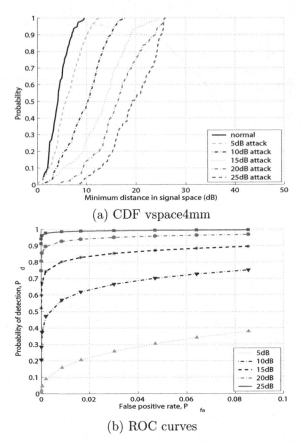

(a) CDF vspace4mm

(b) ROC curves

FIGURE 4.5. CoRE 802.11: (a) Cumulative Distribution Function (CDF) of minimum distance D_s in signal space. (b) Minimum distance D_s: Receiver Operating Characteristic (ROC) curves.

hardware. In this section, we thus derive an attack detection scheme applicable to any signal strength based localization system.

4.5.1 Overview

All of the above algorithms take a vector $\mathbf{s}$ of n RSS readings to (or from) n landmarks for the node to be localized. Note that as described in Chapter 3 $\mathbf{s}$ corresponds to a point in a n-dimensional signal space. Under normal conditions, the RSS vectors obtained from the physical positions in a floor form a surface S in the n-dimensional signal space; we can think of this surface as comprising 'valid' points in signal space. Due to measurement noise, multipath effects, and unknown biases, $\mathbf{s}$ will fluctuate around this idealized RSS surface.

A localization attacker would perturb the correct **s** to produce a corrupted n-dimensional RSS vector **s**′. In signal space, **s**′ will be moved away from the ideal surface constructed by the correct RSS vectors. The stronger the attack, the more likely the vector **s**′ will be distant from the RSS surface. We thus choose the minimum distance to the surface S, i.e. $D_s = \min\{\|\mathbf{s}' - \mathbf{s}^j\| : \text{where } \mathbf{s}^j \in S\}$, as the test statistic for signal strength based attack detection. The key advantage of this approach is that the attack detection is independent of the localization algorithms and can be performed before the localization process.

Although it is possible to devise a statistical model for D_s based on models for normal measurement errors, in this section we shall take a different approach and apply empirical methodologies from training data to determine thresholds for defining the critical region.

4.5.2 Finding Thresholds

Choosing an appropriate threshold τ will allow the detection scheme to be robust to false detections. In order to obtain the thresholds, we don't need to know the exact RSS surface in the signal space (in practice, it is hard to determine and exhibits discontinuities due to wall boundaries). Instead, we can obtain the thresholds through empirical training. During the offline phase, we can collect the RSS vectors for a set of known positions over the floor and construct a radio map. During the localization phase, we get an observed vector $\mathbf{s}^{\mathbf{obs}}$, and we can then determine whether the $\mathbf{s}^{\mathbf{obs}}$ is being attacked by calculating the D_s using the pre-constructed radio map.

We define that if

$$D_s > \tau, \tag{4.16}$$

the signal strength readings are under attack. We use the distribution of the training data to help decide on the thresholds. Figure 4.5 (a) shows the CDF of the D_s in signal space. We found that the curve of D_s shifted to the right under signal strength attacks, especially for larger attacks, thereby suggesting that we can use D_s as a test statistic for detecting attacks, and also that we can use the non-attacked CDF to obtain τ for a given α value.

4.5.3 Experimental Evaluation

We next present the evaluation of the effectiveness of using minimum distance D_s for attack detection. Figure 4.6 presents the Detection Rate under different threshold (TH) levels as a function of signal strength attacks for both the 802.11 and the 802.15.4 networks in CoRE and the 802.11 network in the Industrial Lab. Figure 4.5 (b) is the corresponding ROC curves under signal attenuation attacks for the 802.11 network in CoRE. We found that, in general, the effectiveness of the attack detection scheme is similar across the different networks and buildings. Interestingly, we found that the performance of the attack detection scheme under signal amplification attacks is uniformly better than those for signal attenuation attacks, although the shapes of the DR curves are similar. Because of the higher detection rates under amplification attacks, we do not present additional amplification results in the remainder of the chapter. All these results are highly encouraging because they show our methods are quite general and do not depend on a specific network or environment.

Further, we observed that the DR under the 802.15.4 network in CoRE outperformed the DR under the 802.11 networks in both CoRE and Industrial Lab for the signal attenuation attacks as well as the signal amplification attacks. For attack strengths of 15dB or larger, the DR in the 802.15.4 network is over 95% and equals 100% when attack severity reaches 20dB and larger. We believe that the better landmark placement for localization [73] of the 802.15.4 network can account for its higher detection rates, although further investigation of this effect is required.

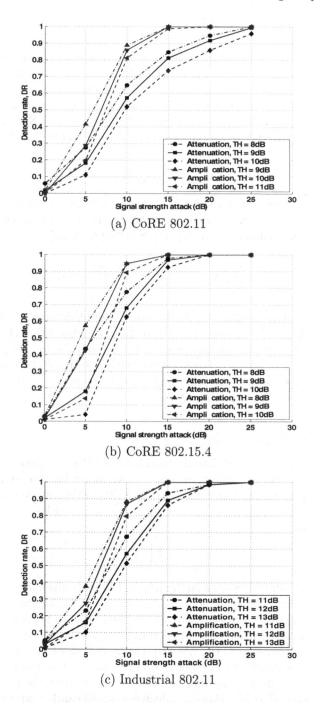

(a) CoRE 802.11

(b) CoRE 802.15.4

(c) Industrial 802.11

FIGURE 4.6. Minimum distance in signal space $\mathbf{D_s}$: attack detection across different networks and buildings.

4.6   Other Test Statistics

In this section, we examine algorithm-specific test statistics, which use properties specific to a particular localization algorithm. We have chosen a representative set of diverse algorithms. For the multilateration category, we investigate the NLS algorithm, while for signal strength based algorithms, we study both Area Based Probability (ABP) and Bayesian Networks (BN) algorithms. Detailed descriptions of these can be found in [14, 15, 73].

4.6.1   Nonlinear Least Squares (NLS)

As presented in Section 4.4, NLS is a multilateration algorithm that tries to satisfy the condition shown in Equation (4.2). The estimated $(\hat{x}, \hat{y})$ is the solution that minimizes the Sum of Squared Errors $\mathcal{E}^2$:

$$\mathcal{E}^2 = \sum_{i=1}^{n} [\sqrt{(x_i - \hat{x})^2 + (y_i - \hat{y})^2} - d_i]^2. \qquad (4.17)$$

We define a test statistic $\mathcal{E} = \sqrt{\mathcal{E}^2}$ because $\mathcal{E}$ will likely increase under the attack. The CDF of $\mathcal{E}$ presented in Figure 4.7 (a) confirms that the $\mathcal{E}$ grows rapidly with the attack severity. Figure 4.7 (b) and Figure 4.8 show that the performance of attack detection when using $\mathcal{E}$ for the 802.11 network in CoRE is comparable to that using residuals in Section 4.4. The thresholds are also obtained from training.

4.6.2   Area Based Probability (ABP)

Turning to signal strength based algorithms, ABP is an area-based algorithm that uses Bayes' Rule to return an area which has the highest likelihood of capturing the true location [15]. ABP divides the floor into a set of tiles. The total likelihood that the wireless node resides at each tile is calculated using:

$$P = \prod_{i=1}^{n} P_i \qquad (4.18)$$

where n is the total number of landmarks and P_i is the likelihood of observing the measured RSS reading at landmark i which is usually modeled as a Gaussian random variable. The total likelihood is calculated at each tile, and the returned location estimation is either a region whose likelihood is above a certain level, or is the tile with the maximum likelihood.

When under attack, the corrupted RSS readings reduce the set of likely positions on the floor to localize a node. We found that the highest tile-likelihood denoted as $likelihood_{max}$ decreases significantly under attack, as well as the sum of the likelihoods over all the tiles, $likelihood_{sum}$. We explored both $likelihood_{sum}$ and $likelihood_{max}$ as test statistics. The thresh-

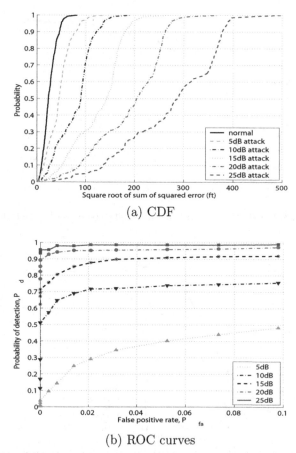

(a) CDF

(b) ROC curves

FIGURE 4.7. CoRE 802.11, NLS: (a) Cumulative Distribution Function (CDF) of $\mathcal{E}$. (b) $\mathcal{E}$: Receiver Operating Characteristic (ROC) curves.

olds are learned from the training data by taking the negative log of the values of the highest likelihood and the sum of the likelihoods.

The effectiveness of using $likelihood_{sum}$ and $likelihood_{max}$ for attack detection in ABP are presented in Figure 4.9 and Figure 4.10. We found that using $likelihood_{sum}$ under threshold equal to 2 had better performance than others in detecting larger attacks, but on the other hand resulted in slightly higher false positive rates around 7%.

4.6.3 Bayesian Networks (BN)

Another representative signal strength based algorithm, BN, utilizes Bayesian networks [14]. With Bayesian statistical inference, BN predicts the probability distribution of the unknown positions. BN uses a Monte-Carlo sampling technique (Gibbs sampling) to compute the full joint-probability distribu-

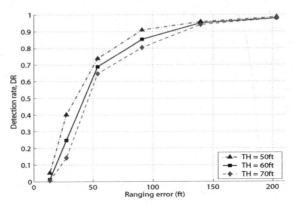

FIGURE 4.8. CoRE 802.11, NLS using $\mathcal{E}$: effectiveness of attack detection

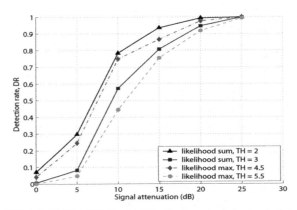

FIGURE 4.9. CoRE 802.11, ABP: effectiveness of attack detection

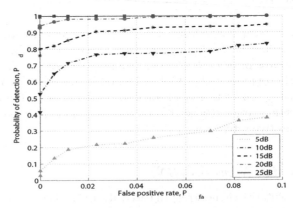

FIGURE 4.10. CoRE 802.11, ABP: Receiver Operating Characteristic (ROC) curves

tion for not just the position coordinates, but also for every random variable in the Bayesian network. Without an attack, the contribution from each landmark to the full joint-probability distribution is almost uniform. Under an attack, we found that the contribution from each landmark can become significantly reduced as the attack severity increases. Thus, we can use the fraction of contribution to the joint probability as a test statistic in BN.

Another method we explored is to use the probability likelihood because the conditional probability distribution of the coordinates in BN relies on the prior and the likelihood. We observed that under an attack, the value of the likelihood became significantly smaller. During the sampling process, the calculation of the likelihood uses the same approach as in Equation (4.18). Because the absolute value of the likelihood is very small, we take the negative log of the likelihood and use it as a test statistic for attack detection in BN.

Figure 4.11 shows the effectiveness of using the fraction of contribution and the likelihood for attack detection in BN. The detection rates are over 90% for attack strength of 20dB or larger. The false positive rates are about 10%. Comparing the absolute performance of these two methods with the other schemes we proposed in this chapter, the performance of these two methods is qualitatively worse.

4.7 Discussion

Comparing all of our detection schemes, Figure 4.12 shows the DR as a function of the signal attenuation attacks for the 802.11 network in the CoRE building. Surprisingly, we found that the performance of all the schemes provided qualitatively similar detection rates, although utilizing the residuals in LLS and the sum of likelihoods in ABP slightly outperformed the others, while using the fraction of contribution and the likelihood in BN underperformed the others.

Based on these similar performance characteristics, it is advantageous to use the minimum distance in the signal space D_s for signal strength based algorithms. Since the attack detection can be performed prior to the localization process and thus results in localization computation cost savings under attack. Additionally, the attack detection performance under the 802.15.4 network when using D_s outperforms the 802.11 network with 100% detection rate for large attacks as shown in Figure 4.6.

Moving to examine the relationship between attack detection and localization error, Figure 4.13 shows the DR when using residuals in LLS for attack detection, and the localization errors under the corresponding signal attacks with different localization algorithms. The figure shows that detection rates are more than 90% for attack strength equal to or greater than

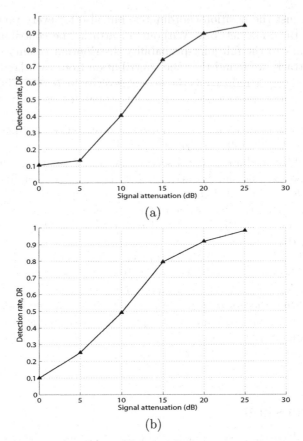

FIGURE 4.11. CoRE 802.11, BN: (a) Using fraction of contribution of each landmark for attack detection with threshold = 0.15. (b) Using likelihood in Bayesian
inference for attack detection with threshold = 0.25.

15dB, and at this attack strength the average localization error is about
35ft.

The above result is quite encouraging, as it shows that an attacker cannot
cause gross localization errors without there being a very high probability
of detection (¿95%). In the case of RSS, with mean errors of 10-15 ft [15],
an attacker can not cause errors of about 2-3 times over the average error
without a very high probability of detection. Even for detection rates as
low as 50%, the attacker's position error is limited to about 20 ft.

4.8   Conclusion

In this chapter, we analyzed the problem of detecting non-cryptographic attacks on wireless localization. We proposed a theoretical foundation by formulating attack detection as a statistical significance testing problem. We

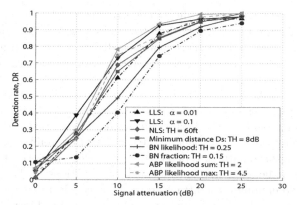

FIGURE 4.12. CoRE 802.11: Comparison between generic and specific test statistics for attack detection.

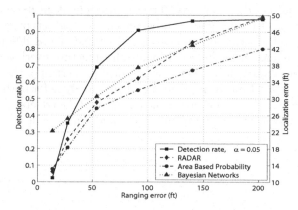

FIGURE 4.13. CoRE 802.11: Relationships among Detection Rate (DR), ranging error, and localization error.

then concentrated on test statistics for two broad localization approaches: multilateration and signal strength. For multilateration that uses Linear Least Squares, we derived a closed-form representation for the attack detector. Further, for localization schemes that employ signal strength, we showed that by utilizing the signal strength as a common feature, the minimum Euclidean distance in the signal space can be used as a test statistic for attack detection independent of the localization process. Further, we derived additional test statistics for a selection of representative localization algorithms.

We studied the effectiveness and generality of our attack detection schemes using a trace-driven evaluation involving both an 802.11 (WiFi) network and an 802.15.4 (ZigBee) network in two real office buildings. We evaluated the performance of our attack detection schemes in terms of detection rates and receiver operating characteristic curves. Our experimental results provide strong evidence of the effectiveness of our attack detection schemes with high detection rates, over 95% and low false positive rates, often below 5%. Also, our approach is generic across a diverse set of algorithms, networks, and buildings. Interestingly, we found that the performance of the different attack detection schemes are more similar than different. This result shows that different localization systems have similar attack detection capabilities, and consequently that system designers can focus on using algorithms that provide the highest localization accuracy rather than having to tradeoff position accuracy against attack detection abilities.

After a localization attack is detected in a wireless network, the next important and challenging step is to localize the positions of the adversaries and further to eliminate the attack from the network. In the next chaper, we illustrate this idea further by examining the applicability of localization methods to locate an adversary participating in a spoofing attack. A spoofing attack is an attack where the attacker forges its identity and masquerades as another device, or even creates multiple illegitimate identities. Although the identity of a node can be verified through cryptographic authentication, authentication is not always desirable or possible because it requires key management and additional infrastructure overhead. We will take a different approach by using the physical properties of the radio signal and propose a scheme using K-means cluster analysis for both detecting spoofing attacks as well as localizing the positions of the adversaries without adding any overhead to the wireless devices and sensor nodes.

5

Robust Statistical Methods for Attack-tolerant Localization

5.1 Introduction

The infrastructure provided by wireless networks promises to have a significant impact on the way computing is performed. Not only will information be available while we are on the go, but new location-aware computing paradigms along with location-sensitive security policies will emerge. Already, many techniques have emerged to provide the ability to localize a communicating device [17, 20, 35, 54, 77].

Enforcement of location-aware security policies (e.g., this laptop should not be taken out of this building, or this file should not be opened outside of a secure room) requires trusted location information. As more of these location-dependent services get deployed, the very mechanisms that provide location information will become the target of misuse and attacks. In particular, the location infrastructure will be subjected to many *localization-specific* threats that cannot be addressed through traditional security services. Therefore, as we move forward with deploying wireless systems that support location services, it is prudent to integrate appropriate mechanisms that protect localization techniques from these new forms of attack.

The purpose of this chapter is to examine the problem of secure localization from a viewpoint different from traditional network security services. In addition to identifying different attacks and misuse faced by wireless localization mechanisms, we take the viewpoint that these vulnerabilities can be mitigated by exploiting the redundancy present in typical wireless deployments. Rather than introducing countermeasures for every possible attack, our approach is to provide *localization-specific, attack-tolerant* mechanisms that shield the localization infrastructure from threats that bypass traditional security defenses. The idea is to live with bad nodes rather than eliminate all possible bad nodes.

In this chapter, to address the localization attacks, we propose the use of robust statistical methods. In Section 5.3 and Section 5.4 we focus our discussion on applying robust mechanisms to two broad classes of localization: triangulation and fingerprinting methods. We introduce the notion of coordinated adversarial attacks on the location infrastructure, and present a strategy for launching a coordinated attack on triangulation-based methods. For triangulation-based localization, we propose the use of least median squares (LMS) as an improvement over least squares (LS) for achieving robustness to attacks. We formulate a linearization of the least squares location estimator in order to reduce the computational complexity of LMS. Since LS outperforms LMS in the absence of aggressive attacks, we devise an online algorithm that can adaptively switch between LS and LMS to ensure that our localization algorithm operates in a desirable regime in the presence of varying adversarial threats. For fingerprinting-based location estimation, we show that the use of traditional Euclidean distance metrics is not robust to intentional attacks launched against the base stations involved in localization. We propose a median-based nearest neighbor scheme that employs a median-based distance metric that is robust to location attacks. The use of median does not require additional computational resources, and in the absence of attacks has performance comparable to existing techniques. Finally, we present conclusions in Section 5.5.

5.2   Robust Localization: Living with Bad Guys

As discussed in the previous section, wireless networks are susceptible to numerous localization-specific attacks. These attacks will be mounted by clever adversaries, and as a result will behave dramatically different from measurement anomalies that arise due to the underlying wireless medium. For example, signal strength measurements may be significantly altered by opening doorways in a hallway, or by the presence of passersby. Although these errors are severe, and can degrade the performance of a localization scheme, they are not intentional, and therefore not likely to provide a persistent bias to any specific localization scheme. However, the attacks mentioned in Section 2.3 will be intelligent and coordinated, causing significant bias to the localization results.

Solutions that can combat some of these localization attacks have been proposed, often involving conventional security techniques [35,54]. However, as noted earlier, it is unlikely that conventional security will be able to remove all threats to wireless localization. We therefore take the viewpoint that instead of coming up with solutions for each attack, it is essential to achieve robustness to unforeseen and non-filterable attacks. Particularly, localization must function properly even in the presence of these attacks.

Our strategy to accomplish this is to take advantage of the redundancy in the deployment of the localization infrastructure to provide stability to contaminated measurements. In particular, we develop statistical tools that may be used to make localization techniques robust to adversarial data. As a byproduct, our techniques will be robust to non-adversarial corruption of measurement data. For the purpose of the discussion, we shall focus our attention on two classes of localization schemes: triangulation, and the method of RF fingerprinting. We have chosen these two methods since they represent a broad survey of the methods used. Our discussion and evaluations will focus on the case where we localize a single device. Localizing multiple nodes involves applying the proposed techniques for each device that is to be localized.

The methods we will propose here make use of the median. Median-based approaches for data aggregation in sensor networks have recently been proposed [78, 79], and use the median as a resilient estimate of the average of aggregated data. On the other hand, localizing a device involves estimating a device's position from physical measurements not directly related to position, such as signal strength. Applying robust techniques to wireless sensor localization is challenging as it involves not only integrating robust statistical methods that estimate position from other types of measurements, but also must consider important issues such as computational overhead.

5.3   Robust Methods for Triangulation

Triangulation methods constitute a large class of localization algorithms that exploit some measurement to estimate distances to anchors, and from these distances an optimization procedure is used to determine the optimal position. The robust methods that we describe can be easily extended to other localization techniques, such as the Centroid method.

Triangulation methods involve gathering a collection of $\{(x, y, d)\}$ values, where d represents an estimated distance from the wireless device to an anchor at (x, y). These distances d may be stem from different types of measurements, such as hop counts in multi-hop networks (as in the case of DV-hop [24]), time of flight (as in the case of CRICKET), or signal strength. For example, in a hop-based scheme like DV-hop, following the flooding of beacons by anchor nodes, hop counts are measured between anchor points and the wireless device, which are then transformed into distance estimates.

In the ideal case, where the distances are not subjected to any measurement noise, these $\{(x, y, d)\}$ values map out a parabolic surface

$$d^2(x, y) = (x - x_0)^2 + (y - y_0)^2, \tag{5.1}$$

whose minimum value (x_0, y_0) is the wireless device location. Gathering several $\{(x_j, y_j, d_j)\}$ values and solving for (x_0, y_0) is a simple least squares problem that accounts for overdetermination of the system and the presence of measurement noise.

However, such an approach is not suitable in the presence of malicious perturbations to the $\{(x, y, d)\}$ values. For example, if an adversary alters the hop count, perhaps through a wormhole attack or jamming attack, the altered hop count may result in significant deviation of the distance measurement d from its true value. The use of a single, significantly incorrect $\{(x, y, d)\}$ value will drive the location estimate significantly away from the true location in spite of the presence of other, correct $\{(x, y, d)\}$ values. This exposes the vulnerability of least squares localization method to attacks, and we would like to find a robust alternative, as discussed below, to reduce the impact of attacks on localization.

5.3.1   Robust Fitting: Least Median of Squares

The vulnerability of the least squares algorithm to attacks is essentially due to its non-robustness to "outliers". The general formulation for the LS algorithm minimizes the cost function

$$J(\theta) = \sum_{i=1}^{N} [u_i - f(v_i, \theta)]^2, \tag{5.2}$$

where θ is the parameter to be estimated, u_i corresponds to the i-th measured data sample, v_i corresponds to the absissas for the parameterized surface $f(v_i, \theta)$, $|y_i - f(x_i, \theta)|$ is the residue for the i-th sample, and N is the total number of samples. Due to the summation in the cost function, a single influential outlier may ruin the estimation.

To increase robustness to outliers, a robust cost function is needed. For example, the method of least median of squares, introduced by Rousseeuw and described in detail in [80], is one of the most commonly used robust fitting algorithms. Instead of minimizing the summation of the residue squares, LMS fitting minimizes the median of the residue squares

$$J(\theta) = \text{med}_i [y_i - f(x_i, \theta)]^2. \tag{5.3}$$

Now a single outlier has little effect on the cost function, and won't bias the estimate significantly. It is known that in absence of noise, LMS tolerates up to 50 percent outliers among N total measurements, and still give the correct estimate [80].

The exact solution for LMS is computationally prohibitive. An efficient and statistically robust alternative [80] is to solve random subsets of $\{(x_i, y_i)\}$ values to get several candidate $\hat{\theta}$. The median of the residue squares for each candidate is then computed, and the one with the least median of residue squares is chosen as a tentative estimate. However, this tentative estimate is obtained from a small subset of samples. It is desirable to include more samples that are not outliers for a better estimation. So, the samples are reweighted based on their residues for the tentative estimate, followed by a reweighted least squares fitting to get the final estimate.

The samples can be reweighted in various ways. A simple thresholding method given by [80] is

$$w_i = \begin{cases} 1, & |\frac{r_i}{s_0}| \leq \gamma \\ 0, & \text{otherwise} \end{cases} \tag{5.4}$$

where γ is a predetermined threshold, r_i is the residue of the i-th sample for the least median subset estimate $\hat{\theta}$, and s_0 is the scale estimate given by [80]

$$s_0 = 1.4826(1 + \frac{5}{N-p})\sqrt{\text{med}_i r_i^2(\hat{\theta})}, \tag{5.5}$$

where p is the dimension of the estimated variable. The term $(1 + \frac{5}{N-p})$ is used to compensate the tendency for a small scale estimate when there are few samples.

Assume we are given a set of N samples, and that we aim to estimate a p-dimensional variable θ from this ensemble. The procedure for implementing the robust LMS algorithm is summarized as follows:

1. Choose an appropriate subset size n, the total number of subsets randomly drawn M, and a threshold γ.

2. Randomly draw M subsets of size n from the data ensemble. Find the estimate $\hat{\theta}_j$ for each subset. Calculate the median of residues r_{ij}^2 for every $\hat{\theta}_j$. Here $i = 1, 2, \cdots, N$ is the index for samples, while $j = 1, 2, \cdots, M$ is the index for the subsets.

3. Define $m = \arg \min_j \text{med}_i \{r_{ij}^2\}$, then $\hat{\theta}_m$ is the subset estimate with the least median of residues, and $\{r_{im}\}$ is the corresponding residues.

4. Calculate $s_0 = 1.4826(1 + \frac{5}{N-p})\sqrt{\text{med}_i r_{im}^2}$.

5. Assign weight w_i to each sample using Equation (5.4).

6. Do a weighted least squares fitting to all data with weights $\{w_i\}$ to get the final estimate $\hat{\theta}$.

5.3.2 Robust Localization with LMS

In the absence of attacks, the device location estimate $(\hat{x}_0, \hat{y}_0)$ can be found by least squares, i.e.

$$(\hat{x}_0, \hat{y}_0) = \arg \min_{(x_0, y_0)} \sum_{i=1}^{N} [\sqrt{(x_i - x_0)^2 + (y_i - y_0)^2} - d_i]^2. \tag{5.6}$$

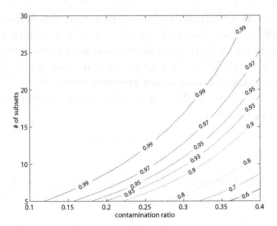

FIGURE 5.1. The contour plot of the equation (5.8): probability to get at least one good subset over contamination ratio and the number of subsets when $n = 4$.

In presence of attacks, however, the adversary produces "outliers" in the measurements. Instead of identifying this misinformation, we would like to live with them and still get a reasonable location estimate (identification of misinformation will come out as a byproduct naturally). To achieve this goal, we use LMS instead of least squares to estimate the location. That is, we can find $(\hat{x}_0, \hat{y}_0)$ such that

$$(\hat{x}_0, \hat{y}_0) = \arg \min_{(x_0, y_0)} \operatorname{med}_i [\sqrt{(x_i - x_0)^2 + (y_i - y_0)^2} - d_i]^2. \qquad (5.7)$$

Then the above LMS procedure can be used.

However, before using the algorithm, we need to consider two issues: First, how to choose the appropriate n and M for LMS-based localization? Second, how to get an estimate from the samples efficiently? The answers depend on the required performance and the affordable computational complexity. Considering that power is limited for sensor networks, and that the computational complexity of LMS depends on both the parameters and algorithmic implementation, we would like to gain the robustness of LMS with minimal additional computation compared to least squares, while exhibiting only negligible performance degradation. These two issues are now addressed.

1) How to choose the appropriate n and M?

The basic idea of the LMS implementation is that, hopefully, at least one subset among all subsets does not contain any contaminated samples, and the estimate from this good subset will thus fit the inlier (non-corrupted) data well. Since the inlier data are the majority ($> 50\%$) of the data, the median of residues corresponding to this estimate will be smaller than that from the bad subsets.

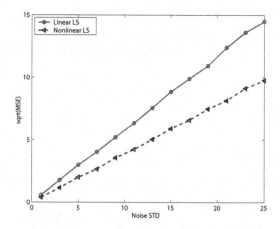

FIGURE 5.2. The comparison between linear LS, and nonlinear LS starting from the linear estimate.

We now calculate the probability P to get at least one good subset without contamination. Assuming the contamination ratio is ϵ, i.e, ϵN samples are outliers, it is easy to get that

$$P = 1 - (1 - (1 - \epsilon)^n)^M. \tag{5.8}$$

For a fixed M and ϵ, the larger n, the smaller is P. So the size of a subset n is often chosen such that it's just enough to get an estimate. In our case, although the minimum number of samples needed to decide a location is 3, we have chosen $n = 4$ to reduce the chance that the samples are too close to each other to produce a numerically stable position estimate.

Once n is chosen, we can decide the value of P for a given pair of M and ϵ. A contour plot of P over a grid of M and ϵ is shown in Figure 5.1. For larger ϵ, a larger M is needed to obtain a satisfactory probability of at least one good subset. Depending on how much contamination the network localization system is required to tolerate and how much computation the system can afford, M can be chosen correspondingly. Because the energy budget of the sensors is limited, and the functionality of the sensor network may be ruined when the contamination ratio is high, we chose $M = 20$ in our simulations, so that the system is resistant up to 30 percent contamination with $P \geq 0.99$.

2) How to get a location estimate from the samples efficiently?

To estimate the device location (x_0, y_0) from the measurements $\{x_i, y_i, d_i\}$, we can use the least squares solution specified by equation (5.6). This is a nonlinear least squares problem, and usually involves some iterative searching technique, such as gradient descent or Newton method, to get the solution. Moreover, to avoid local minimum, it is necessary to rerun the algorithm using several initial starting points, and as a result the computation is relatively expensive. Considering that sensors have limited power, and

LMS finds estimates for M subsets, we may want to have a suboptimal but more computationally efficient algorithm.

Recall that equation (5.6) is equivalent to solving the following equations when $N \geq 2$:

$$
\begin{aligned}
(x_1 - x_0)^2 + (y_1 - y_0)^2 &= d_1^2 \\
(x_2 - x_0)^2 + (y_2 - y_0)^2 &= d_2^2
\end{aligned}
\tag{5.9}
$$

$$
\vdots
$$

$$
(x_N - x_0)^2 + (y_N - y_0)^2 = d_N^2
$$

Averaging all the left parts and right parts respectively, we get

$$
\frac{1}{N} \sum_{i=1}^{N} [(x_i - x_0)^2 + (y_i - y_0)^2] = \frac{1}{N} \sum_{i=1}^{N} d_i^2.
\tag{5.10}
$$

Subtracting each side of the equation above from equation (5.9), we linearize to get the new equations

$$
(x_1 - \frac{1}{N} \sum_{i=1}^{N} x_i)x_0 + (y_1 - \frac{1}{N} \sum_{i=1}^{N} y_i)y_0 =
$$

$$
\frac{1}{2}[(x_1^2 - \frac{1}{N} \sum_{i=1}^{N} x_i^2) + (y_1^2 - \frac{1}{N} \sum_{i=1}^{N} y_i^2) - (d_1^2 - \frac{1}{N} \sum_{i=1}^{N} d_i^2)]
$$

$$
\vdots
\tag{5.11}
$$

$$
(x_N - \frac{1}{N} \sum_{i=1}^{N} x_i)x_0 + (y_N - \frac{1}{N} \sum_{i=1}^{N} y_i)y_0 =
$$

$$
\frac{1}{2}[(x_N^2 - \frac{1}{N} \sum_{i=1}^{N} x_i^2) + (y_N^2 - \frac{1}{N} \sum_{i=1}^{N} y_i^2) - (d_N^2 - \frac{1}{N} \sum_{i=1}^{N} d_i^2)],
$$

which can be easily solved using linear least squares.

Due to the subtraction, the optimum solution of the linear equations (5.11) is not exactly the same as the optimum solution of the nonlinear equations (5.9), or equivalently equation (5.6). However, it can save computation and also serve as the starting point for the nonlinear LS problem . We noticed that there is a non-negligible probability of falling into a local minimum of the error surface when a random initial value is used with Matlab's *fminsearch* function to find the solution to equation (5.6). We observed that initiating the nonlinear LS from the linear estimate does not get trapped in a local minimum. In other words, the linear estimate is close to the global minimum of the error surface. A comparison of the performance of the linear LS technique , and the nonlinear LS searching starting from the linear estimate is presented in Figure 5.2. Nonlinear searching from the linear estimate performs better than the linear method at the price of a higher computational complexity. Here, we only used 30 samples, and that the performance difference between the linear and nonlinear methods should decrease as the number of samples increases.

5.3.3 Simulation

To test the performance of localization using LMS, we need to build a threat model first. In this work, we assume that the adversary successfully gains the ability to arbitrarily modify the distance measurements for a fraction ϵ of the total anchor nodes. The contamination ratio ϵ should be less than 50 percent, the highest contamination ratio LMS can tolerate. The goal of the adversary is to drive the location estimate as far away from the true location as possible. Rather than randomly perturbing the measurements of these contaminated devices, the adversary should *coordinate* his corruption of the measurements so that they will push the localization toward the same wrong direction. The adversary will thus tamper measurements so they lie on the parabolic surface $d_a^2(x, y)$ with a minimum at (x_a, y_a). As a result the localization estimate will be pushed toward (x_a, y_a) from the true position (x_0, y_0) in the absence of robust countermeasures. The larger distance between (x_a, y_a) and (x_0, y_0), the larger the estimate deviates from (x_0, y_0). So we use the distance $d_a = \sqrt{(x_a - x_0)^2 + (y_a - y_0)^2}$ as a measurement of the strength of the attack.

In our simulation, in addition to the underlying sensor network, we had a localization infrastructure with $N = 30$ anchor nodes that were randomly deployed in a $500 \times 500 m^2$ region. We assume that the sensor to be localized gets a set of $\{x_i, y_i, d_i\}$ observations by either DV-hop or another distance measurement scheme. In other words, the d_j may come from multi-hop measurements. The measurement noise obeys a Gaussian distribution with mean 0 and variance σ_n^2. The adversary tampers $N\epsilon$ measurements such that they all "vote" for (x_a, y_a). LS and LMS localization algorithms are applied to the data to obtain the estimates $(\hat{x}_0, \hat{y}_0)$. For computational simplicity, we use linear least squares to get location estimates, realizing that a nonlinear least squares approach will improve the performance a little, but won't change the other features of the algorithms. The distance between the estimate and the true location is the corresponding estimation error.

For each contamination ratio ϵ and measurement noise level σ_n, we observed the change of the square root of mean square error (MSE) with the distance $d_a = \sqrt{(x_a - x_0)^2 + (y_a - y_0)^2}$. As an example, the performances at two different pairs of σ_n and ϵ are presented in Figure 5.3, where the value at each point is the average over 2000 trials. As expected, the estimation error of ordinary LS increases as d_a increases due to the non-robustness of the least squares to outliers. The estimation error of LMS increases first until it reaches the maximum at some critical value of d_a. After this critical value, the error decreases slightly and then stabilizes. In other words, *if LMS is used in localization, it's useless or even harmful for the adversary to attempt to conduct too powerful of an attack.*

The performance of the LS and LMS algorithms are affected by both the contamination ratio and the noise level. Figure 5.4 (a) illustrates the

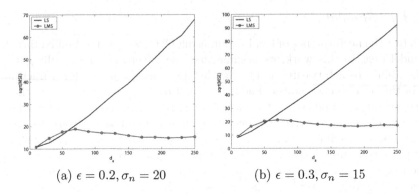

(a) $\epsilon = 0.2, \sigma_n = 20$ (b) $\epsilon = 0.3, \sigma_n = 15$

FIGURE 5.3. The performance comparison between LS and LMS for localization in presence of attack.

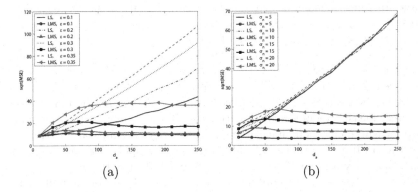

(a)   (b)

FIGURE 5.4. (a) The impact of ϵ on the performance of LS and LMS algorithms at $\sigma_n = 15$. (b) The impact of σ_n on the performance of LS and LMS algorithms at $\epsilon = 0.2$.

degradation of the performance as ϵ increases at a fixed $\sigma_n = 15$, while Figure 5.4(b) illustrates the impact of measurement noise σ_n on the performance at a fixed $\epsilon = 0.2$. Not surprisingly, the higher the contamination ratio, the larger the measurement noise, the larger is the estimation error. Also, since we chose n and M so the system would be robust up to 30 percent contamination, 35 percent contamination results in severe performance degradation as shown in Figure 5.4(a). More computations might improve the performance at high contamination ratio, but as noted earlier, due to the limitation of the power in sensor network, we trade the performance for reduced complexity.

We also noticed from Figure 5.3 and Figure 5.4 (b) that at low attacking strength, the performance of LS is actually better than LMS. In order to elucidate the reason for this behavior, let us look the simpler problem of fitting a line through data. In Figure 5.5, we present the line-fitting scenario

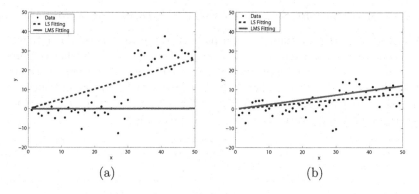

FIGURE 5.5. Example linear regression demonstrating that LMS performs worse than LS when the inlier and outlier data are too close.

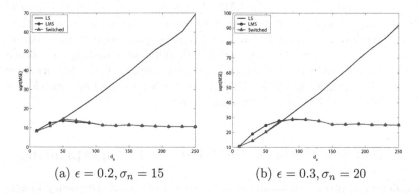

FIGURE 5.6. The performance of the switching algorithm comparing to LS and LMS algorithms.

using an artificial data set with 40 percent contamination. We generated 50 samples, among which 20 samples with $x = 31, \cdots, 50$ are the contaminated outliers. When the outlier data are well separated from the inlier data, LMS can detect this and fit the inlier data only, which gives a better fitting than LS. However, when the outlier data are close to the inlier data, it's hard for LMS to tell the difference, so it may fit part of the inlier data and part of the outlier data, thus giving a worse estimate than LS.

Therefore, when the attack strength is low, LS performs better than LMS. Further, in this case, LS also has a lower computational cost. Since power consumption is an important concern for sensor networks, we do not want to use LMS when not necessary. We have developed an algorithm, discussed below, where we may switch between LS and LMS estimation and achieve the performance advantages of each.

5.3.4   An Efficient Switched LS-LMS Localization Scheme

We use the observation that when outliers exist, the variance of the data will be larger than that when no outlier exists. Moreover, the farther outliers are from the inliers, the larger the variance. This suggests that the variance of the data can be used to indicate the distance between inliers and outliers. Therefore, we can do a LS estimate over the data first, and use the residues to estimate the data variance $\hat{\sigma}_n$ from the residuals r_i, i.e.

$$\hat{\sigma}_n = \sqrt{\frac{\sum_{i=1}^{N} r_i^2}{N-2}}.$$

Then the ratio $\frac{\hat{\sigma}_n}{\sigma_n}$ represents the variance expansion due to possible outliers. If the normal measurement noise level σ_n is known, which is a reasonable assumption in practice, we can compare the $\frac{\hat{\sigma}_n}{\sigma_n}$ to some threshold T. If $\frac{\hat{\sigma}_n}{\sigma_n} > T$, we choose to apply the LMS algorithm; otherwise, we just use the LS estimate we have calculated. We refer to this as the switched algorithm. In our simulation, we found that $T = 1.5$ gives quite good results over all tested ϵ and σ_n pairs. Two examples with different ϵ and σ_n are shown in Figure 5.6. After the switching strategy is deployed, the performance curves (the triangles in Figure 5.6) are very close to the lower envelop of the performance of LS and LMS algorithms.

5.4   Robust Methods for RF-Based Fingerprinting

A different approach to localization is based upon radio-frequency finger-printing. One of the first implementations was the RADAR system [16,81]. The system was shown to have good performance in an office building. In this section, we will show how robustness can be applied to such a RF-based system to obtain attack-tolerant localization.

In RADAR, multiple base stations are deployed to provide overlapping coverage of an area, such as a floor in an office building. During set up, a mobile host with known position broadcasts beacons periodically. The signal strengths at each base station are measured and stored. Each record has the format of $\{x, y, ss_1, \cdots, ss_N\}$, where (x, y) is the mobile position, and ss_i is the received signal strength in dBm at the i-th base station. N, the total number of base stations, should be at least 3 to provide good localization performance. To reduce the noise effect, each ss_i is usually the average of multiple measurements collected over a time period. The collection of all measurements forms a radio map that consists of the featured signal strengths, or fingerprints, at each sampled position.

Following setup, a mobile may be localized by broadcasting beacons and using the signal strengths measured at each base station. To localize the mobile user, we search the radio map collected in the setup phase, and find the fingerprint that best matches the signal strengths observed. That is, the central base station compares the observed signal energy $\{ss'_1, \cdots, ss'_N\}$ with the recorded $\{x, y, ss_1, \cdots, ss_N\}$, and pick the location (x, y) that minimizes the Euclidean distance $\sqrt{\sum_{i=1}^{N}(ss_i - ss'_i)^2}$ as the location estimate of the mobile user. This technique is called *nearest neighbor in signal space (NNSS)*. A slight variant of the technique involves finding the k nearest neighbors in signal space, and averaging their coordinates to get the location estimate. It was shown in [16] that averaging 2 to 4 nearest neighbors improves the location accuracy significantly.

The location estimation method described above is not robust to possible attacks. If the reading of signal strength at one base station is corrupted, the estimate can be dramatically different from the true location. Such an attack can be easily launched by inserting an absorbing barrier between the mobile host and the base station. Sudden change of local environment, such as turning on a microwave near one base station, can also cause incorrect signal strength readings. To obtain reasonable location estimates, in spite of attacks or sudden environmental changes, we propose to deploy more base stations and use a robust estimation method to utilize the redundancy introduced. In particular, instead of minimizing the Euclidean distance $\sqrt{\sum_{i=1}^{N}(ss_i - ss'_i)^2}$ to find nearest neighbors in signal space, we can minimize the median of the distances in all dimensions, i.e. minimize $med_{i=1}^{N}(ss_i - ss'_i)^2$ to get the "nearest" neighbor. In this way, a corrupted estimate won't bias the neighbor searching significantly.

We tested the proposed method through simulations. As pointed out in [16], the radio map can be generated either by empirical measurements, or by signal propagation modeling. Although the modeling method is less accurate than the empirical method, it still captures the data fairly well and provides good localization. In [16] a *wall attenuation factor* model was used to fit the collected empirical data and, after compensating for attenuation due to intervening walls, it was found that the signal strength varies with the distance in a trend similar to the generic exponential path loss [82]. In our simulation, we use the model, which we adopted from [16],

$$P(d)[dBm] = P(d_0)[dBm] - 10\gamma log(\frac{d}{d_0}), \qquad (5.12)$$

to generate signal strength data. We used the parameter $d_0 = 1m$, $P(d_0) = 58.48$ and $\gamma = 1.523$, which were obtained in [16] when fitting the model with the empirical data. We emphasize that the trends shown in our results are not affected by the selection of the parameters. We also added random zero-mean Gaussian noise with variance 1dBm so that the received signal strengths at a distance have a similar amount of variation as was observed in [16].

The rectangular area we simulated was similar to the region used in [16], and had a size $45m \times 25m$, which is a reasonable size for a large

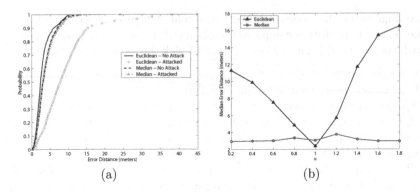

(a)   (b)

FIGURE 5.7. (a) The CDF of the error distance for the NNSS method in Euclidean distance and in median distance, with and without an attack (one reading is modified to $\alpha \cdot ss_i$, where $\alpha = 0.6$). (b) Median of the error distance vs. the attacking strength α (one reading is modified to $\alpha \cdot ss_i$).

indoor environment. Instead of three base stations, we employed six to provide redundancy for robust localization. We collected samples on a grid of 50 regularly spaced positions in order to form the radio map. During localization, a mobile sends beacons, and the signal strengths at the base stations are recorded. The nearest neighbors in signal space in terms of Euclidean distance and median distance are each found. The coordinates of the four nearest neighbors are averaged to get the final location estimate of the mobile user.

To simulate the attack, we randomly choose one reading ss_i and modify it to $\alpha \cdot ss_i$, where α indicates the attacking strength. $\alpha = 1$ means no attack. Figure 5.7 (a) shows the cumulative distribution function (CDF) of the error distance for the NNSS method in Euclidean distance and in median distance, with and without an attack. In presence of an attack with $\alpha = 0.6$, which is very easy to launch from a practical point of view, the Euclidean-based NNSS method shows significantly larger error than when there is no attack, while for the median-based NSSS approach there is little change (the curves with and without attack almost completely overlap in Figure 5.7 (a)). Although its performance is slightly worse than Euclidean-NNSS in the absence of attacks, median-NNSS is much more robust to possible attacks. In Figure 5.7 (b), we plot the 50th percentile value of the error distance for a series of α from 0.2 to 1.8. NNSS in median distance shows good performance across all α's.

With six base stations, the system can tolerate attacks on up to two readings. For simplicity, we assume the adversary randomly selects two readings and modifies them to $\alpha \cdot ss_i$. We note that such an approach is not a coordinated attack, and there may be better attack strategies able to produce larger localization error. Figure 5.8 (a) shows the CDF of the error distance at $\alpha = 0.6$, and Figure 5.8 (b) shows the change of median

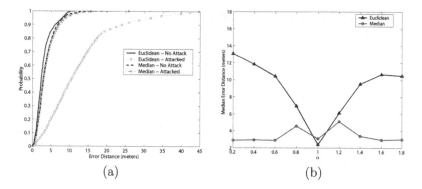

FIGURE 5.8. (a) The CDF of the error distance for the NNSS method in Euclidean distance and in median distance, with and without an attack (two readings are modified to $\alpha \cdot ss_i$, where $\alpha = 0.6$). (b) Median of the error distance vs. the attacking strength α (two readings are modified to $\alpha \cdot ss_i$).

error distance with α. Again, the median-NNSS exhibits better resistance to attacks. We observed the same phenomenon as that in the triangulation method: it is better for the adversary to not be too greedy when attacking the localization scheme. Finally, we note that the computational requirements for Euclidean-NNSS and median-NNSS are comparable. The fact that there is only marginal performance improvement for Euclidean-NNSS when there are no attacks suggests that a switched algorithm is not critical for fingerprinting-based localization.

5.5   Conclusion

As wireless networks are increasingly deployed for location-based services, these networks are becoming more vulnerable to misuses and attacks that can lead to false location calculation. Towards the goal of securing localization, this chapter has made two main contributions. It first enumerates a list of novel attacks that are unique to wireless localization algorithms. Further, this chapter proposes the idea of tolerating attacks, instead of eliminating them, by exploiting redundancies at various levels within wireless networks. We explored robust statistical methods to make localization attack-tolerant. We examined two broad classes of localization: triangulation and RF-based fingerprinting methods. For triangulation-based localization, we examined the use of a least median squares estimator for estimating position. We provided analysis for selecting system parameters. We then proposed an adaptive least squares and least median squares position estimator that has the computational advantages of least squares in the absence of attacks and switches to a robust mode when being attacked. For fingerprinting-based localization, we introduced robustness through the use of a median-based distance metric.

6

Spatio-Temporal Access Control by Dual-using Sensor Networks

6.1 Introduction

Historically, wireless networks have freed users from the confines of static, wired networks, and traditionally access control mechanisms are not based upon the geographic properties associated with the wireless user. The fact that wireless networks are becoming increasingly ubiquitous, however, suggests that it is not necessary to restrict access to services based solely on conventional identity-based authenticators. Rather, the wireless infrastructure can facilitate location-aware computing paradigms, where services are only accessible if the user is in the right place at the right time. For example, location-aware security services, such as ensuring that a file can only be accessed within a specific secure room, or that a laptop no longer functions when it is taken outside of a building, are not only desirable but will soon become feasible.

Wireless sensor networks are traditionally used to monitor phenomena and record data corresponding to the underlying physical information field. The small form factor and low-power design associated with many sensor architectures, suggests that this inherently pull-oriented use of sensor networks can be turned upside down to push information into the physical environment over extended periods of time. By using sensor nodes themselves to transmit signals into the wireless medium, it is possible actuate the physical environment to convey information useful for new types of mobile services. This notion of *inverted sensor networks* can be used to facilitate new forms of access control that can now be based upon whether an entity is at the right place at the right time in order to witness the information needed to access content.

In this chapter, we propose the use of inverted sensor networks to provide access control based upon a user's spatio-temporal information. Access control systems involve two main components: the security policy, which is a statement of who is allowed to access which service; and security mechanisms, which are the methods by which the security policies are enforced. Therefore, in this chapter, we will explore both of these aspects in the context of using inverted sensor networks for spatio-temporal access control (STAC) . This chapter explores both of these components by providing a formalism for spatio-temporal access control (STAC) models, as well as key distribution issues associated with spatio-temporal access control utilizing inverted sensor networks.

We begin the chapter in Section 6.2 by providing a brief overview of inverted sensor networks and their use in achieving spatio-temporal access control. Following this high-level description, we will explore both the specification and enforcement of STAC policies in more detail. In Section 6.3, we describe the components involved involved in STAC and provide several different means to describe STAC policies. In particular, we show that it is possible to achieve more flexible representations of STAC policies through the use of finite automata. We briefly examine how STAC can be achieved using a centralized entity, and propose a location-oriented challenge response protocol to support STAC. However, we believe that centralized schemes can expose a user to a severe privacy risk, and therefore, in order to achieve better user privacy, in Section 6.5, we turn our focus to the problem of using the inverted sensor network to enforce STAC policies. By having sensor networks appropriately announce keys at the correct time, it becomes possible to confine the access to content objects to specific spatio-temporal regions. Finally, we conclude the chapter and discuss further directions we are currently investigating in Section 6.7.

6.2   Overview of Inverted Sensor Networks

Traditionally, wireless sensor networks have been used for a variety of monitoring applications, ranging from military sensing to industry automation and traffic control. A common feature to all of these different sensor applications is that they *pull* data from the environment, i.e. they collect data and route this data (or process locally) for external applications to utilize. This pull-oriented formulation is the original, intuitive usage of sensors nodes. However, since sensor nodes consist of radios that they use for communication, it is also possible to turn the the role of the sensor node upside down and make the sensor *push* information into the environment.

This notion of an *inverted sensor network* is contrary to the traditional application of sensor networks where the sensors themselves merely gather data that another device can use to actuate and make decisions with. In the inverted model, however, the environment is changed using the radio, and a new information field is created in RF space that may be monitored using other devices. Due to the low-power nature of sensor node transmissions, the information injected into the environment by a sensor node is inherently localized.

The localized effect that a sensor node can have on the RF environment makes it possible to support access control based on the spatio-temporal location of a node that monitors the environment. Spatio-temporal access control allows for objects to be accessed only if the accessing entity is in the right place at the right time. STAC may thus be supported by having the environment locally convey the cryptographic information (keys) needed by the entity in order to access enciphered objects. To achieve STAC, then, a sufficiently dense network of low-power radios, as exists in traditional sensor nodes, is needed, and each sensor node can transmit a time-varying schedule of assigned cryptographic information in support of STAC.

While conventional access control is based on the user's identity, there are many scenarios where identity-based authentication is not only inconvenient but also unnecessary, and instead user spatio-temporal contexts are more desirable for basing access control upon:

1. A company may restrict its commercially confidential documents so that they can only be accessed while inside a building during normal business hours.

2. Network connectivity may be provided only to users who are located in a specific room (e.g. a conference room) during a specified meeting time.

3. Devices, such as corporate laptops, can be made to cease functioning if it leaves the building.

4. During sporting events, a sports service may transmit value-add information, such as live scores and player information, during the game, but only want those within the stadium to be able to access this information.

5. Movies or entertainment may be made to be accessible only to vehicles that are located on a specific road.

The above examples illustrate cases where both objects and services may be restricted based on location. Although the implementation of STAC to services, such as network connectivity, might not necessitate use of encipherment, we shall take the viewpoint in this paper that all objects/services that we wish to control access to can be suitably protected through encryption and appropriate key management. The extension of the ideas provided in this paper to more general cases, such as controlling access to network

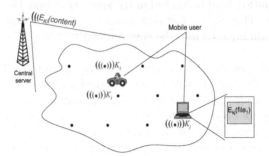

FIGURE 6.1. Our basic architecture for spatio-temporal access control consists of a central entity that supplies encrypted content, an auxiliary network of sensor nodes that emit keys, and mobile users that desire to access content based on their spatio-temporal context.

connectivity, can be done through simple resource allocation mechanisms, and we thus consider general cases a straight-forward application of the techniques detailed in this paper.

Throughout the rest of this paper, we shall refer to the STAC communication model depicted in Figure 6.1. In this model, we have an object that is protected through encryption with a set of keys. These keys may, for example, encrypt different portions of the object. Assisting in the STAC process is a broad array of sensor nodes, spread out over the region of interest, that each hold a set of decryption keys. These sensors can be configured to emit specific keys at specific times with specific power levels. As a result, only users that are near the right sensor at the right time can witness keys and access objects. The schedule of key transmissions, and their corresponding power levels, can either be pre-loaded, or controlled by an external entity connected to the sensor network, as depicted in the figure. Additionally, in the STAC model depicted in the figure, we present two different means by which the user can obtain a protected object: first, it can be broadcasted via a central entity (much like a television broadcast); or it can downloaded, locally stored on the user's device, and then accessed when the device witnesses the appropriate keys.

6.3   Spatio-Temporal Access Control Model

In this section, we capture the essential features of a STAC system and present a basic STAC model. The pieces described here serve as an un-

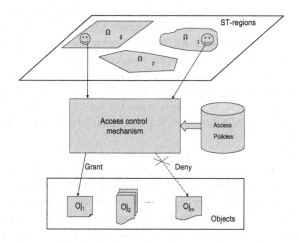

FIGURE 6.2. An conceptual picture of STAC model

derlying framework for our later discussion where we use inverted sensor networks to achieve STAC.

6.3.1 STAC Components

Analogous to the primary features of the RBAC model, as described in [53], STAC involves five basic elements: *users*(USERS), *objects*(OBS), *operations*(OPS), *spatio-temporal regions*(STRGNS), and *permissions*(PRMS). Figure 6.2 illustrates the core components of the STAC model. A set of STAC *access policies* describes rules that specify what permissions/privileges may be granted/rejected based upon a user's access attempt. The central idea of STAC is to control the *permission* or rejection of a *user*'s *operation* on an *object* based on the *spatio-temporal region* the user is in, according to these *access policies*. *Access control mechanisms* are the collection of techniques (e.g. encipherment) used to enforce this process. We now specify each of these components in more detail.

User. A *user* is generically defined as any entity, such as a process or person, that seeks access to objects.

Operation. An *operation* is a function that a user can execute on an object. The types of operations depend on the types of applications and systems. For example, when considering access to a file, the operations could either be simply access/no-access, or they can be multilevel such as read, write or execute. For the simplicity of discussion, we shall restrict our attention to binary operations. We note, however, that multilevel cases can be converted to the binary cases by defining a set of new objects, one for each operation, that substitute for the original one. For example, given an object *f1* with 3 operations read, write or execute, we define 3 new objects *f1read, f1write, f1exec* that substitute for the original object *f1*, and each of these new objects becomes a binary case. Throughout this paper, we shall represent access privileges for binary operations by 1 corresponding to access approval, while 0 corresponds to access rejection.

Permission. The set of *permission* is defined as PRMS $\subseteq 2^{(OPS \times OBS)}$, where OPS is the set of operations and OBS is the set of objects. A permission is an approval of an operation on an object. Since the operations discussed here are binary, each permission *prms* then has the form $p \in \{0,1\} \times OBS$.

Object. An *object* can be an information container, such as a file or an exhaustible system resource (e.g. a wireless service or CPU cycles). As stated in the last section, we assume the objects can be protected through encryption and appropriate key management. When an object is involved in the STAC system, it is endowed with temporal character, and can be generally categorized into static and streaming cases. A streaming object continually evolves with time, such as a movie being broadcast to an entire network of users, or live scores from a sporting event being transmitted to the audiences in a sports arena. On the other hand, a static object does not change with time (except for the possible exception of version revision, as often occurs for software objects). Since streaming objects are time-varying, their corresponding access control policies will inherently become more complicated to express.

As an example, consider an object O_j that is a streaming object. We may partition this object into subobjects according to time intervals, e.g. $O_j = \{O_j[t_0, t_1), O_j[t_1, t_2), O_j[t_2, t_3)\}$. Here O_j has been broken down into three pieces according to the time intervals $[t_0, t_1)$, $[t_1, t_2)$, and $[t_2, t_3)$ respectively. These subobjects may be further decomposed, and such a decomposition naturally raises the issue of the maximum amount that an object can be decomposed. We refer to the smallest constituent piece of a larger object as the *object atom*. The size of an object atom is determined by the *temporal resolution* of a STAC system.

Spatio-temporal region. A useful concept for specifying STAC for streaming objects is the notion of a *spatio-temporal region*. A spatio-temporal region, denoted by Ω is defined as a set of 3-tuple, $\Omega = \{(x, y, t) :$ valid areas in space and time$\}$, where (x, y) represents a spatial location and t represents an arbitrary time instance, and hence each (x, y, t) is a point in 3-dimensional spatio-temporal space. The spatio-temporal regions, which we shall refer to as ST-regions for brevity, are set up by the system according to the access policies, such as which object can be accessed where at what time period. Thus, it is often useful to visualize a ST-region as a continuous region in 3-dimensional space, instead of a set of discrete points. Figure 6.3 shows two example ST-regions Ω_1 and Ω_2. Ω_1 indicates a spatial region that is constantly specified from time 0 to time t; whereas Ω_2 indicates a spatial region that varies with time and, hence, if associated

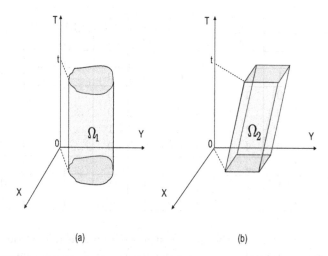

FIGURE 6.3. Examples of two spatio-temporal regions Ω_1 and Ω_2.

with an object's spatio-temporal access policy, would require that a user must move in a specific manner in order to maintain access privileges to an object. For an object ob and an operation op, a ST-region is called the *secure ST-region of (ob, op)* if the operation op is allowed to be performed on the object ob at this ST-region. In the case that the operation is binary, as we focus on in this paper, we simply refer to such a region as the *secure ST-region of ob*.

6.3.2 Access policies and their representations

Basic access policies and their representations by an access control matrix

Access policies outline the rules and regulations for appropriately accessing the objects. In STAC, a basic access policy is a 3-tuple $A = \{(\Omega; op; O_j;)\}$, where $O_j \in OBS$, $op \in OPS$, $\Omega \in STRGNS$, which is interpreted to mean that within the ST-region Ω, the operation op on object O_j is approved. Access control matrices can be used to represent such access policies. In the access control matrix, the columns correspond to objects, the rows to ST-regions, and the cell where the column and row intersect specifies the access privilege. Table 1 shows an example of an access control matrix for STAC, where the operations on objects are binary.

In this table, for example, object $S1$ can be accessed in the spatio-temporal location Ω_1, but not in Ω_2 (which has been further decomposed into smaller spatio-temporal regions). As a more involved example, we can decompose object Mv (which might correspond to a movie) down into sub objects (e.g. first 10 minutes, second 10 minutes, etc.). By similarly decomposing region Ω_2 into smaller regions Ω_{21}, Ω_{22}, Ω_{23} and Ω_{24}, we may now describe a more refined STAC policy, where Mv_1 is accessible at location

TABLE 6.1. Access Control Matrix of STAC

		S1	S2	F1	Mv			...	Oj_m	
		-	-	-	Mv_1	Mv_2	Mv_3	...	Oj_{m_1}	Oj_{m_2}
Ω_1	-	1	1	0	0	0	0	...	0	0
Ω_2	Ω_{21}	0	1	1	1	0	0	...	0	0
	Ω_{22}	0	1	1	0	1	0	...	0	0
	Ω_{23}	0	1	1	0	0	1	...	0	0
	Ω_{24}	0	1	1	0	0	0	...	0	0
...	...	...	...	...	...	...	...	...	...	...
Ω_n	Ω_{n1}	1	0	0	1	1	1	...	1	0
	Ω_{n2}	1	0	0	1	1	1	...	0	1

Ω_{21} but not Ω_{22}, and thus the user must move its physical location with an appropriate time period to Ω_{22} in order to access Mv_2, and hence resume access to Mv.

Complex access policies and their representations by FA

The example of object Mv just described gives insight into the power of STAC. In particular, a STAC system can perform complex access control by decomposing objects into *object atoms* and suitably associating smaller *region atoms* with these objects. By doing so, it is possible to grant or deny a user access to an object not only based on his current location, but also based on his previous spatio-temporal behavior. For example, we might require (for additional security), that a user have the ability to access object ob_1 at location l_1 and time t_1, and then be at location l_2 at time t_2 in order to access object ob_2. That is, without having been to (l_1, t_1) and having accessed ob_1, the user would not have the requisite access control information needed in order to access ob_2 at location (l_2, t_2).

Such a form of access control is stateful, and is unwieldy for representing with access control matrices. Notice in Table 6.1, for example, that although object Mv_1 can be accessed at Ω_{21}, Mv_2 can be accessed at Ω_{22}, and Mv_3 can be accessed at Ω_{23}, there is no information contained in the matrix that specifies that the user had to be at Ω_{21} before proceeding to Ω_{22} in order to access Mv_2.

In order to represent these more advanced forms of STAC policies, we must employ a syntactical framework that facilitates the representation of state. We now show how automata theory can be employed to describe such complex polices. First, though, we provide a brief review of automata theory to facilitate our later discussion. A *finite automaton* is denoted by a 5-tuple $(Q, \Sigma, \delta, q_0, F)$, where Q is a finite set of *states*, Σ is a finite *input alphabet*, q_0 in Q is the *initial* state, $F \subseteq Q$ is the set of *final* states, and δ is the *transition function* mapping $Q \times \Sigma \to Q$ [83]. A string x is said to be *accepted* by a finite automaton $M = (Q, \Sigma, \delta, q_0, F)$ if $\delta(q_0, x) = p$

for some $p \in F$. The *language accepted by* M, designated $L(M)$, is the set $\{x|\delta(q_0, x)\}$.

We may represent an access policy using an automaton $M = (Q, \Sigma, \delta, q_0, F)$ and capture the user's history for consideration in the access policy. To do so, we let the input alphabet $\Sigma = STRGNS \cup OBS$. A string x that is accepted by M is a sequence mixed by ST-regions that a user is required to locate and the objects that the user is required to access. By properly designing the δ and the set of states Q, the desired access policy can be expressed.

In order to illustrate how, we now provide an example. Suppose that we have a movie that is divided into 3 parts—Mv_1, Mv_2 and Mv_3. Further, suppose that the access policy we want to describe is that in order for a user to be able to view a later part of the movie, he must have finished viewing the part(s) that came before that part. Further, suppose that we have the additional spatio-temporal requirement that Mv_1 can only be accessed at location l_1 at time t_1, Mv_2 can only be accessed at location l_2 at time t_2, and Mv_3 can only be accessed at location l_3 at time t_3.

A finite automaton M_1 for this policy is defined as follows. $M_1 = (Q, \Sigma, \delta, q_0, F)$, where $Q = \{q_0, q_1, q_2, q_3, p_1, p_2, p_3\}$; $\Sigma = \{\Omega_1, \Omega_2, \Omega_3, Mv_1, Mv_2, Mv_3\}$; $\Omega_1 = (l_1, t_1)$, $\Omega_2 = (l_2, t_2)$, $\Omega_3 = (l_3, t_3)$; $F = \{p_1, p_2, p_3\}$. Here, in our specification, there are two classes of states q_j and p_j. The states q_j correspond to a description of the user being at the spatio-temporal contextual state needed to access the movie object Mv_j, where as state p_j corresponds to a description that the user has been in state q_j, and then performed the required accessing of the movie object Mv_j (and consequently, can move to the next ST-region Ω_{j+1}). If we examine the transition diagram depicted in Figure 6.4, we can describe the transition mapping as follows:

$$\delta(q_0, \Omega_1) = q_1$$

$$\delta(q_1, Mv_1) = p_1$$

$$\delta(p_1, \Omega_2) = q_2$$

$$\delta(q_2, Mv_2) = p_2$$

$$\delta(q_2, \Omega_3) = q_3$$

$$\delta(q_3, Mv_3) = p_3.$$

We now walk through this transition mapping. The user starts in a null state q_0, and if it has moved to location Ω_1 at time t_1, it is described as being in state q_1, and hence has the ability to access Mv_1. Once the user has watched the movie portion Mv_1, its state transitions to p_1, and the user may now move to location Ω_2 (by time t_2). The process continues, as the user moves to a new location, its state changes so it can access the content, then having accessed the content the user can now move to the next location. Formally, we may state the accepted language of this automaton

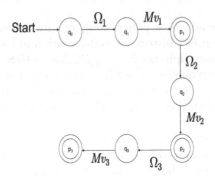

FIGURE 6.4. The transition diagram for M_1

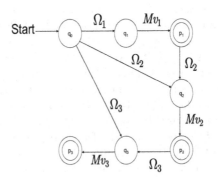

FIGURE 6.5. The transition diagram for M_1'

M_1 as the collection of valid strings: $L(M_1) = \{\Omega_1 M v_1, \ \Omega_1 M v_1 \Omega_2 M v_2,$
$\Omega_1 M v_1 \Omega_2 M v_2 \Omega_3 M v_3\}$.

Finite automata are fairly flexible and able to represent stateful access control policies rather easily. As another example, the finite automaton M_1 can be modified easily to support similar but slightly different policies. For example:

- If we add two transition functions to M_1. $\delta(q_0, \Omega_2) = q_2$, $\delta(q_0, \Omega_3) = q_3$, as shown in the transition diagram for the finite automaton M_1' in Figure 6.5, then the STAC policy takes on a different interpretation. Now, the STAC policy does not explicitly require that the user must have accessed the prior content $M v_{j-1}$ before accessing $M v_j$. Instead, all that is required is that the user is at Ω_j in order to access $M v_j$. This scheme corresponds to the movie access policy specified in the access control matrix in Table 6.1.

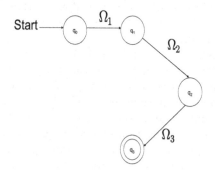

FIGURE 6.6. The transition diagram for M_1''

- We note that it is also possible to specify policies that have no involvement with the object, but instead only require the user to go through a particular spatio-temporal path. Such policies, as depicted in the transition diagram in Figure 6.6 for an automaton M_1'', might not be directly interesting from an access control point of view, but they can be useful as building blocks for more complicated STAC policies. For example, one can easily envision requiring that a user go to a succession of different locations, prior to being able to access content.

6.4 Centralized Mechanisms for STAC

Access control mechanisms are the means to enforce access policies. There are usually two steps necessary for supporting spatio-temporal access control: encipher objects to prevent unauthorized access to objects; and ensure that only users at an appropriate spatio-temporal location can acquire the keys needed to decipher objects.

Before delving into using inverted sensor networks to achieve STAC, we note that it is possible to provide access control to objects by holding the objects and then require that an entity prove that it is at a specific location before distributing the object to that entity. Or, as another centralized approach, an entity can have an enciphered object and, upon proving that it is in a specific location, the central server can deliver the set of keys needed to decipher and access the object.

In order to facilitate such centralized approaches for STAC, it is necessary to have trustworthy location information. Conventional localization schemes [16, 20, 24, 81] can provide location information, but recent efforts by many researchers have revealed that these localization algorithms can be attacked/subverted by non-cryptographic mechanisms. In response to this weakness, there has been a concerted effort to develop secure positioning techniques, where the localization algorithm is robust to measurement attacks or impersonation [47, 54].

Although these techniques can reliably localize an entity, we feel that such methods don't reflect the natural operation of an authentication protocol used in access control. Instead, we now present a different form of the locationing problem in which we have a mobile node *claiming* that it is at a particular location, and we desire the centralized entity to *authenticate* that claim [46]. Our location verification scheme, depicted in Fig. 6.7, involves a *challenge-response* protocol , and uses variable power configurations for the underlying wireless network to corroborate the claimed location.

Suppose we have an infrastructure of anchor points AP_j of known locations (x_j, y_j), where $j = 1, 2, ..., K$, capable of emitting localization beacons. Suppose that a mobile device contacts the infrastructure, claiming that it is at a location (x, y). It is the task of the infrastructure to validate this claim. To do so, it will issue a *challenge* to the mobile by creating a random test power configuration intended to verify the claimed location. This power configuration corresponds to the powers used by the different access points when transmitting locationing beacons. The power configuration will involve a power of 0 for some access points, meaning that these APs do not transmit, while a power of P_j is chosen for other APs. The powers P_j are chosen to define a radio region about AP_j so that the node *should* be able to witness the beacon from its claimed position (x, y). The determination of a radio region Ω_j is done using a propagation model.

The infrastructure now sends the challenge "Which APs do you hear?" to the mobile. The power levels of the APs are temporarily adjusted and location beacons are issued. The mobile then responds with a list of the APs it was able to witness, and the infrastructure checks this response. If a device incorrectly reports that it heard an AP that was not present, then this is clear evidence that the device's truthfulness, and hence its position, is false. However, if a device reports some APs correctly, but fails to report an AP that it should have heard, then we do not conclude the device's location is false. Rather, it may be that the beacon was simply missed due to poor propagation. We can assert the likelihood that a device misses a beacon using the underlying propagation model, and incorporate this confidence measure into verifying the device's location. In order to enhance the confidence levels of the claimed location, the challenge-response process may be repeated several times with different configurations.

We now provide a basic analysis of the performance of the power modulated challenge response protocol under two scenarios: the adversary is not able to witness any AP during the challenge, and a legitimate device is truly where it claims to be located. We will suppose that there are K APs in the WLAN, and that each round of the protocol we randomly choose k APs and set their power levels to provide a 95% coverage guarantee of

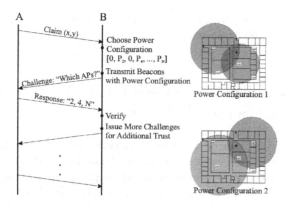

FIGURE 6.7. Location verification based upon inclusion principle and the notion of power modulated challenge-response.

the claimed (x, y) location. If the adversary knows that k APs were used but cannot witness any APs, then there is a probability of $\binom{K}{k}^{-1}$ of correctly guessing the APs. If the procedure runs N protocol rounds, then the probability of the adversary incorrectly being authenticated is $p_a = \binom{K}{k}^{-N}$. On the other hand, if the device is legitimate and at the correct location, the probability that it successfully witnesses all k APs is $(0.95)^k$. Thus, the probability of a legitimate device failing to authenticate in N protocol rounds is $p_d = 1 - (0.95)^{kN}$. Ideally, both p_a and p_d should be small. This implies that there is a fundamental tradeoff between K, k, and N in such a scheme. We note, however, that the above analysis only captures two extreme cases, and does reveal the true complexities associated with the authentication problem. In particular, the analysis does not reflect the possibility of an adversary capable of witnessing some APs and hence who possesses partial knowledge of the challenge pattern. The primary focus of this paper, though, is on our proposed decentralized mechanisms for STAC, and we will thus provide further analysis of centralized challenge-response location verification techniques in an expanded follow-up work.

One drawback of centralized STAC techniques is that the interaction between a user and the infrastructure inherently introduces an issue related to the user's privacy. In particular, a user's location information becomes exposed in centralized STAC mechanisms, and consequently users may be tracked by the infrastructure [62]. Although there have been some efforts recently that have examined location privacy, e.g. [84], these efforts are primarily focused on services that provide location traces to other services, and not on privacy issues associated with the measurement of a transmitter's location.

We will revisit the privacy issue in a later discussion. However, in the next session, we present our inverted sensor network construction, which achieves user location privacy by not requiring interaction between a user and the network infrastructure.

6.5   Decentralized Approach for STAC through Inverted Sensor Networks

In this section, we describe the use of inverted sensor networks to support spatio-temporal access control. In the basic inverted sensor network scheme, we assume that sensor networks are deployed in a regular (lattice) pattern, and that the sensors employ constant power levels when transmitting keys. This leads to a coverage issue, and further raises the issue of how precisely we may cover a specified spatio-temporal region Ω using the basic configuration. Since the general sensor network will not be deployed in a regular pattern, and since sensor nodes can employ variable power levels across the network, we next examine the issue of adjusting the coverage region to support a spatial region. Finally, we examine issues related to key deployment/management and the frequency of key announcement in order to support a desired time resolution for the region Ω.

6.5.1   Inverted sensor network infrastructure

There are three basic components involved in the inverted sensor network approach to STAC, as shown in Figure 6.1. The first component is a centralized content distributor that does not need to have any interaction with the user. Instead, the content distributor might broadcast or supply objects for download that have been enciphered with a set of keys known by the central server.

The second component is the auxiliary network (the inverted sensor network), which consists of sensor motes that have been deployed to cover a region of interest. These sensor motes will transmit a schedule of encryption keys that vary with time. Here, we assume that time is broken down into intervals, and that during any given interval, the corresponding key will be repeatedly transmitted at regular intervals. Any entity within the radio range of the sensor mote can, should it wish, acquire the key that is announced by that sensor at that time. We note that the keys that are transmitted by the sensor motes must be initially distributed to these nodes for use in supporting STAC.

Finally, the third component is the mobile user itself, which must move around the region of interest in an appropriate manner in order to acquire the keys transmitted by the sensor nodes.

In order to explore the properties of the inverted sensor network, we assume that sensor nodes are deployed in a regular, hexagonal fashion across the region of interest. It is well known that a regular hexagonal lattice is more efficient than either a rectangular or quincunx sampling in two-dimensions. Hence, we assume the coverage area is partitioned into hexagon cells of the same size, and that a single sensor node is placed at the center of each hexagon. Prior to initiating the STAC enforcement, we assume that each sensor has been assigned a schedule of keys (for simplicity, we shall say that the keys have been distributed by a centralized key distribution center). During STAC operation, each sensor emits a key according to its schedule, which can be observed by any other entity within radio range of the sensor.

If we let a be the length of an edge of one of the regular hexagons, and assume isotropic radio propagation then we may assume that each radio coverage is a circle with radius r. There are two natural choices for how the hexagonal lattice is deployed: either we deploy the sensor nodes so that the radio ranges do not share any overlap (and hence the circular regions would touch each other only at tangent points), or we can assume that we have deployed the sensors so that there is some overlap between the radio regions. Since the first option implies gaps in the radio coverage of the sensor grid, we assume that the deployment is of the second type. In this case, we have a deployment such as the one depicted in Figure 6.8, and hence the radius of radio coverage is $r = a$.

Given a ST-region of an object that needs to be protected, we assume that the content distributor knows the key schedule of all of the sensor nodes (perhaps it is also the key distributor and generated the key schedule in the first place), and that the object has been encrypted with the symmetric encryption key corresponding to the key that is being transmitted by a specific sensor node at a specific time. Exploring this idea further, reveals a some system requirements. If we need a STAC region that is larger than a single radio region of a sensor node, then we need all sensor nodes within the ST-region to transmit the same key. This requires that either the key distribution center has a priori knowledge of the ST-regions (perhaps corresponding to one or more objects that need to be protected) that need to be enforced for spatio-temporal access control, or that the key distribution center has a means to adapt the key transmission schedule of the sensor network. As an example, Figure 6.8 illustrates the key distribution scheme for two secure ST-regions Ω_1 for object O_1 and Ω_2 for object O_2. Here, we suppose k_1 and k_2 are the decryption keys for O_1 and O_2 respectively at a particular time. Therefore, it is necessary that k_1 has been assigned to all sensors whose radio discs are inside the rectangle Ω_1, and k_2 to the sensors whose radio discs are inside the rectangle Ω_2.

6.5.2 Improving the coverage

As seen in Figure 6.8, the regions of Ω_1 and Ω_2 are not fully covered by the keys sent by sensors. In particular, the concept of an *approximating ST-region* naturally arises.

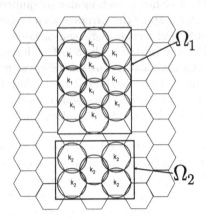

FIGURE 6.8. An example of how the keys may be assigned in order to cover two ST regions Ω_1 and Ω_2 for an inverted sensor network support spatio-temporal access control.

Definition 1: An *approximating ST-region* $\overline{\Omega}$ of a ST-region Ω given an STAC mechanism Σ is the spatio-temporal region that the object is actually accessible under Σ.

In practice, the approximating ST-region is not likely to be the same as the original, desired ST-region, and in fact we are only able to protect an approximation of the original ST-region. If we restrict our attention to just the spatial portion of a ST-region (which we shall denote by R), then we want the approximating ST-region to cover as large of a portion of the desired ST-region as possible. Before we proceed onto exploring how to optimize the approximating ST-region, we note that our discussion has centered around approximating a ST-region from the inside, and that it is possible to consider approximating regions that are *larger* than the ST-region that they are meant to represent. In such a case, we include sensors whose radio regions share any overlap with the desired ST-region. For this case, we would aim to reduce the amount of extra area covered by the ST-region.

For this work, though, we focus on approximating a ST-region from within, and hence we would like to minimize the amount of blank area between a desired ST-region and the approximating ST-region. Although we have deployed our sensors in a regular fashion, we may adjust the transmission powers of the sensor nodes in such a way so as to improve the amount of area covered by the radio regions.

To formalize this, let us define the region of interest to be G, and define $S = \{s_j\} \in G$ to be a set of sensor nodes. Here, we use the notation s_j to refer to the spatial position of node j. In the discussion that follows, we do not require that the sensor positions fall on a lattice, but instead the positions can be more general. For a given region $R \subseteq G$ we may specify what it means to cover (or fill) R from within.

Definition 2: A *cover of R from inside*, denoted C, is a set of circles C_j centered at s_j such that the union of the circles is fully within the inside R, that is $C = \bigcup(C_j) \subseteq R$. A cover from inside C is a function of a subset of the sensor nodes that are selected, and the corresponding transmission powers assigned to each of these sensor nodes. Hence, if we denote $\underline{P} = \{p_1, p_2, ...\}$ to be the power allocation vector for nodes $\{s_j\}$, then we may represent the cover as $C(\underline{P})$.

A cover from the inside will typically not completely cover the region R in the formal topological sense, and thus we are interested in measuring how accurately a cover from the inside approximates R. To capture this notion, we introduce the blank region and its corresponding area.

Definition 3: For a cover from inside C of a region R, a *Blank Region* is the set $B(C, R) = R \setminus \bigcup(C_j')$. A measure of the magnitude of the blank region is the *Blank Area*, $BA(C, R) = Area(R \setminus \bigcup(C_j'))$.

In an access control system, it is desirable to minimize the blank area. We note that it is easily possible to minimize the blank area by covering areas outside of R. This, however, implies that users who are not in the restricted area can access protected content, and hence should be considered as security weakness. Consequently, we take the viewpoint, that it is desirable to from a security point-of-view to sacrifice some area between the boundary of the STAC region and the actual approximation region so long as we do not have any key information being leaked outside the desired access control region.

In order to accomplish this, we may adjust the transmission powers in such a way to best cover from the inside a particular region R. We now present Algorithm 1, which describes a greedy algorithm for constructing an approximation of a region R from within, with the objective of minimizing the blank area $BA(C, R)$. In this algorithm, the input is the collection of sensor node locations $\{s_j\}$, as well as the desired region R. Additionally, we provide a constraint m which describes the maximum allowed transmission radius for each sensor node. For example, m might be determined by policy or by hardware restrictions. In this algorithm, we assume that there is a direct way to relate the actual transmission power to a corresponding coverage radius p_i (for example, by employing a propagation model). Hence, rather than explicitly define the algorithm in terms of assigning wattage to different nodes, we are instead formulating the problem in an equivalent manner using distances. We use the notation $d(s_i, s_j)$ to denote the distance between the nodes s_i and s_j.

The algorithm basically starts by assigning powers to ensure that the maximal coverage region for each sensor node is as large as possible, while remaining inside of the region R. Then, the algorithm proceeds to remove redundant nodes or power assignments. Using Algorithm 1, it is possible to achieve a more efficient coverage pattern for an arbitrary region R than

---

Data: Spatial Region R, the set of sensor node
locations $s_1, s_2, ..., s_n$ and the maximal radius constraint m
Result: An cover from inside C, such that $BA(C, R)$ is minimal.
for *Every s_i inside of R* **do**
   | Draw an inscribed circle centered at s_i, r_i denotes the
   | radius of the inscribed circle.;
   | $p_i = \min(m, r_i)$;
end
Sort the sensors by the decreasing of their inscribed circle's radius.
$s_{(1)}, s_{(2)}, ..., s_{(n)}$;
for $i = 1$ *to* n **do**
   | **if** $p_{(i)} != 0$ **then**
      | **for** $j = i + 1$ *to* n **do**
         | **if** $r_{(j)} + d(s_{(i)}, s_{(j)}) < r_{(i)}$ **then**
            | $p_{(i)} == 0$;
         | **end**
      | **end**
   | **end**
end

Algorithm 1: An algorithm for finding a near minimal blank area given a set of sensor locations $\{s_j\}$ and a desired region R to cover from the inside.

using a default deployment pattern where every node has the same power assignment. We present an example power configuration that results in Figure 6.9.

In order to further illustrate the advantages of adapting the power levels, we conducted a simulation study where the ST-region to be protected is a square with sides of length d, and deployed sensors in a uniform hexagonal tiling where the distance between sensor nodes is r. We varied the ratio d/r, and measured the blank area for both the uniform coverage and adaptive coverage resulting from Algorithm 1 using Monte Carlo sampling techniques. We found that at small d/r, it is not possible for the uniform deployment to cover the square (going beyond the boundary of the square is not allowed), but the power allocations assigned by Algorithm 1 adapts the transmission power to cover square without going outside the square. As we increase d/r, we allow more sensor nodes to fall within the square, and we see that the adaptive power allocation algorithm consistently results in less blank area than uniform power allocation.

Finally, we note that Algorithm 1 is flexible and can be applied to address the power allocation for any placement of sensor nodes beyond the regular lattice patterns that we have focused our discussions on.

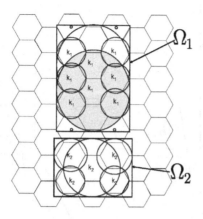

FIGURE 6.9. An example of the key distribution coverage pattern after the power allocations of each sensor node have been adjusted using Algorithm 1.

6.5.3 Dynamic Encryption and Key Updating

Spatio-temporal access control not only involves access based on an entity's spatial location, but also implies that there might be important temporal contexts that affect the ability of a user to access content. There are many cases where we can specify a STAC policy for an object (such as an entire movie) that has the requirement that access changes with time. One particularly important example of an object that would have such a policy is a streaming object that varies with time. For this case, it is necessary to decompose an object into smaller object atoms (as defined earlier), and then treat each of these smaller object atoms as individual objects that are protected over a spatial region that is fixed over a smaller time interval. As an example, we can consider an object with a formal ST-region Ω_2, but would have to approximate Ω_2 via a collection of smaller and simpler ST-regions with temporal resolution τ, as depicted in Figure 6.10. Here, in this figure, region Ω_2 is thus approximated as the union

$$\Omega_2 = \bigcup_{j=1}^{4} \Omega_{2j}.$$

Each of these ST-atoms Ω_{2j} will be enciphered by a corresponding key k_j, and hence the encryption of such an object will be dynamic in the sense that the key will vary with time. For streaming objects, if we were to not employ dynamic encryption, then it would be possible for an adversary to observe the key in a valid ST-region (e.g. Ω_{21}) at a valid time, and then access the rest of the content from a region not allowed by the spatio-temporal access control policy.

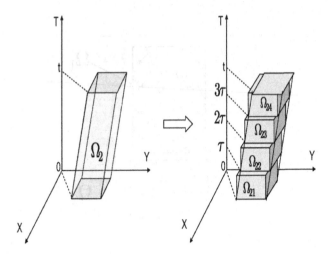

FIGURE 6.10. A ST-region Ω_2 is originally specified in a continuous, smooth manner. However, in practice it is necessary to decompose the region into ST region atoms Ω_{2j}, and correspondingly decompose the object into smaller object atoms.

As a further issue, we note that dynamic encryption is only meant to protect the content during the initial access period for that content. By this we mean, that once a user has recorded a STAC-protected object and has satisfied the spatio-temporal requirements to access the object, that user has effectively unlocked the object for him/her to access at any later time. Essentially, once a user has the file and the keys, access is granted from that point on. We note that dynamic encryption is only meant to protect dynamically evolving content/objects. If it is desired to strictly limit a user to accessing content during a specific time and not after that time, then it is necessary to employ the use of additional security mechanisms, such as trusted operating systems and secure containers which would guarantee that key information is stored and accessible to the user only during a specified time.

Dynamic encryption and the decomposition of objects into smaller, more refined ST-regions requires that the each ST-region is associated with different keys, and hence the problem of managing the keying information in an inverted sensor network becomes important. Although one could envision that a key distribution center might be able to manage and frequently update keys within the sensor network by issuing updates to each sensor node (e.g. through a gateway between the sensor network and the broader Internet) every time the key needs to change, such a scheme is impractical. Rather, we should reduce the frequency at which the key distribution center interacts with each sensor node by having the center distribute a single set of keys corresponding to a keying schedule.

In order to do so, the KDC must either initially install a large set of keys prior to deployment, or the KDC can communicate as needed with the sensor nodes through a set of keys shared between the KDC and each sensor, i.e. $K_{s_i,KDC}$. There is significant overhead associated with frequent updates of keys, and in order to reduce this overhead, we make use of a chain of one-way hash functions to generate and store keys.

A one-way chain $(V_0, ..., V_n)$ is a collection of values such that each value V_i (except the last value V_n) is a one-way function of the next value V_{i+1}. In particular, we have that $V_i = H(V_{i+1})$ for $0 \leq i < N$. Here H is a one-way function, and is often selected as a cryptographic hash function. For setup of the one-way chain, the generator chooses at random the root or seed of the chain, i.e., the value V_n, and derives all previous values V_i by iteratively applying the hash function H as described above, yielding a chain as in:

$$V_0 \leftarrow V_1 \leftarrow V_2 \leftarrow \cdots \leftarrow V_{n-1} \leftarrow V_n.$$

By employing the hash chain, the entity responsible for key distribution need only send the anchor seed and the times at which the sensor node should change keys. For example, if we let the last key be the key seed, i.e. $V_n = K_n$, then the KDC simply performs

$$KDC \rightarrow SN : E_{K_{(s_i,KDC)}}(K_n, t_0, t_1, ...t_n).$$

The sensor then can derive K_1, K_2, all the way up to K_{n-1} locally by applying H. When the keys are used up, the central server will repeat the process.

One necessary system requirement for STAC, though, is that all sensor nodes maintain synchronization with each other and the server so as to guarantee that keys are transmitted during the correct time period.

6.6   Discussion on the operation of inverted sensor networks

The use of the inverted sensor network for spatio-temporal access control achieves several advantages when compared to a centralized scheme: first, it reduces the risk of a privacy breach; second, it is naturally resistant to location spoofing attacks; and third, it facilitates new classes of applications that can easily be implemented.

6.6.1   Reduced Contextual Privacy Risk

Privacy is the guarantee that information, in its general sense, is observable or decipherable by only those who are intentionally meant to observe or decipher it. The phrase "in its general sense is meant to imply that there may be types of information besides the content of a message that are associated with a message transmission. When you access an ATM at a bank, this action is observable, and some contextual information regarding your actions is revealed to anyone observing you. An adversary that witnesses you going to a bank should naturally conclude that you likely have withdrawn money, and he does not need to launch any sophisticated cryptographic attacks to acquire your money (he simply robs you as you walk away from the bank).

In this bank example, a user's contextual information is revealed. More generally, the issue of contextual privacy has come up in other scenarios, such as database access and location-services. Generally speaking, there is a risk of a privacy breach when any entity A contacts another entity B asking for some service. For example, in the case of database privacy, when an entity A requests information from B, B's sole function should be to provide the service or answer to the query. It might not be desirable for B to know the details of the query or the specific answer [85–87].

In access control systems, there is always the risk of a privacy breach. When the user requests a service, this request can be logged by the entity providing the access control. In Kerberos, for example, all of the emphasis is placed on secure, authenticated exchanges but it is possible for the servers that administer service granting tickets to record which user has made a request for which service, thereby tracking a user's usage patterns or preferences. Similarly, in a centralized spatio-temporal access control, when the user attempts to prove that it is in a specific location, this can be recorded by the centralized entity, and used to infer the user's activities—it becomes possible to not only infer the user's current position, but also by accumulating the user's position over time it is possible to discover the user's habits.

In order to avoid this privacy risk, what is needed is a technique for access control that does not pass through a centralized entity. In our use of inverted sensor networks, the user (or users) are supplied content through some external means. For example, the content may be streaming content that is broadcast, and any entity that wants to access the content is free to do so by simply being at the right place at the right time to acquire the keys transmitted by the inverted sensor network. Since the users don't have to interact explicitly with any entity, there is no information revealed as to whether a specific user is accessing specific content, and there is no information revealed about the user's spatio-temporal profile.

6.6.2   Resistant to Positioning Spoofing

Any scheme that requires that a user prove that it is in a specific location becomes susceptible to attacks that might be launched against the localization infrastructure [47] (such as an adversary trying to prove that it is in a location that it is not, which would allow the adversary to access content he is not intended to access). Although as described in Chapter 5 secure localization schemes can mitigate this threat, there might be a limit to the effectiveness of these techniques against non-cryptographic attacks. By using inverted sensor networks, however, this issue is bypassed. There is no reliance on an entity proving its location to another entity. We do note, however, that inverted sensor networks can be physically attacked by adversaries destroying nodes, capturing and reprogramming nodes, or by simply covering sensors so that their transmissions are blocked. These issues however, are common to all wireless networks, and must be addressed through careful deployment of the sensor nodes (e.g. placing the nodes on the ceiling in a building might make it harder for vandalism).

6.6.3 Support of Applications with Little Effort

Taking advantage of a sensor network in an inverted fashion in order to facilitate spatio-temporal access control represents what we feel is an exciting opportunity to develop new classes of location-based applications. Programming STAC applications becomes relatively easy with the assistance of an inverted sensor network: sensor nodes can be deployed in support of STAC and then loaded with their key schedule; while a user's application simply needs to receive broadcast content, listen for the appropriate key, and then decrypt the content.

One promising new style of application that we envision is a spatio-temporal scavenger hunt, which might be an interesting paradigm for educational applications. In the scavenger hunt application, the user receives content and can only access it at the right place and right time. This content, once opened, might give the user a puzzle to solve describing where the user must next go to in order to advance to the next stage of the scavenger hunt. If the user solves the puzzle and makes it to the next location within a specified time limit, then the user would get the next puzzle. The process can continue until the user achieves the final objective.

There are many variations that are possible to the basic scavenger hunt. The objective of the scavenger hunt game can be specified by constructing a suitable transition diagram, which would detail the rules and paths allowed for a user to traverse the game. For example, we may create a cut through in the transition diagram (similar to the cut throughs in Figure 6.5) which would allow for a user to bypass the requirement of going to a certain area and accessing a certain object. This could correspond, for example, to a user solving a more challenging puzzle and being allowed to advance further in the game. Or, as another variation, if a user solves a puzzle and moves to the next location in a short amount of time, then the user might receive a different decryption key than if it had taken longer to solve the puzzle, thus allowing the user to access a different puzzle when it is in this location. Overall, we believe that inverted sensor networks can easily facilitate new classes of applications.

6.7 Conclusion

As wireless networks become increasingly prevalent, they will provide the means to support new classes of location-based services. One type of location-oriented service that can be deployed are those that make use of spatio-temporal location-information to control access to objects or services. In this paper we have examined the problem of location-based access control by exploring how access control policies can be formally specified, and presenting two different mechanisms for supporting spatio-temporal access control. We first showed how it is possible for a centralized entity, with the assistance of a localization infrastructure, can verify a claimed location. Once a location has been verified, an object can be distributed for access. The second approach that we explored, which was the primary focus of this paper, involved inverting the role of a sensor network. Specifically, an appropriately deployed sensor network can consist of nodes that locally transmit a schedule of keys, thereby facilitating access to enciphered objects. We then examined issues of optimizing the region covered by the sensor network in support of spatio-temporal access control by providing an algorithm for optimizing sensor node power allocation.

We believe that spatio-temporal access control in general, and inverted sensor networks in particular, represent a promising paradigm for the development of new location-oriented applications. The techniques outlined in this paper represent the beginning of a larger effort to develop spatio-temporal applications using inverted sensor networks, and we are currently implementing the system outlined in this paper.

Part II

Defending Against Wireless Spoofing Attacks

7

Relationship-based Detection of Spoofing-related Anomalous Traffic

7.1 Introduction

One serious threat facing wireless networks is spoofing, whereby one radio device can alter its network identifiers to that of another network device. Spoofing attacks are very easy to launch in many wireless networks. For example, in an 802.11 network, a device can alter its MAC address by simply issuing an `ifconfig` command. This weakness is serious, and there are numerous attacks, ranging from denial of service attacks [66] to session hijacking [88] to attacks on access control lists [89], that are facilitated by spoofing.

Although full-scale authentication mechanisms are the natural solution for coping with issues of identity verification, there are several reasons why it is desirable to explore other, complementary security measures for wireless networks. First, the use of authentication requires the existence of reliable key management/maintenance. This is often not possible. For example, WLAN hotspots are often deployed in local coffee shops or book stores without any means to check the identities of their constantly evolving user base and without any means to distribute authentication keys. Further, many wireless devices can be easily pilfered, their memory scanned, and their programming altered. This can lead to authentication keys becoming compromised and, in order to maintain the integrity of authentication checks, it is necessary to periodically refresh these keys through methods that are not subject to compromise [41–44]. Altogether, full-scale authentication [38–40] runs the risk of authentication keys being compromised and requires an extensive infrastructure to maintain the integrity of authentication methods.

In this paper, we take the viewpoint that it is desirable to have a light-weight security layer that is separate from conventional network authentication methods. Towards this objective, we propose a suite of light-weight security solutions for wireless networks that complement the use of conventional authentication services and are intended to detect multiple devices using the same network identity. Our approach involves introducing detectors of anomalous behavior within the medium access control (MAC) layer. If no authentication service is available, then our approach can serve as an effective anti-spoofing mechanism, while if authentication is available, then our approach can further mitigate the risk of compromised keys while lightening the load on authentication buffers, which are often the target of resource consumption attacks. Our approach does not replace authentication, as it does not rely on the explicit use of authentication keys to identify entities. Instead, our strategy involves the verification of forge-resistant relationships between packets coming from a claimed network identity. Our methods are generic, operate locally and are suitable for a broad array of wireless networks.

The rest of the paper is organized as follows. We first overview the basic strategy by describing a formal model for relationship-based security in Section 7.2. There are two varieties of relationships that we explore in this paper: first, relationships that are introduced through auxiliary fields in packets (Section 7.3); and second, relationships that result from the use of intrinsic properties associated with the transmission and reception of packets (Section 7.4). We support our strategies with theoretical analysis and provide guidelines for appropriately selecting associated parameters. We then provide an example of how our schemes can activate security responses through multi-level classification in Section 7.5. We validate our detection methods in Section 7.6, where we have conducted experiments using the ORBIT wireless testbed. Finally, Section 7.7 concludes the paper.

7.2   Strategy Overview

Typically, binding approaches are employed in order to defend against network identity spoofing [68, 69, 90]. The general scenario we are concerned with involves two or more devices claiming a particular network identity u_j. We refer to these devices as A, B, C, and so on. In general, we will only need to refer to three devices: A, B, and X. Device X is a monitoring device and is responsible for detecting anomalous network behavior. The monitor X may be another wireless network participant or it may be part of a separate security infrastructure. Associated with identity u_j will be a particular sequence of packets $\{P_j(k)\}$. The packet $P_j(k)$ consists of the payload, which comes from a higher-layer service. In addition to the payload, each packet has a state $S_j(k)$ associated with it. The state $S_j(k)$ may itself be a field contained within $P_j(k)$ or may be a property measured at the receiver.

For each identity u_j, we require that there is a rule, $\mathcal{R}$, that specifies the relationship between a set of observed $S_j(k)$ states. Suppose that we define an observation $\omega_j(k)$ to be a collection of N states corresponding to the kth packet from u_j. For example, one choice for $\omega_j(k)$ might be to take $\omega_j(k) = \{S_j(k), S_j(k-1)\}$, which is simply two consecutive states. A relationship consistency check (RCC) is a binary function $R_j(\omega_j(k))$ that returns 1 if the states in $\omega_j(k)$ obey the rule $\mathcal{R}$ with respect to each other (i.e. they are consistent), or returns 0 if they do not (i.e. they are inconsistent and suspicious). As we shall explore later, one possible relationship that we might require is that $S_j(k) = S_j(k-1) + 1 \pmod{M}$ for some modulus M. A potential RCC would return 1 if $S_j(k)$ is close to $S_j(k-1)$ or return 0 if they are far apart.

Simply using any relationship $\mathcal{R}$ and checking the corresponding RCC at a monitoring device is not enough to provide reliable security. We must have the guarantee that an adversary cannot easily forge a set ω_j that would pass the RCC. Therefore, just as it is necessary to add security properties to hash functions, in order to have cryptographically useful hash functions, we need to add forgeability requirements to the relationship $\mathcal{R}$. This leads to the following definition.

Definition 1. *An ϵ-forge-resistant relationship $\mathcal{R}$ is a rule governing the relationship between a set of states for a node j for which there is a small probability ϵ of another device being able to forge a set of states $\tilde{\omega}$ such that a monitoring device would evaluate the corresponding RCC as $R(\tilde{\omega}) = 1$.*

The definition of a forge-resistant RCC (RRCC) then corresponds to defining a detector capable of identifying anomalous network traffic resulting from the spoofing of another node's identity. The output $R_j(\omega_j(k))$ is either 1 or 0, and is a declaration by a monitoring device of whether suspicious network behavior has occurred. The output of the RRCC may be viewed as deciding between two different hypotheses based upon an observation vector $\omega_j(k)$. Hypothesis testing is thus the appropriate framework for defining a RRCC [91]. Suppose the observation vectors $\omega_j(k)$ for identity u_j are N-dimensional, then $R_j(\omega_j(k))$ partitions N-dimensional space into two critical regions Ω_0 and Ω_1. The region Ω_0 corresponds to those observation vectors that will be classified as non-suspicious data (the null hypothesis $\mathcal{H}_0$), while Ω_1 corresponds to those observation vectors that are suspicious or anomalous (the alternate hypothesis $\mathcal{H}_1$).

There are several measures for quantifying the effectiveness of an R. The probability of false alarm $P_{FA} = Pr(\mathcal{H}_1; \mathcal{H}_0)$ is the probability that we decide $\omega_j(k)$ is suspicious when it was legitimately created. The probability of missed detection $P_{MD} = Pr(\mathcal{H}_0; \mathcal{H}_1)$ is the probability of deciding the vector $\omega_j(k)$ is legitimate when it was actually created by an adversary. The power, or probability of detection, of R is simply $P_D = 1 - P_{MD}$. An

RRCC with a forge-resistance of ϵ therefore corresponds to an RRCC with $P_{MD} = \epsilon$. We note that it is often useful to characterize an RRCC via the pair $(\epsilon, \delta) = (P_{MD}, P_{FA})$, as there might be several choices for R that have equivalent probability of missed detections, yet drastically different false alarm characteristics. We therefore have the following definition:

Definition 2. *An RCC, R, corresponding to an ϵ-forge-resistant relationship $\mathcal{R}$, is an (ϵ, δ)-resistant RCC, or (ϵ, δ)-RRCC if $P_{FA} = \delta$.*

The (ϵ, δ)-RRCC will operate at a monitoring device to detect anomalous behavior. Based upon the choice of ϵ and δ, anomaly detection might be used to drive various security responses. For example, in a WLAN, if an access point detects suspicious network behavior with low δ values, the network might automatically switch to activating higher-layer authentication services. Or, if an RRCC with moderate δ values is used, the network might only respond by issuing a message to the network administrator, warning of a *potential network intrusion*.

7.3   Forge-resistant Relationships via Auxiliary Fields

There are several sources for forge-resistant relationships. We now look at two approaches that involve the use of auxiliary state fields contained in the packet to detect identity spoofing in wireless networks. We note that other strategies are possible, such as might arise from using specific signatures associated with the RF transmitter itself, which will be described in Chapter 15.

7.3.1   Anomaly Detection via Sequence Number Monotonicity

The first family of rules that we explore is based upon requiring packet sequence numbers to follow a monotonic relationship. This property is motivated by an observed behavior of the firmware of many 802.11 devices, and has been proposed for detecting spoofing [92,93]. In the discussion that follows, we shall formalize these earlier works by placing the sequence number method in the context of hypothesis testing, and deriving forge-resistance properties under different network conditions.

In 802.11 wireless networks (WLAN or adhoc), before transmitting each packet, an 802.11 header is appended. One field of the 802.11 header is the sequence control field, which is inserted directly by the firmware into the header. There are two parts that comprise the sequence control field: the fragmentation control, and the sequence number. The sequence number is a 12-bit field, providing sequence numbers with the range between 0 and 4095. Following each packet transmission, the firmware increments the sequence number field by 1. This monotonic relationship allows us to define a rule $\mathcal{R}_{seq}$. Except where noted, the remaining discussion regarding $\mathcal{R}_{seq}$ shall focus on the case of a 12-bit sequence number field, as is used in 802.11. Extension to arbitrary cases is straight-forward.

Before we discuss forge-resistance, we will examine the behavior of the sequence number field for two important cases: first, a single node using a specified MAC address transmitting packets to a receiver; second, two nodes using the same MAC address (one spoofing the other) to transmit packets. We present the results of a simple experiment using 802.11 devices in Figure 7.1, where the sequence number of consecutive received packets at the receiver is presented for these two cases[1]. When there is only one source, the sequence number is a monotonic function of the number of packets sent, and wraps back to 0 after the sequence number reaches 4095, as shown in part(a) of Figure 7.1. We provide a magnified depiction of the monotonicity in the lower part of Figure 7.1(a). The jump near packet number 3055 is due to packetloss. When there are two sources sending packets to the receiver, each device follows its own monotonic relationship, and the observed sequence number pattern is a mix of these two patterns, as shown in Figure 7.1 (b). The mixing results in a discontinuous trace, which an anomalous behavior we wish to detect.

The forge-resistance of $\mathcal{R}_{seq}$ is not dependent on whether another device can alter its sequence number. Although it is true that it is *somewhat* difficult for an adversary to set the sequence number field to an arbitrary value, it is possible for an adversary to bypass firmware and set an arbitrary value for the sequence number field. There have been several efforts recently to develop a software MAC layer for 802.11 that bypasses firmware restrictions and facilitate arbitrary packet generation [94, 95]. Instead, forge-resistance follows from the fact that, even if an adversary changes its sequence number based upon the last observed sequence number used by the legitimate device, the legitimate device will follow its own sequence and hence duplicate sequence numbers will be transmitted. As we will show, this duplication is detectable, and forge-resistance follows from the fact that an adversary cannot stop the legitimate device from transmitting.

Our discussions will consider the following traffic patterns :

- *Normal Operations with Packetloss:* Under benign conditions, a single node will transmit a stream of packets, where an occasional packet will be lost due to packetloss. This is not considered anomalous traffic.

- *Non-Adaptive Adversaries:* One or more adversaries will blindly spoof the MAC address of a device without adjusting their sequence num-

---

[1] We note that this behavior reflects both data and management packets. Even during transmission of management packets, the sequence number field is incremented. Additionally, we note that all sequence number experiments have retransmission disabled, as would correspond to multicast/broadcast mode.

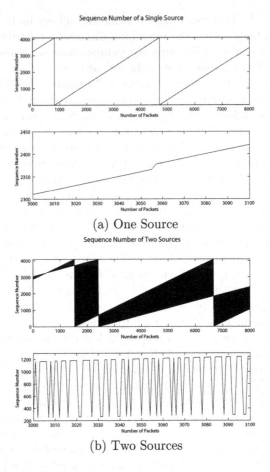

(a) One Source

(b) Two Sources

FIGURE 7.1. Sequence number of consecutive received packets at a receiver in case of different number of sources. In part (a), only one source is present in the network, while in part (b), there are two sources sending out packets to the same receiver. The lower figures in both part (a) and (b) are the zoomed-in versions of the corresponding upper figures.

bers. Highly discontinuous traces, such as depicted in Figure 7.1 (b), will result.

However, we note that more complex adversarial models are possible, where the adversary not only alters its MAC address, but also adapts its sequence number based on what it witnesses from the legitimate device. We note, however, that there is generally little advantage to the adversary in this case as, even if the adversary chooses to transmit a specific sequence number to pass a short term check, the legitimate device will eventually choose to transmit its own packet with that sequence number, thereby revealing an anomaly. Based on this observation, we focus our discussion on the two traffic cases above.

We now examine the Normal Operations with Packetloss case. When there is only one source sending out packets to the receiver, the sequence numbers associated with the source's MAC address should increase by precisely one (modulo 4096) for consecutively received packets under ideal network conditions. In practice, packetloss will cause the increment between successively received packets to be more than 1. Further, there is no case where the increment should be 0. The likelihood of the difference being greater than 1 depends on the packetloss rate of the link. Let $\{x_n, x_{n+1}, x_{n+2}, \cdots\}$ denote the sequence numbers of consecutive received packets, and $x_n \in [0, 4095]$. We assume that the packetloss occurs with a rate p, and that packet loss is independent between successive packets. We now note that the difference $x_{n+1} - x_n = \tau$ will lie between 1 and 4096 (A difference of 0 is considered an anomalous behavior and, hence, mapped to 4096). Hence, the probability of $x_{n+1} - x_n = \tau$ is $\frac{(1-p)p^{\tau-1}}{1-p^{4096}} \approx (1-p)p^{\tau-1}$ for $1 \leq \tau \leq 4096$. The approximation $(1-p)p^{\tau-1}$ is extremely close when p^{4096} is very close to 0, which is valid for any link of interest (e.g. packetloss $p < 0.5$). The approximation $(1-p)p^{\tau-1}$ gives $E[\tau] = \frac{1}{1-p}$, while the variance of τ is $\sigma_\tau^2 = E[\tau^2] - E[\tau]^2 = \frac{p}{(1-p)^2}$. By examining the formula for $E[\tau]$ and σ_τ^2, it is clear that the mean and standard deviation are relatively small (compared to 4096), even for low-quality links. For example, for 50% packetloss, $E[\tau] = 2$ and $\sigma_\tau = 1.41$. In general, we can infer that, even for wireless networks with poor connectivity, the difference between successive packets will not be large.

We now turn to the case of one or more devices spoofing another's identity and discuss appropriate classification policies. We start with the Non-Adaptive Adversaries case and look at the case of only one attacker. Let y denote the sequence number from the real source, and x the sequence number of the attacker. Assuming these two random variables are independent and uniformly distributed in $[0, 4095]$, then their difference $z = x - y$ follows a triangular distribution from -4095 to 4095 (the convolution of the distributions of x and y). If we consider the gap, $\tau = z \pmod{4096}$, and map a difference of 0 to 4096, then τ is a uniform distribution over $[1, 4096]$, and we have $E[\tau] = 2048.5$. In this case, the standard deviation is $\sigma = 1182$. Comparing the statistical behavior of the gap for the dual-source case with the gap for a single-source case, we see a large difference between the τ values, suggesting that the gap is a powerful statistical discriminant between normal network behavior and the behavior when the network is under a spoofing attack. Since τ is calculated using consecutive sequence numbers, the case of more than one adversary will appear similar to a dual-source case as it will involve at least two sources following their own sequence number progression. We may thus consider the multiple-source case as analogous to the dual-source case.

A natural question that arises is the issue: What if the adversarial device B transmitted first? Then, in this case the detection rule will follow B's sequence numbers, and when the legitimate device A finally communicates, the discontinuity will be detected. On the other hand, if the legitimate device never communicates, then the sequence number detector will track the adversarial sequence number and never identify an anomaly. Thus, anomalous traffic detection is only possible in the presence of heterogeneous sources: we must have the legitimate device transmit! This is a reasonable assumption as the legitimate device will at least transmit periodic control packets, such as beacons. For example, in 802.11 WLANs beacon frames are periodically broadcast by access points to facilitate AP discovery by client NICs.

We now turn to the issue of building a detector using sequence numbers, and quantifying its probability of detection and false alarm. We shall first focus our discussion on the more general Non-Adaptive case. Using $\mathcal{R}_{seq}$, we may define an RRCC detection scheme as follows. Rather than operate strictly on two consecutive packets, the detection scheme should operate on a window of packets coming from a specific MAC address u_j. Our detector uses a window $\omega_j(k) = \{S_j(k), S_j(k-1), \cdots, S_j(k-L-1)\}$, consisting of L consecutive sequence number state fields $S_j(k)$. The detector calculates the $L-1$ sequence number differences, $\{\tau_1, \tau_2, \cdots, \tau_{L-1}\}$, where $\tau_l = S_j(k-l+1) - S_j(k-l)$ (mod 4096). A difference of 0 is considered an anomalous behavior and, hence, mapped to 4096. By using a window of data points, we may define a family of detectors with varying sensitivity levels. The basic detector determines that anomalous behavior has occurred if $max_{l=1}^{L-1}\{\tau_l\} > \gamma$, where γ corresponds to a threshold that governs the probability of false alarm and missed detection. Alternative detector strategies might declare anomalous behavior if two or more difference τ_l are greater than a threshold.

Based on the probability calculations earlier and the window size L, we may calculate the ϵ and δ values for this RRCC. The null hypothesis $\mathcal{H}_0$ for this problem corresponds to a single source using a specific network identity. In order to calculate the probability of false alarm $P_{FA} = Pr(\mathcal{H}_1; \mathcal{H}_0)$ we must calculate the probability $Pr(\tau_l < \gamma; \mathcal{H}_0)$ for a threshold γ. Using the approximation, it is easy to see that $Pr(\tau_l < \gamma; \mathcal{H}_0) = (1-p^\gamma)$. Since we assumed that packetloss applies independently to each packet transmitted, we get $\delta = P_{FA} = 1 - (1-p^\gamma)^{L-1}$ for a window of size L. The calculation for $\epsilon = P_{MD} = Pr(\mathcal{H}_0; \mathcal{H}_1)$ involves calculating $Pr(\tau_l < \gamma; \mathcal{H}_1)$ for a threshold γ. As discussed earlier, under the adversarial hypothesis $\mathcal{H}_1$ the distribution of τ_l is uniform over $[1, 4096]$, and hence $Pr(\tau_l < \gamma; \mathcal{H}_1) = \frac{\gamma-1}{4096}$. The probability of missed detection is thus $\epsilon = Pr(\mathcal{H}_0; \mathcal{H}_1) = \left(\frac{\gamma-1}{4096}\right)^{L-1}$. The choice of γ may be determined by setting a desired probability of false alarm rate δ and solving for γ from $\delta = 1 - (1-p^\gamma)^{L-1}$. In order to illustrate the performance of the sequence number monotonicity detector, we plot receiver operating characteristic (ROC) curves for $p \in \{0.3, 0.5, 0.8\}$ and for $L = 2$ in Figure 7.2. In particular, we note that there is a rapid improvement in

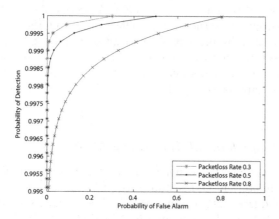

FIGURE 7.2. Receiver operating curve for sequence number monotonicity detector while $L = 2$.

probability of detection (i.e. $1 - \epsilon$) as packetloss p decreases. In fact, since most operational networks have packetloss smaller than 0.1, we can infer that the probability of detection will be very close to 1. Additionally, by referring to the equations for ϵ, it is clear that increasing L also rapidly increases the probability of detection.

The monitor should not observe any out of order sequence number packets. However, throughout the experimental validation, we witnessed that roughly 0.01% of the time packets came out of order, and that these were always beacon packets yielding a gap 4095. The reason lies in the fact that the device driver tries it's best to send out periodic beacons on time. The detection algorithm can be easily extended to include the out of order transmission situation by accepting an occasional gap of 4095 as a "single" source pattern. We now briefly discuss the performance of the detector in the presence of an Adaptive Adversary. We note that, for a window of packets, the Adaptive Adversary is likely to produce a cluster of sequence number gaps of 4096, which are easily detectable. As long as the adversary is not preventing the legitimate device from transmitting, the only factor that could prevent the detection of anomalous traffic in the Adaptive Adversary model is the ambient packetloss. However, the packetloss will affect both the legitimate sender and the attacker, equally and unpredictably for both! This packetloss implies that the adversary cannot control which packets it sends will get through. Further, since the adversary cannot predict which of the source's packets will successfully arrive, the adversary cannot do better than forging a packet with a sequence number that is one greater than the previous packet it has witnessed. This observation also implies that the adversary is most detectable if it is transmitting at roughly the same rate as the source and that, in order to avoid detection, the adversary should either inject very few packets, or flood the network with packets. For the first case, this has the effect of lessening the severity of the attack, while the second case would have the ultimate effect of making the adversary the only transmitter, thereby preventing the legitimate sender from transmitting. Later, in Section 7.5, we shall examine a detector that is quantifies the severity of a spoofing attack scenario by quantifying both the size of the sequence number gaps as well as the number of gaps present in a window.

7.3.2   One-Way Chain of Temporary Identifiers

In the previous section, the sequence numbers followed a publicly known pattern, and ultimately forge-resistance required a guarantee that a legitimate device would frequently communicate. We now explore an alternative, where the state field is also a temporary identifier that changes for each transmitted packet. However, unlike the sequence number case, for this method the temporary identifiers are difficult for the adversary to predict– the adversary must solve a cryptographic puzzle in order to figure out the next value of the temporary identifier. Thus, the puzzle must be solved before the adversary can transmit a single packet with a state field that would pass the RRCC. For every packet that the adversary wishes to send, he must solve another puzzle or randomly guess the state field value.

The advantage of this strategy is that it places extra burden on the adversary in order to prevent him from injecting traffic. Even if the legitimate source does not communicate, the adversary must solve the cryptographic puzzle for each packet it wishes to create in order to be able to not be detected.

In the One-Way Chain of Temporary Identifiers method a temporary identifier field (TIF) serves as the state field for anomaly detection. For identity u_j, the relationship between TIFs $S_j(k)$ will follow a reverse one-way function chain F [40, 96–98]. The source first chooses a number n that is larger than the number of packets that will be sent during the total period of communication. The source chooses a final TIF $S_j(k)$, and then calculates earlier TIFs via the reverse one-way chain $S_j(k-1) = F(S_j(k))$ for $1 \leq k \leq n$.

Suppose, at an arbitrary time during the operation of the protocol, a monitoring node associates a TIF $S_j(k)$ with identity u_j and that the next packet the monitor receives claiming to come from node u_j has TIF S'_j. We would like to test to see if S'_j could have produced $S_j(k)$ via the one-way function. Ideally, the verification process would check whether $S_j(k) = F(S'_j)$. If this fails, then the RRCC would declare that spoofing had occurred. However, in practice, there is packetloss, and this verification might fail for legitimate reasons. Therefore, much like is done in TESLA [97, 98], we use the loss-tolerant property of one-way chains. We declare that spoofing has occurred if the test $S_j(k) = F^l(S'_j)$ fails for all $1 \leq l \leq L$, where $F^l(x) = F(F(\cdots(x)\cdots))$ is l compositions of F applied to

x. Under this strategy, we test the chain up to L times in case there was severe packetloss.

We now examine the issue of forge-resistance. First, we examine the complexity associated with an adversary successfully creating a valid sequence of L TIFS. In order to succeed in this task, the adversary uses a previous observed TIF value, $S_j(k)$ and must solve for the L consecutive $S_j(k+1), S_j(k+2), \cdots, S_j(k+L)$ via $S_j(k+j) = F^{-1}(S_j(k+j-1))$. Even though the function F is publicly known, its cryptographic one-way pseudo-random properties implies that an adversary must essentially "brute-force" search for the inverse. Suppose that $F : \{0,1\}^b \rightarrow \{0,1\}^b$ is a one-way pseudo-random function then, on average, finding a single inverse will require 2^{b-1} applications of F, and similarly finding L inverses will, on average, require $L2^{b-1}$ applications of F. Since, by definition, F is itself a polynomial (in b) time algorithm, the $L2^{b-1}$ factor dominates the adversary's complexity. In particular, even modest values of b are sufficient. In practice, such pseudo-random functions can be created using encryption functions, such as AES.

Next, we examine the forge-resistance of this method by calculating P_{MD} and P_{FA} assuming that the adversary does not have the computational capabilities to invert the one-way function. Let us suppose that S'_j has been witnessed. The case $\mathcal{H}_0$ is that S'_j is legitimate, while the case $\mathcal{H}_1$ is that S'_j is forged. The probability of missed detection $P_{MD} = Pr(\mathcal{H}_0; \mathcal{H}_1)$ is the probability that the adversary was able to make an S'_j such that $S_j(k) = F^l(S'_j)$ for at least one l. The probability $P_{FA} = Pr(\mathcal{H}_1; \mathcal{H}_0)$ corresponds to the likelihood that at least L consecutive legitimate packets were lost due to packetloss. P_{FA} may be calculated using the packetloss rate p to be $\delta = P_{FA} = p^L/(1-p)$. Calculating P_{MD}, however, involves the properties of F. Again, suppose that $F : \{0,1\}^b \rightarrow \{0,1\}^b$ is a one-way pseudo-random function mapping b input bits into b output bits. Following the properties of cryptographic one-way functions, with good pseudo-randomness, the output distribution will be uniformly distributed over the 2^b possible outputs. Under this assumption, $\epsilon = P_{MD} = 1 - (1 - 2^{-b})^L$. Thus, we have an (ϵ, δ)-RRCC that is parameterized by the network packetloss p, the bitlength b of the TIF, and the length L of the verification window.

We illustrate the performance of the TIF detector by examining ROC curves for $p \in \{0.3, 0.5, 0.8\}$ and for $b = 10$ and $b = 16$ in Figure 7.3. From these curves, we see that the probability of detection increases as p decreases, and that the probability of detection increases as b increases. The result is intuitive since, as b increases, it becomes harder for the adversary invert the one-way function or do an exhaustive search to find next TIF. The TIF detector achieves high probability of detection with small probability of false alarm. By comparing the ROC curves for TIF and the sequential detector, we see that even a small value of b can be quite powerful. Moving

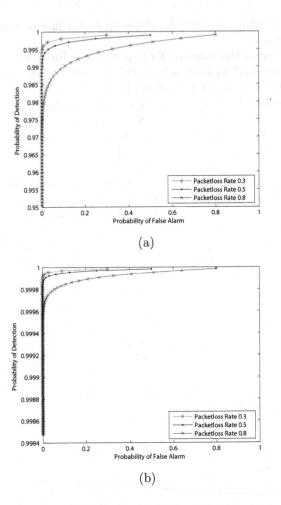

FIGURE 7.3. (a) Receiver operating curve for One-Way Chain of Temporary
Identifiers detector with $b = 10$. (b) Receiver operating curve for One-Way Chain
of Temporary Identifiers detector with $b = 16$

from $b = 10$ to $b = 16$ leads to the TIF ROC curves having better detection
performance than the sequential ROC curves.

Finally, we note that, for our purpose, the initial TIF $S_j(0)$ need not be
distributed to the monitoring nodes. If $S_j(0)$ is distributed to the monitor using an authenticated channel, then we will be able to perform entity
authentication, similar to the strategy proposed in LHAP [38, 39]. This
bootstrapping incurs extra overhead associated with authentication, and
is contrary to our objective of detecting whether there is more than one
source present. We emphasize that our objective is not conventional authentication, but rather to detect whether there is more than one source

present. This means that, if the adversary is spoofing another device's identity, and the other entity never communicates, then we will not detect this. However, if the legitimate device (or any other device attempting to spoof that identity) communicates, the TIF strategy will capture this anomalous traffic scenario without requiring authentication!

Finally, we discuss the storage overhead of TIFs. If the number of packets within a communication session is large, then the required one-way key chain could be long. One may employ efficient one-way function chain constructions to reduce the storage requirements [99]. Further, the sources may use a single TIF for a fixed amount of packets at a time. If an adversary uses the same TIF for the transmission of its forged packets, the monitor can detect anomalous traffic by observing that the amount of packets sharing the same TIF is larger than the maximum amount allowed. The detection of anomalous traffic follows from the fact that an adversary cannot stop the legitimate device from using the same TIF. We note, though, that for network scenarios with packet loss, the receivers may receive fewer packets with the same TIF than the threshold amount, but this is a situation that an adversary cannot control or seek to exploit. In this case, the monitor can transition to the next TIF (even though not all packets for a particular TIF have been used) by verifying the next TIF according to the one-way chain property.

7.4 Forge-resistant Relationships via Intrinsic Properties

In the previous section, we introduced an extra field into the packets to create relationships for anomaly detection. A potential drawback of those methods, though, is that they introduce additional communication overhead. In this section, we seek to exploit properties that are inherent in the transmission and reception of packets, and hence do not require additional communication overhead.

7.4.1 Traffic Arrival Consistency Checks

Our basic strategy is to use traffic shaping to control the single-hop interarrival times observed by a monitoring device within the radio range of the source. The interarrival statistics are then used to discriminate between spoofing and non-spoofing scenarios.

Suppose we have a limited network scenario involving only two nodes A and B using the same network identity, and a monitoring device X that records the times at which it witnesses packets coming from that network identity. In particular, this network scenario does not involve any background traffic and, hence, the traffic that will be monitored by X is only a result of A and B's behavior.

Now suppose, without loss of generality, that device A is the initial device using a particular network identity. For the traffic arrival consistency check, device A will send out packets such that the time between packets being transmitted follows a fixed distribution. For example, A might act as a Poisson source with rate λ, or a constant rate (CBR) source with a fixed period. Throughout the operation of this consistency check, A will maintain this behavior. Regardless of A's behavior, we assume that the monitor X knows the interarrival times τ follow a specified distribution $f_T(\tau)$.

Now, when the second source B starts communicating, a new distribution will be observed. The traffic arrival relationship $\mathcal{R}_{tr}$ corresponds to the distribution $f_T(\tau)$ that X witnesses for A. When the RRCC corresponding to $\mathcal{R}_{tr}$ is used, a test distribution $f_E(\tau)$ is measured and the objective is to decide whether $f_E(\tau)$ corresponds to $f_T(\tau)$ or not.

For the purpose of our discussion, we shall use the χ^2-test for goodness of fit. We note, though, that other tests, like the Kolmogorov-Smirnov test [100], are also suitable. Let us suppose the distribution $f_T(\tau)$ is divided into M adjacent intervals to yield the discrete distribution f_i for $1 \le i \le M$. The test distribution $f_E(\tau)$ will then be measured as a histogram using the same partitions. For n samples of the test distribution, we will denote N_i to be the amount of occurrences of the test events in the ith partition, and thus $N_1 + N_2 + \cdots N_M = n$. Then the χ^2-test statistic , is

$$\chi^2 = \sum_{i=1}^{M} \frac{(N_i - nf_i)^2}{nf_i}. \tag{7.1}$$

Our null hypothesis for this problem $\mathcal{H}_0$ is that the N_i follow the same distribution as f_i. If the new observations N_i match the original distribution f_i, then we expect the test statistic χ^2 to be small. We reject $\mathcal{H}_0$ if $\chi^2 > \chi^2_{M-1,1-\delta}$, where $\chi^2_{M-1,1-\delta}$ is the $1 - \delta$ critical point for the chi-squared distribution with $(M-1)$ degrees of freedom. This threshold corresponds yields a probability δ of false alarm, i.e. rejecting $\mathcal{H}_0$ when the measurements in fact do follow f_i.

We now explore the basic behavior that we can expect regarding $\mathcal{R}_{tr}$. In particular, suppose that A produces $f_A(\tau)$ at X. Then, if B starts communicating, it will increase the amount of traffic that X witnesses coming from that network identity. As a result, it should be expected that the interarrival times will decrease on average. One might expect that an anomalous $f_E(\tau)$ will be shifted to the left compared to $f_A(\tau)$. As we shall see later in the experimental section, we do witness an average decrease in the interarrival times. We have depicted this in Figure 7.4. However, the characteristics of an anomalous $f_E(\tau)$ can exhibit complicated phenomena compared to $f_A(\tau)$. As in Figure 7.4, the combination of two CBR sources can create a bimodal distribution.

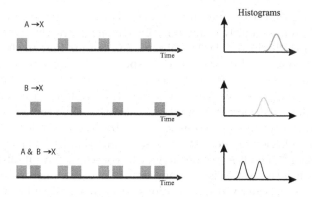

FIGURE 7.4. The basic phenomena underlying traffic shaping and traffic relationship methods for anomaly detection. Here A transmits with a specific traffic
distribution. When A and B both communicate, the resulting histogram will appear quite different from either distribution.

The traffic arrival consistency check can be a very powerful method for
detecting anomalous traffic. It should be emphasized, however, that this
powerful detection method is possible because the device A follows a fixed
traffic pattern. This necessitates that A perform traffic shaping and control the time at which packets are transmitted. When higher-layer services
supply packets to the network stack, it is necessary that the network stack
perform traffic shaping at the various buffers internal to the network software in order to control the transmission rate of packets. Ultimately, this
detection strategy is appropriate when delay-tolerant services are running
on the wireless network.

7.4.2 Joint Traffic Load and Interarrival Time Detector

Merely using the traffic arrival statistics alone in deciding whether spoofing
is present does not yield a reliable detector as there are other factors besides spoofing that can affect the traffic pattern. In particular, for wireless
networks that employ carrier sensing as a basis for medium access control,
the presence of background traffic from other communicators can affect the
observed traffic arrival pattern. Although we may attempt to use judicious
buffering strategies to control the release of packets to conform to a desired
interarrival distribution, for carrier-sensing based wireless networks we are
nonetheless forced to delay transmission until an existing transmission com-
pletes. This introduces extra delay to the observed interarrival times, and
this amount of delay increases as the background traffic levels increase.
Additionally, packet losses will also cause the traffic interarrival increase.

Therefore, using just the interarrival distribution alone is not suitable
for detecting spoofing when there is large amounts of background traffic.
Rather, we should take into account the level of background traffic and

how this might affect the interarrival times. For this more general scenario, we instead propose jointly examining the average interarrival time and the background traffic load. The key observation here is that in the presence of no background traffic, we expect a specific average interarrival time, which can be directly determined from the original traffic distribution that A transmits, while this average interarrival time will increase as the background traffic levels increase. Suppose we define $\bar{\tau}$ to be the observed average interarrival time, and Λ to denote the observed traffic load. Following the methods of anomaly detection, we may partition $(\Lambda, \bar{\tau})$ space into two distinct regions. The first region, Region I, corresponds to normal system behavior (when the system is not under attack). Region I may be defined either using formally derived specifications for normal system behavior (as is done in specification-based anomaly detection [101]) or empirically through a training phase (such as is done in network intrusion detection [102, 103]). In our work, we shall focus on empirical methods in which, during a training phase, $(\Lambda, \bar{\tau})$ data is collected from non-adversarial traffic scenarios. The region of normal system behavior may then be defined by binning the data according to Λ values and calculating a prediction interval for a specified percentile level for observed $\bar{\tau}$ data. For example, we may calculate the 99% prediction interval for the observed data. Region I is then defined as the region falling above the horizon connecting the percentile levels from different Λ bins. Region II is defined as the complement of Region I, and is the region that falls below Region I. Region II corresponds to non-normal system behavior and thus represents anomalous scenarios.

Later, in Section 7.6, we will revisit this classifier by presenting case studies where we illustrate how these regions may be defined through an example using on experimental data.

7.5   Enhanced Detectors using Multi-Level Classification

The previous two sections described two practical methods for detecting anomalous traffic. These methods were binary classifiers. In practice, however, it is desirable to attach a measurement of the severity of a threat with an assessment. For example, if the detector is to send a message to the network administrator, it is highly desirable (from the administrator's point of view) that the message should measure the severity of the potential security breach, thereby allowing the administrator to prioritize and gauge his response. In this section, we will show how to take an RRCC and associate with it an additional measurement quantifying the severity of the violation. Due to space limitations, we present our multi-level classifier for only the $\mathcal{R}_{seq}$ case and note that the extension to both of the traffic analysis methods of Section 7.4 is straight-forward.

To begin, we note that for $\mathcal{R}_{seq}$ described earlier, using shorter time windows makes it harder to draw a conclusion about a long-term threat pattern (perhaps a single abnormal gap value is not considered a severe threat). Further, in the discussion of $\mathcal{R}_{seq}$, there was no discrimination based on the size of a sequence number gap: a gap of 10 is weighted the same as a gap of 1000, though clearly a gap of 1000 is a more anomalous event.

When assessing the severity of a threat, it is desirable to utilize the distribution of the observed gap statistics over a sufficiently long time window, and to weight larger gaps more heavily. In order to meet these two criteria, we propose the use of a *weighted severity metric* . In describing the weighted severity metric, suppose that the monitor has knowledge of the normal (single-source for a single identity) distribution f_j, for $j \in [1, 4096]$, corresponding to the sequence number gaps of devices in the ad hoc network (e.g. this could be measured a priori empirically). During operation of the network, the monitor will observe the sequence number gaps for windows of L packets (here L should be large enough to characterize current traffic pattern). From this window, the monitor obtains the frequency for each possible gap value, i.e. q_j for $j \in [1, 4096]$. The weighted severity metric is defined as

$$D(f, q) = \sum_{j=1}^{4096} w_j |f_j - q_j|, \tag{7.2}$$

where w_j is the weight for the jth sequence number gap. There are various ways to define w_j, and the definition should satisfy the following properties: (1) all w_j should be nonnegative; and (2) w_j should be monotonic with j, as a large sequence number gap should contribute more to the severity metric than a small sequence number gap. In this work, we define $w_j = -\log(p^{j-1}) = -(j-1)\log p$, where p is the average network packetloss rate. Hence, a sequence number of gap 1, which indicates single source traffic, does not contribute to the severity metric, as $w_1 = 0$. It is clear that the severity metric not only describes the degree of the deviation of the gaps, but also the frequency of large deviations. A few occurrences of small sequence number gaps will not produce a large value, while either frequent small gaps or a few large gaps will yield larger severity values.

Using the severity metric, we now show how it is possible to define a multi-level classifier that classifies traffic into three different categories: benign, low-threat, and severe-threat. Our multi-level classifier uses an L-packet time window, and the severity $D = D(f, q)$ for that window. Under the normal (single-source) scenario, the average and standard deviation of sequence number gaps are small. For example, for a packetloss rate of 50%, the average and standard deviation are 2 and 1.41, respectively. We thus can map out a benign region (Region I), corresponding to small D values. For this region, the detector concludes that no spoofing is present. For the dual-source case the average and standard deviation of sequence number gap were 2048.5 and 1182, and hence larger w_j values will factor into the calculation of D. We may thus define a severe-threat region, Region III, which corresponds to large D values. A D measurement falling in Region III would indicate significant amounts of spoofing activity, and hence trigger a severe-threat alarm. Region II corresponds to an intermediate case (e.g. the adversary was not persistent and only spoofed for a short period of time), and falls in between Region I and Region III. In this case, a warning message would state that a low-threat violation occurred. The specification of the boundaries corresponds to selecting the threshold levels, s_1 and s_2, and depends on the link quality and the definition of w_j. These values may be adjusted according to the network administrator's definition of benign, low-threat, and high-threat events, and is a matter of the administration security policy. Later, in Section 7.6, we will explore a more detailed example of this scheme.

7.6   Experimental Validation on the ORBIT Wireless Testbed

Although we have provided a theoretical framework for detecting anomalous traffic associated with device spoofing, we believe that it is necessary to evaluate wireless protocols in actual systems implementations. In particular, it is recognized that theoretical and simulation methods often fall short of capturing actual system behavior, especially for wireless networks [104, 105]. Notably, theory and simulations simplify assumptions about the physical layer and the complex inter-layer interactions associated with wireless networking.

In this section, we evaluate our proposed methods using the ORBIT wireless testbed. The ORBIT testbed is an open-access resarch facility intended to provide a flexible platform for the evaluation of future wireless networking protocols, middleware, and applications [106]. ORBIT consists of a grid of wireless nodes, with the nodes separated by 1m, as depicted in Figure 7.5 (a). Each node, depicted in Figure 7.5 (b), has a 1GHz VIA C3 processor with 512MB of RAM, a 20GB local hard disk, two 802.11a/b/g interfaces. Data is collected via a library of measurement tools [107] and backhauled via Ethernet for databasing and storage. Currently, both Atheros and Intel 802.11a/b/g cards are available on ORBIT. In our experiments, it was not necessary to use the full testbed, and we selected a subset of the Atheros nodes in our tests.

In the discussions that follow, we have chosen not to include results on $\mathcal{R}_{TIF}$ as the use of a cryptographic field in the $\mathcal{R}_{TIF}$ method makes the challenge of detecting spoofing entirely dependent on an adversary's ability to invert the one-way function chain, and not dependent on any properties of the wireless network. Hence, the performance of $\mathcal{R}_{TIF}$ can be completely

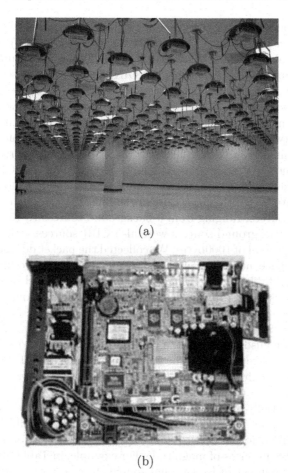

(a)

(b)

FIGURE 7.5. (a) The ORBIT wireless testbed. (b) An ORBIT node.

evaluated using the pseudorandom properties of the underlying one-way function.

7.6.1   Validation of Detection using Sequence Numbers

From the discussion in Section 7.3, the average sequence number gap for consecutively received packets depends on the link packet loss rate. Further, the average value and standard deviation are very small for packets from the same source, while there is large variation when more than one source uses the same identity.

We performed several experiments to validate the utility of sequence numbers for detecting more than one entity using a device. Our experiments consisted of two sources A (which acted as the legitimate source) and B

(which acted as a non-Adaptive spoofing source) that shared the same MAC address, and a monitoring device X. A was located at $(3, 2)$ on ORBIT, while B was at $(4, 5)$, and X was at $(5, 4)$. In order to capture the effect of different link quality conditions on ORBIT, we selected several of the other ORBIT nodes to act as background traffic sources. The background traffic sources used their own MAC addresses and only served to adjust packetloss. The first experiment we conducted was to quantify the mean difference between consecutive packets from the same MAC address when there was a single source A, and when there was spoofing present (i.e. B using the same MAC address as A). Our sources operated as CBR sources, emitting packets with an interarrival rate of 10msec and a payload length of 1000 bytes. We varied the packetloss in the links by introducing additional background sources, ranging from 0 background sources to 15 background sources. Our background sources were also CBR sources with a period of 50msec and payload of 1000bytes. We collected the packet delivery statistics and the sequence number gap statistics and report the results in Table 7.1. From this table, we see that, as we increase the level of background traffic, the packetloss rate increases for both the single source and dual-source experiments. When there is only a single source, the mean sequence number gap τ is small (less than 3) and the standard deviation σ of the sequence gap is also small (on the order of 0.5). This trend exists for the single-source case regardless of the background traffic level. In contrast, the dual-source case exhibits significantly higher τ and σ across all background traffic levels.

Finally, we briefly make a qualitative comparison between the experimental results of Table 7.1 and Section 7.3.A. Examining the table shows that the general trend for theoretical and experimental values for $E[\tau]$ and σ_τ are comparable in terms of magnitude. For example, in Table I(a), when we increased the number of background sources to 4 to get a packetloss rate of $p = 0.011$, we observed an experimental $E[\tau] = 1.1$, while theoretical

TABLE 7.1. Sequence number gap statistics for 802.11 sources with varying levels of background traffic.

# background sources	0	1	4	7	15
Packetloss Rate	0	0.001	0.011	0.062	0.440
$E[\tau]$	1.0	1.1	1.1	1.2	2.0
σ_τ	0	0.2	0.3	0.5	0.2

(a) Single-source case

# background sources	0	1	4	7	15
Packetloss Rate	0.011	0.030	0.060	0.080	0.498
$E[\tau]$	1955	1985	1916	1964	515
σ_τ	2037	393	1531	599	1083

(b) Dual-source case

TABLE 7.2. Sequence number gap statistics for the dual-source case with 10 background sources versus the adversary's sending rate.

Time between packets sent by B	∞msec	10msec	50msec	100msec	200msec
Packetloss Rate	0.018	0.048	0.037	0.036	0.002
$E[\tau]$	1.4	676	1878	1419	1007
σ_τ	0.5	1175	1140	1346	1340

results would yield $E[\tau] = 1.011$. For the standard deviation, we observed experimental values of $\sigma_\tau = 0.3$ while theoretical values would yield 0.106. Similarly, as noted earlier, for the dual source (adversarial) case, Table I(b) shows the dramatic increase in sequence number gap $E[\tau]$ and σ_τ when compared to the single source case, which was noted in the theoretical discussion. We note that although the general trends between theory and experimental results are consistent, mismatch between the precise values is expected. This mismatch arises as a result that the theoretical calculations are based on several independence assumptions: packetloss is independent across packets, and further that the adversary's packets are sent independently of each other. Neither assumption can be controlled in a realistic experimental scenario.

The second experiment examined the effect of an increased adversarial rate of transmission on the sequence number gap. In this experiment A was a 10msec CBR source, while we varied the period at which B sent packets. In the experiment, B varied from sending as a 200msec CBR source to a 10msec CBR source. The average gap along with the standard deviation error bars are reported in Table 7.2. With the exception of the ∞msec case, the gap values are very large. It should be noted that the ∞msec case corresponds to B being absent (single-source only).

Taken together, these two experiments conclusively indicate that there is a large deviation in behavior between the behavior of the sequence number gap when there is a single source and when there are two devices sharing the same identity. This suggests that the sequence number field is very promising for detecting spoofing.

We built an anomalous traffic detection algorithm using the window of sequence numbers method described in Section 7.3. In this experiment, we looked at two different window sizes of $L = 2$ and $L = 10$ packets. In both cases, the source A was Poisson source with average interarrival time 40msec. We introduced an adversary that acted as a Poisson source with average interarrival time λ, and varied λ from ∞ (no adversary) to 10msec to 200msec. Additionally, we introduced 15 background sources on ORBIT to control packetloss. These background sources transmitted packets as Poisson sources, each with an interarrival rate of 40msec. In all cases, the packets transmitted were 1000bytes. We calculated P_D and P_{FA} for a threshold of $\gamma = 4$ and $\gamma = 6$, and present the results in Table 7.3. The use of a longer window significantly reduces the probability of missed

TABLE 7.3. P_{MD} and P_{FA} for $\mathcal{R}_{seq}$ RRCC with different thresholds γ when A is Poisson with interarrival time 40msec, B is Poisson with interarrival time λ, and 15 background Poisson sources of average interarrival time 40msec.

| λ for B | $L = 2$ packets | | | | $L = 10$ packets | | | |
| | $\gamma = 4$ | | $\gamma = 6$ | | $\gamma = 4$ | | $\gamma = 6$ | |
	P_{MD}	P_{FA}	P_{MD}	P_{FA}	P_{MD}	P_{FA}	P_{MD}	P_{FA}
∞	0	0.001	0	0.0004	0	0.008	0	0.001
200	0.228	0.001	0.228	0.0008	0.018	0.001	0.019	0.0007
100	0.248	0.0001	0.249	0	0.013	0.0006	0.015	0
20	0.230	0	0.230	0	0.016	0	0.018	0
10	0.180	0.0001	0.180	0	0.032	0	0.032	0

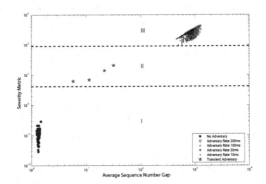

FIGURE 7.6. Severity metric versus average sequence number gap with a window size of $L = 100$ packets, when A is Poisson with interarrival time 40msec. Several cases for the adversary B are presented: Poisson with interarrival time ∞ (no adversary), 200msec, 100msec, 20msec and 10msec; and a transient adversary that interjects only a few packets. For all cases, there are 15 background Poisson sources of average interarrival rate 40msec.

detection. In some cases, when a longer window is used, there is an slight increase in likelihood of false alarm.

We next examined the multi-level classifier described in Section 7.5. We processed the same set of data used in Table 7.3 with a window size of $L = 100$ packets. The packet loss p was estimated from the collected data for the single-source case and used to define w_j's. In order to assist in the visualization of our data, we plot the severity versus the average sequence number gap for different adversarial rates in Fig. 7.6. In this figure, there are three visible "clusters," which will correspond to the three regions of Section 7.5.

When there is no adversary, the severity metric is small, yielding values below 0.5 (and the average sequence number gap is 1.6). When two persistent sources exist, in spite of the adversary's rate, the severity metric significantly increases to 300 (here, the average gap is roughly 900), as shown in the upper right corner. Since our objective was to study a multi-level classifier, we next looked at the case of a "transient" adversary, one that only spoofed a small amount of packets in the time window. Specifically, we had an adversary B that would imitate A by randomly sending between 1 to 3 packets during each 100 packet window, with gap values from 450 to 1000 for these packets. For the transient adversary, the D values were between the values observed for the benign and full adversarial cases, as shown in the figure. We note that adjusting the amount of spoofed packets in a 100 packet window, or the gap values would lead to larger or smaller D values that would continuously fall between the two extremes of benign and full adversarial cases.

In practice, the administrator should define the boundaries between these regions. To accomplish this, the regions may be defined heuristically, or based upon an organizationally-defined security policy. A quantitative approach to defining these regions would involve a training phase, where benign and full adversarial cases are run, and the data is used to define Region I and Region III directly (Region II would follow directly as the region in between these two regions) For example, in the data presented in Fig. 7.6, we used the D-values to find the (upper) 99.9% prediction interval thresholds for no adversary case, and the (lower) 99.9% prediction interval thresholds for the full adversary cases. The resulting thresholds were $d_1 = 4.20$ and $d_2 = 90.13$ (calculated as the minimum of the 99.9% prediction intervals from each of the adversarial data sets). We have indicated these thresholds using the dashed line in Fig. 7.6. For this choice of thresholds, we are employing a conservative strategy for defining normal traffic (as represented by the large gap between the non-adversarial data and the threshold d_1). Yet, in spite of this conservative threshold, the transient adversary would result in a "low-threat" warning.

7.6.2 Validation of Detection using Traffic Statistics

In order to validate the $\mathcal{R}_{tr}$ method, we conducted an experiment on OR-BIT where we controlled the rate at which data packets were released to the card at the sources, and witnessed the corresponding distributions at the monitoring node. The experimental layout of the traffic statistics tests was slightly different from ORBIT experiments used for $\mathcal{R}_{seq}$ and is presented in Figure 7.7 (a). In this set up, we had two different source nodes, and one monitoring node. Node A was located at $(3, 2)$, node B at $(4, 5)$, while the monitor X was at $(5, 4)$.

In these experiments, we used A as the original source and had A release packets at a constant rate of 200msec. In addition to our source, we had 1 background traffic source that transmitted as a CBR source with a period of 2msec between packets. The background traffic did not use the same MAC address as A or the adversarial node B. The purpose of this background source was to provide a channel where occasional medium access contention would occur. Our monitor X recorded the time at which packets were received and passed up the network stack. The time stamping was conducted at the MAC layer. These readings were used to calculate the interarrival times between packets, and the resulting histogram for the interarrival times when A transmitted with the background traffic is presented in Figure 7.7 (b).

We then looked at two dual-source cases. The first dual-source case involved node A communicating as a 200msec source, while B also communicated as a 200msec constant rate source. The resulting histogram is presented in Figure 7.7 (c). The second dual-source case that we performed an experiment for involved A communicating as a 200msec source, while B communicated as a 50msec constant rate source. The histogram for the second dual-source case is presented in Figure 7.7 (d). By examining these histograms, it is clear that neither of the dual source cases looks similar to the single-source case. In the first dual-source case in Figure 7.7 (c), we see that the resulting distribution is bimodal. In order to see why this occurs, we refer the reader back to Figure 7.4, where the time between an A and B packet is less than the time between a B and A packet. In Figure 7.7 (d), we do not see a bimodal distribution, but instead see a dominant mode at roughly 50msec interarrival time, and the rest of the probability spread out between 0 and 50msec. This is due to the fact that the 50msec rate is so much faster than the 200msec rate, which causes only a few occurrences of an interarrival time less than 50msec.

Although it is clear that these two dual-source distributions do not look anything like the single source distribution, we nonetheless performed the Chi-Squared test . We broke the data from the two dual-source cases into data windows of $n = 250$ data readings, and calculated the χ^2 statistic for each data window for each of the dual-source data sets. We then calculated the mean and variance of the χ^2 statistics. For the first dual-source case, the average χ^2 statistic was $\overline{\chi}^2 = 139386$, and had a standard deviation of $\sigma_{\chi^2} = 2073$. The second dual-source case had an average χ^2 value of $\overline{\chi}^2 = 143073$, with a standard deviation of $\sigma_{\chi^2} = 6 \times 10^{-6}$. The small value for σ_{χ^2} for the second case is due to the fact that the dual-source distribution did not overlap the single source distribution. For these tests, the 1% critical value was 9.21. Since the χ^2 statistics were significantly larger than the 1% critical value in both cases, we may numerically conclude that the detector would discriminate between single and dual-source cases.

The second set of traffic experiments we conducted involved A acting as a source that transmitted packets with an interarrival time that was uniformly drawn from 150 to 200msec. During these experiments, we had a single background traffic source operating as a CBR source with interarrival time 2msec. The resulting histogram of interarrival times witnessed at X is presented in Figure 7.8 (b). We then performed two dual-source cases. The first involved an adversary B that transmitted packets with an interarrival time of $Unif(150, 200)$msec, while the second case involved an adversary B that transmitted packets at an interarrival rate of $Unif(0, 50)$msec. The resulting interarrival distribution observed at X are presented in Figures

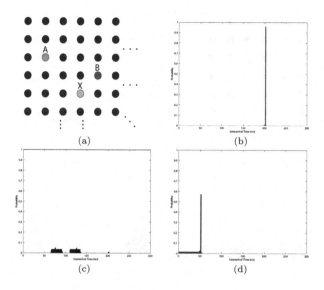

FIGURE 7.7. (a) Experimental setup on the ORBIT grid, node separation is 1 meter. Node A is $(3, 2)$, B is $(4, 5)$. The monitor X is at $(5, 4)$. (b) Interarrival time distribution when A communicates as a CBR with packet rate 200msec to X. (c) Interarrival time distribution when both A and B communicate as CBR with packet rate 200msec. (d) Interarrival time distribution when A is a 200msec constant rate source, and B is a CBR of 50msec. All experiments had 1 background CBR source with period 2msec.

7.8 (c) and (d). From these figures, it is again clear that there is significant deviation between the single source and dual-source cases. To corroborate this claim, we report the Chi-squared test when we broke the data gathered into data windows of $n = 250$ records. The average χ^2 statistic for the first dual-source case was $\overline{\chi}^2 = 7535$, and had a standard deviation of $\sigma_{\chi^2} = 749$. The average χ^2 statistic for the second dual-source case was $\overline{\chi}^2 = 9973$, and had a standard deviation of $\sigma_{\chi^2} = 5 \times 10^{-12}$. The low standard deviation of the second dual-source experiment is due to the lack of overlap between the $A \rightarrow X$ and $A + B \rightarrow X$ distributions for this case. For both dual-source cases, the 1% critical value for the Chi-squared test was 37.57, and thus conclusively show that the traffic discrimination test is quite powerful at differentiating between single and dual-source cases under more arbitrary source distributions.

7.6.3 Validation of the Joint Traffic Arrival and Traffic Load Detector

The background traffic loads used in the previous set of experiments is relatively light, and we were interested in how traffic arrival statistics would change as we increased the levels of the background traffic. In order to vali-

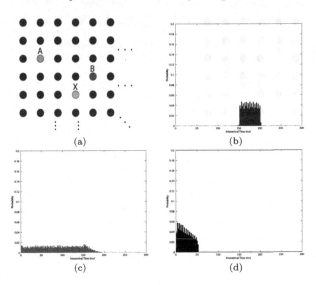

FIGURE 7.8. (a) Experimental setup for uniform source A. (b) Interarrival time distribution observed at X when A is a $Unif(150, 200)$msec source and 1 background CBR source with interarrival time 2msec, (c) Interarrival time distribution observed at X when A is a $Unif(150, 200)$msec source, B is $Unif(150, 200)$msec, and 1 background CBR source with interarrival time 2msec, (d) Interarrival time distribution observed at X when A is a $Unif(150, 200)$msec source, B is $Unif(0, 50)$msec and 1 background CBR source with interarrival time 2msec.

date our detector of Section 7.4 B, which jointly used a measurement of the local channel traffic load and the average interarrival times, we conducted a series of experiments on ORBIT where we controlled the rate at which the packets are released from the sources. Our experimental layout was the same as that in Figure 7.7 (a), where node A was located at (3,2), node B at (4,5) (which claims the same MAC address with node A), and the monitor X was at (5,4). Further, we varied the amount of background traffic sources up to a maximum of 4 sources (each background source used its own MAC address). It should be noted that the background sources used in this experiment more heavily occupy the channel than the background used in the experiments of Section 7.6.2. Typical with most wireless LAN deployments, all nodes were within radio range of every other node.

The first set of experiments was chosen to illustrate how the traffic interarrival distribution changes with the total amount of traffic load in the networks. In the first experiment, we used A as a source that released packets at a constant rate of 200msec. The traffic interarrival distribution is shown is Fig. 7.9 (a). In addition to the source, we added up to 4 background constant rate traffic sources, each transmitting packets with a constant rate of 2msec, into the network. The traffic interarrival distributions for packets from A with 1, 2, and 4 background sources are shown in Fig. 7.9 (b), (c)

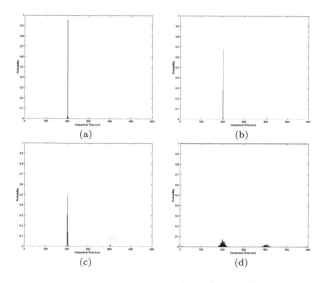

FIGURE 7.9. (a) Interarrival time distribution when A communicates as a CBR with packet rate 200msec to X. (b) Interarrival time distribution when A communicates as a CBR with packet rate 200msec to X, with 1 background 2msec CBR traffic source.(c) Interarrival time distribution when A communicates as a CBR with packet rate 200msec to X, with 2 background 2msec CBR traffic sources. (d) Interarrival time distribution when A communicates as a CBR with packet rate 200msec to X, with 4 background 2msec CBR traffic sources.

and (d), respectively. The network becomes more and more congested as more background traffic sources are put into the network. The packetloss rates for 1, 2 and 4 background sources are 3%, 6% and 45%, respectively. It is clear that the traffic interarrival distribution of congested networks does not follow the distribution corresponding to no background traffic.

We conducted the same experiments for a uniform distribution source, where node A released packets as a uniform distribution of $Unif(150, 200)$msec. Similarly, we collected the traffic interarrival distribution for different degrees of background traffic, as shown in Figure 7.10. We drew the same conclusion that the traffic interarrival distribution for congested networks does not follow the distribution without any background sources even no adversary is involved.

In order to validate the use of $(\Lambda, \bar{\tau})$ to differentiate between anomaly and congestion, we conducted a second series of experiments on the ORBIT wireless testbed. We used the same network topology as the first series of the experiments. We collected the total amount of traffic load in units of bytes and interarrival time of packets from node A's MAC address. As before, we look at cases where the source is a CBR source with rate 200msec, and is a uniform source with rate $Unif(150, 200)$msec. In both experiments, we collected data for different levels of background traffic, ranging from light

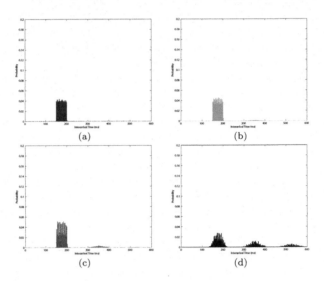

FIGURE 7.10. (a) Interarrival time distribution when A communicates as a uniform with packet rate $Unif(150, 200)$msec to X. (b) Interarrival time distribution when A communicates as a uniform with packet rate $Unif(150, 200)$msec to X, with 1 background 2msec CBR traffic source. (c) Interarrival time distribution when A communicates as a uniform with packet rate $Unif(150, 200)$msec to X, with 2 background 2msec CBR traffic sources. (d) Interarrival time distribution when A communicates as a uniform with packet rate $Unif(150, 200)$msec to X, with 4 background 2msec CBR traffic sources.

weight background with packetloss rate 0%, to a severely congested network scenario with packetloss rate at around 45%.

The total network traffic load in units of *kbps* versus the average interarrival time in a window size of 40sec for the constant rate source is shown in Figure 7.11. The comparison of the effect of the same amount of background traffic and anomalous traffic is shown in Figure 7.11 (a). The average interarrival time remains constant only when light weight background traffic is present for non-spoofed scenarios. When the adversary sends anomalous traffic at the same rate as the background traffic sources, we observed roughly the same amount of total traffic load, but a smaller average interarrival time. When the network becomes increasingly congested, the total amount of traffic and average interarrival time increase for both spoofed and non-spoofed scenario, as shown in Figure 7.11 (b). In all cases, the average interarrival time involving spoofing is always smaller than those of non-spoofed under the same traffic load. In these figures, we have used the 99.5% prediction interval for the observed benign data in order to define the boundary between Region I and Region II. We have depicted the depiction intervals in the figure for each cluster of benign data, and have connected the lower frontier of these regions using a dashed line. Region II is defined as the region that falls below this dashed line.

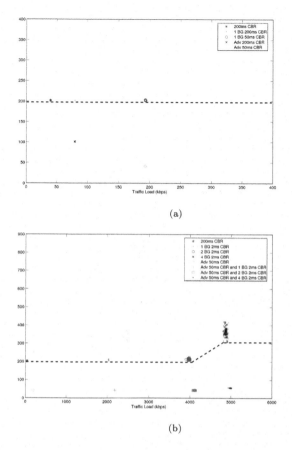

(a)

(b)

FIGURE 7.11. (a) Total network traffic load in kbps versus the average inter-arrival time for light background traffic. (b) Total network traffic load in kbps versus the average interarrival time for medium to heavy background traffic. In both figures, we have defined the boundary between Region I and Region II using the 99.5% prediction interval estimates.

We also collected the total amount of network traffic load versus average interarrival time for the uniform source. The data is collected with a window size of 40sec and is shown in Figure 7.12. Similar to the experiments for the CBR source, we present the same $(\Lambda, \bar{\tau})$ diagrams in Figure 7.12 (a) and (b), where we have varied the traffic from light in Figure 7.12 (a) to heavier in Figure 7.12 (b). As before, we have divided the two dimensional space into two regions using the 99.5% prediction intervals from the benign data. From both of these sets of experiments, we can conclude the validity of using $(\Lambda, \bar{\tau})$ to differentiate between anomalous (spoofed) traffic scenarios and congested traffic scenarios.

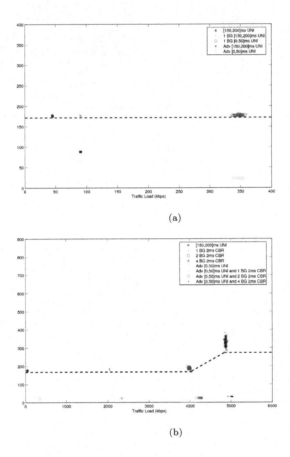

(a)

(b)

FIGURE 7.12. (a) Total network traffic load in kbps versus the average inter-arrival time for light background traffic. (b) Total network traffic load in kbps versus the average interarrival time for medium to heavy background traffic. In both figures, we have defined the boundary between Region I and Region II using the 99.5% prediction interval estimates.

7.7 Conclusion

In this paper we have presented an alternative to traditional identity-oriented authentication methods for detecting device spoofing on a wireless network. Our strategy uses relationships that exist within a stream of packets coming from an individual network identity. Whenever an adversary spoofs a particular identity, the existence of multiple sources causes these relationships to be difficult for an adversary to forge. As a result, it becomes likely that the adversary will reveal its presence. We proposed two different families of relationships that are suitable for wireless networks. The first family involves introducing additional fields in transmitted packets, while the second family involves the implicit properties associated with the transmission and reception of packets. Specifically, we proposed the use of the monotonicity of the sequence number field and the use of temporary identifier fields that evolve according to a one-way function chain. Further, for relationships based upon implicit packet-related properties, we propose that traffic interarrival statistics may be used for detecting anomalous traffic scenarios. We illustrated how these RRCC can be augmented with a measurement of the threat severity in order to facilitate multi-level classification. In all cases, we use these relationships to build forge resistant consistency checks (RRCC) to detect anomalous behavior. Our RRCC do not require the explicit use or establishment of cryptographic keys, and thus RRCC are suitable for application scenarios where the maintenance of keying material is not practical. We supported the validity of our proposed methods through experiments conducted on the ORBIT 802.11 wireless network testbed.

8
Detecting and Localizing Wireless Spoofing Attacks

8.1 Introduction

Due to the openness of wireless and sensor networks, they are especially vulnerable to spoofing attacks where an attacker forges its identity to masquerade as another device, or even creates multiple illegitimate identities. Spoofing attacks are a serious threat as they represent a form of identity compromise and can facilitate a variety of traffic injection attacks, such as evil twin access point attacks. It is thus desirable to detect the presence of spoofing and eliminate them from the network.

The traditional approach to address spoofing attacks is to apply cryptographic authentication. However, authentication requires additional infrastructural overhead and computational power associated with distributing, and maintaining cryptographic keys. Due to the limited power and resources available to the wireless devices and sensor nodes, it is not always possible to deploy authentication. In addition, key management often incurs significant human management costs on the network. In this chapter, we take a different approach by using the physical properties associated with wireless transmissions to detect spoofing. Specifically, we propose a scheme for both detecting spoofing attacks, as well as localizing the positions of the adversaries performing the attacks. Our approach utilizes the Received Signal Strength (RSS) measured across a set of access points to perform spoofing detection and localization. Our scheme does not add any overhead to the wireless devices and sensor nodes.

By analyzing the RSS from each MAC address using K-means cluster algorithm, we have found that the distance between the centroids in signal space is a good test statistic for effective attack detection. We then describe how we integrated our K-means spoofing detector into a real-time indoor localization system. Our K-means approach is general in that it can be applied to almost all RSS-based localization algorithms. For two sample algorithms, we show that using the centroids of the clusters in signal space as the input to the localization system, the positions of the attackers can be localized with the same relative estimation errors as under normal conditions.

To evaluate the effectiveness of our spoofing detector and attack localizer, we conducted experiments using both an 802.11 network as well as an 802.15.4 network in a real office building environment. In particular, we have built an indoor localization system that can localize any transmitting devices on the floor in real-time. We evaluated the performance of the K-means spoofing detector using detection rates and receiver operating characteristic curve. We have found that our spoofing detector is highly effective with over 95% detection rates and under 5% false positive rates.

Further, we observed that, when using the centroids in signal space, a broad family of localization algorithms achieve the same performance as when they use the averaged RSS in traditional localization attempts. Our experimental results show that the distance between the localized results of the spoofing node and the original node is directly proportional to the true distance between the two nodes, thereby providing strong evidence of the effectiveness of both our spoofing detection scheme as well as our approach of localizing the positions of the adversaries.

The rest of the chapter is organized as follows. In Section 8.2, we study the feasibility of spoofing attacks and their impacts, and discuss our experimental methodologies. We formulate the spoofing attack detection problem and propose K-means spoofing detector in Section 8.3. We introduce the real-time localization system and present how to find the positions of the attackers in Section 8.4. Further, we provide a discussion in Section 8.6. Finally, we conclude our work in Section 8.7.

8.2   Feasibility of Attacks

In this section we provide a brief overview of spoofing attacks and their impact. We then discuss the experimental methodology that we use to evaluate our approach of spoofing detection.

8.2.1   Spoofing Attacks

Due to the open-nature of the wireless medium, it is easy for adversaries to monitor communications to find the layer-2 Media Access Control (MAC) addresses of the other entities. Recall that the MAC address is typically used as a unique identifier for all the nodes on the network. Further, for most commodity wireless devices, attackers can easily forge their MAC address in order to masquerade as another transmitter. As a result, these attackers appear to the network as if they are a different device. Such spoofing attacks can have a serious impact on the network performance as well as facilitate many forms of security weaknesses, such as attacks on access control mechanisms in access points [89], and denial-of-service through a deauthentication attack [66]. An overview of possible spoofing attacks can be found in Chapter 7 and [108].

To address potential spoofing attacks, the conventional approach uses authentication. However, the application of authentication requires reliable key distribution, management, and maintenance mechanisms. It is not always desirable to apply authentication because of its infrastructural, computational, and management overhead. Further, cryptographic methods are susceptible to node compromise– a serious concern as most wireless nodes are easily accessible, allowing their memory to be easily scanned.

It is desirable to use properties that cannot be undermined even when nodes are compromised. We propose to use received signal strength (RSS), a property associated with the transmission and reception of communication (and hence not reliant on cryptography), as the basis for detecting spoofing. Employing RSS as a means to detect spoofing will not require any additional cost to the wireless devices themselves– they will merely use their existing communication methods, while the wireless network will use a collection of base stations to monitor received signal strength for the potential of spoofing.

8.2.2 Experimental Methodology

In order to evaluate the effectiveness of our spoofing detection mechanisms, which we describe in the next section, we have conducted experiments using both an 802.11 (WiFi) network as well as an 802.15.4 (ZigBee) network on the 3rd floor of the Computer Science Department at Rutgers University. The floor size is 200x80ft (16000 ft^2). Figure 8.1 (a) shows the 802.11 (WiFi) network with 4 landmarks deployed to maximize signal strength coverage, as shown in red squares. The 802.15.4 (ZigBee) network is presented in Figure 4.2 (b) with 4 landmarks distributed in a squared setup in order to achieve optimal landmark placement [73] as shown in red triangles. The small blue dots in the floor map are the locations used for spoofing and localization tests.

For the 802.15.4 network, we used 300 packet-level RSS samples for each of the 100 locations. We utilized the actual RSS values attached to each packet. We have 286 locations in the 802.11 deployment. Unlike the 802.15.4 data, the RSS values are partially synthetic. We had access to only the mean RSS at each location, but to perform our experiments we needed an RSS value per packet. To generate such data for 200 simulated packets at each location, we used random draws from a normal distribution. We

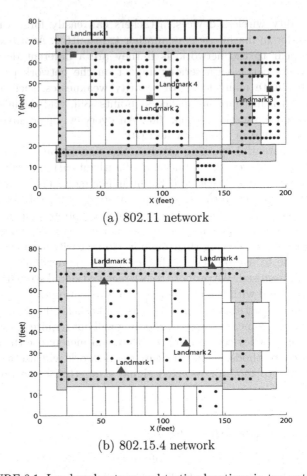

(a) 802.11 network

(b) 802.15.4 network

FIGURE 8.1. Landmark setups and testing locations in two networks.

used the measured RSS mean for the mean of the distribution. For the standard deviation, we computed the difference in the RSS from a fitted signal to distance function, and then calculated the standard deviation of the distribution from these differences over all locations. To keep our results conservative, we took the maximum deviation over all landmarks, which we found to be 5 dB.

Much work has gone into characterizing the distributions of RSS readings indoors. It has been shown that characterizing the per-location RSS distributions as normal, although not often the most accurate characterization, still results in the best balance between algorithmic usability and the resulting localization error [15, 109].

In addition, we built a real-time localization system to estimate the positions of both the original nodes and the spoofing nodes. We randomly selected points out of the above locations as the training data for use by the localization algorithms. For the 802.11 network, the size of the training data is 115 locations, while for the 802.15.4 network, the size of the training data is 70 locations. The detailed description of our localization system is presented in Section 8.4.

To test our approach's ability to detect spoofing, we randomly chose a point pair on the floor and treated one point as the position of the original node, and the other as the position of the spoofing node. We ran the spoofing test through all the possible combinations of point pairs on the floor using all the testing locations in both networks. There are total 14535 pairs for the 802.11 network and 4371 pairs for the 802.15.4 network. The experimental results will be presented in the following sections for the spoofing detector and the attack localizer.

8.3   Attack Detector

In this section we propose our spoofing attack detector. We first formulate the spoofing attack detection problem as one using classical statistical testing. Next, we describe the test statistic for spoofing detection. We then introduce the metrics to evaluate the effectiveness of our approach. Finally, we present our experimental results.

8.3.1   Formulation of Spoofing Attack Detection

RSS is widely available in deployed wireless communication networks and its values are closely correlated with location in physical space. In addition, RSS is a common physical property used by a widely diverse set of localization algorithms [12, 15, 16, 110]. In spite of its several meter-level localization accuracy, using RSS is an attractive approach because it can re-use the existing wireless infrastructure. We thus derive a spoofing attack detector utilizing properties of the RSS.

The goal of the spoofing detector is to identify the presence of a spoofing attack. We formulate the spoofing attack detection as a statistical significance test, where the null hypothesis is:

$$\mathcal{H}_0 : \text{normal (no attack)}.$$

In significance testing, a test statistic $\mathbf{T}$ is used to evaluate whether observed data belongs to the null-hypothesis or not. If the observed test statistic $\mathbf{T^{obs}}$ differs significantly from the hypothesized values, the null hypothesis is rejected and we claim the presence of a spoofing attack.

8.3.2   Test Statistic for Spoofing Detection

Although affected by random noise, environmental bias, and multipath effects, the RSS value vector, $\mathbf{s} = \{s_1, s_2, ... s_n\}$ (n is the number of landmarks/access points(APs)), is closely related with the transmitter's physical location and is determined by the distance to the landmarks [15]. The RSS readings at different locations in physical space are distinctive. As described in Chapter 3 each vector s corresponds to a point in a n-dimensional signal space. When there is no spoofing, for each MAC address, the sequence of RSS sample vectors will be close to each other, and will fluctuate around a mean vector. However, under a spoofing attack, there is more than one node at different physical locations claiming the same MAC address. As a result, the RSS sample readings from the attacked MAC address will be mixed with RSS readings from at least one different location. Based on the properties of the signal strength, the RSS readings from the same physical location will belong to the same cluster points in the n-dimensional signal space, while the RSS readings from different locations in the physical space should form different clusters in signal space.

This observation suggests that we may conduct K-means cluster analysis [111] on the RSS readings from each MAC address in order to identify spoofing. If there are M RSS sample readings for a MAC address, the K-means clustering algorithm partitions M sample points into K disjoint subsets S_j containing M_j sample points so as to minimize the sum-of-squares criterion:

$$J_{min} = \sum_{j=1}^{K} \sum_{s_m \in S_j} \|s_m - \mu_j\|^2 \qquad (8.1)$$

where s_m is a RSS vector representing the mth sample point and μ_j is the geometric centroid of the sample points for S_j in signal space. Under normal conditions, the distance between the centroids should be close to each other since there is basically only one cluster. Under a spoofing attack, however, the distance between the centroids is larger as the centroids are derived from the different RSS clusters associated with different locations in physical space. We thus choose the distance between two centroids as the test statistic **T** for spoofing detection,

$$D_c = \|\mu_i - \mu_j\| \qquad (8.2)$$

with $i, j \in \{1, 2..K\}$. Next, we will use empirical methodologies from the collected data set to determine thresholds for defining the critical region for the significance testing. To illustrate, we use the following definitions, *an original node P_{org}* is referred to as the wireless device with the legitimate MAC address, while *a spoofing node P_{spoof}* is referred to as the wireless device that is forging its identity and masquerading as another device. There can be multiple spoofing nodes of the same MAC address.

Note that our K-means spoofing detector can handle packets from different transmission power levels. If an attacker sends packets at a different transmission power level from the original node with the same MAC address, there will be two distinct RSS clusters in signal space. Thus, the spoofing attack will be detected based on the distance of the two centroids obtained from the RSS clusters.

8.3.3 Determining Thresholds

The appropriate threshold τ will allow the spoofing detector to be robust to false detections. We can determine the thresholds through empirical training. During the off line phase, we can collect the RSS readings for a set of known locations over the floor and obtain the distance between two centroids in signal space for each point pair. We use the distribution of the training information to determine the threshold τ. At run time, based on the RSS sample readings for a MAC address, we can calculate the observed value D_c^{obs}. Our condition for declaring that a MAC address is under a spoofing attack is:

$$D_c^{obs} > \tau. \tag{8.3}$$

Figure 8.2 (a) and (b) show the CDF of the D_c in signal space for both the 802.11 network and the 802.15.4 network. We found that the curve of D_c shifted greatly to the right under spoofing attacks, thereby suggesting that using D_c as a test statistic is an effective way for detecting spoofing attacks.

8.3.4 Performance Metrics

In order to evaluate the performance of our spoofing attack detector using K-means cluster analysis, we use the following metrics:

Detection Rate and False Positive Rate: A spoofing attack will cause the significance test to reject $\mathcal{H}_0$. We are thus interested in the statistical characterization of the attack detection attempts over all the possible spoofing attacks on the floor. The detection rate is defined as the percentage of spoofing attack attempts that are determined to be under attack. Note that, when the spoofing attack is present, the detection rate corresponds to the probability of detection P_d, while under normal (non-attack) conditions it corresponds to the probability of declaring a false positive P_{fa}. The detection rate and false positive rate vary under different thresholds.

Receiver Operating Characteristic (ROC) curve: To evaluate an attack detection scheme we want to study the false positive rate P_{fa} and probability of detection P_d together. The ROC curve is a plot of attack detection accuracy against the false positive rate. It can be obtained by varying the detection thresholds. The ROC curve provides a direct means to measure the trade off between false-positives and correct detections.

8.3.5 Experimental Evaluation

In this section we present the evaluation results of the effectiveness of the spoofing attack detector. Table 8.1 presents the detection rate and false

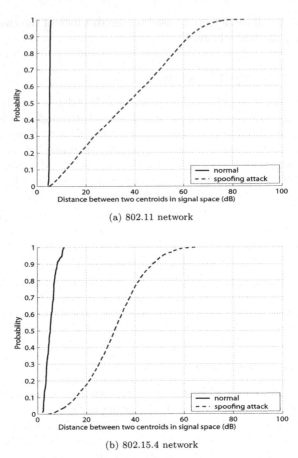

(a) 802.11 network

(b) 802.15.4 network

FIGURE 8.2. Cumulative Distribution Function (CDF) of D_c in signal space

positive rate for both the 802.11 network and the 802.15.4 network under different threshold settings. The corresponding ROC curves are displayed in Figure 8.3. The results are encouraging showing that for false positive rates less than 10%, the detection rates are above 95%. Even when the false positive rate goes to zero, the detection rate is still more than 95% for both 802.11 and 802.15.4 networks.

We further study how likely a spoofing node can be detected by our spoofing attack detector when it is at varying distances from the original node in physical space. Figure 8.4 presents the detection rate as a function of the distance between the spoofing node and the original node. We found that the further away P_{spoof} is from P_{org}, the higher the detection rate becomes. For the 802.11 network, the detection rate goes to over 90% when P_{spoof} is about 13 feet away from P_{org} under τ equals to 5.5dB. While for the 802.15.4 network, the detection rate is above 90% when the distance between P_{spoof} and P_{org} is about 20 feet by setting threshold τ to 9dB.

TABLE 8.1. Detection rate and false positive rate of the spoofing attack detector.

Network, Threshold	Detection Rate	False Positive Rate
802.11, $\tau = 5.5$dB	0.9937	0.0819
802.11, $\tau = 5.7$dB	0.9920	0.0351
802.11, $\tau = 6$dB	0.9884	0
802.15.4, $\tau = 8.2$dB	0.9806	0.0957
802.15.4, $\tau = 10$dB	0.9664	0.0426
802.15.4, $\tau = 11$dB	0.9577	0

This is in line with the average localization estimation errors using RSS [15] which are about 10-15 feet. When the nodes are less than 10-15 feet apart, they have a high likelihood of generating similar RSS readings, and thus the spoofing detection rate falls below 90%, but still greater than 60%. However, when P_{spoof} moves closer to P_{org}, the attacker also increases the probability to expose itself. The detection rate goes to 100% when the spoofing node is about 45-50 feet away from the original node.

8.4   Localizing Adversaries

If the spoofing attack is determined to be present by the spoofing attack detector, we want to localize the adversaries and further to eliminate the attackers from the network. In this section we present a real-time localization system that can be used to locate the positions of the attackers. We then describe the localization algorithms used to estimate the adversaries' position. The experimental results are presented to evaluate the effectiveness of our approach.

8.4.1   Localization System

We have developed a general-purpose localization system to perform real-time indoor positioning. The detailed system architecture is presented in [112] Here we provide a brief system overview. This system is designed with fully distributed functionality and easy to plug-in localization algorithms. It is built around 4 logical components: Transmitter, Landmark, Server, and Solver. The system architecture is shown in Figure 8.5.

8.5   Architecture Design

Transmitter: Any device that transmits packets can be localized. Often the application code does not need to be altered on a sensor node in order to localize it.

Landmark: The Landmark component listens to the packet traffic and extracts the RSS reading for each transmitter. It then forwards the RSS information to the Server component. The Landmark component is stateless and is usually deployed on each landmark or access point with known locations.

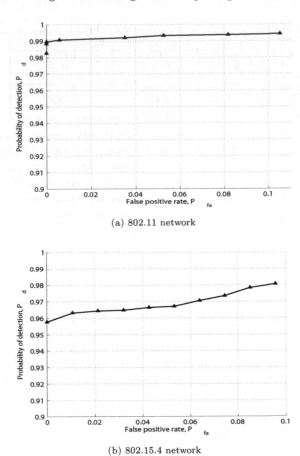

(a) 802.11 network

(b) 802.15.4 network

FIGURE 8.3. Receiver Operating Characteristic (ROC) curves

Server: A centralized server collects RSS information from all the Landmark components. The spoofing detection is performed at the Server component. The Server summarizes the RSS information such as averaging or clustering, then forwards the information to the Solver component for localization estimation.

Solver: A Solver takes the input from the Server, performs the localization task by utilizing the localization algorithms plugged in, and returns the localization results back to the Server. There are multiple Solver instances available and each Solver can localize multiple transmitters simultaneously.

During the localization process, the following steps will take place:

1. A Transmitter sends a packet. Some number of Landmarks observe the packet and record the RSS.

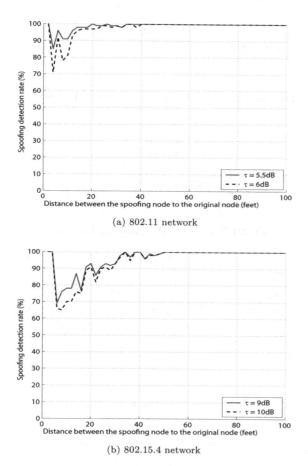

(a) 802.11 network

(b) 802.15.4 network

FIGURE 8.4. Detection rate as a function of the distance between the spoofing node and the original node.

2. Each Landmark forwards the observed RSS from the transmitter to the Server.

3. The Server collects the complete RSS vector for the transmitter and sends the information to a Solver instance for location estimation.

4. The Solver instance performs localization and returns the coordinates of the transmitter back to the Server.

If there is a need to localize hundreds of transmitters at the same time, the server can perform load balancing among the different solver instances. This centralized localization solution also makes enforcing contracts and privacy policies more tractable.

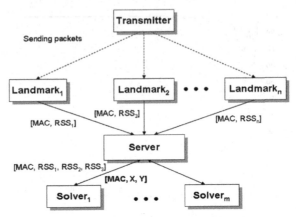

FIGURE 8.5. GRAIL system architecture

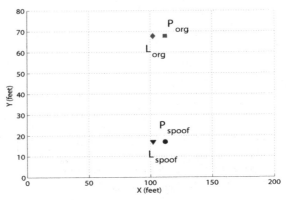

FIGURE 8.6. Relationships among the original node, the spoofing node, and their location estimation through localization system.

8.5.1 Attack Localizer

When our spoofing detector has identified an attack for a MAC address, the centroids returned by the K-means clustering analysis in signal space can be used by the server and sent to the solver for location estimation. The returned positions should be the location estimate for the original node and the spoofing nodes in physical space. Using a location on the testing floor as an example, Figure 8.6 shows the relationship among the original node P_{org}, the location estimation of the original node L_{org}, the spoofing node P_{spoof}, and the localized spoofing node position L_{spoof}.

In order to show the generality of our localization system for locating the spoofing nodes, we have chosen two representative localization algorithms using signal strength from point-based algorithms and area-based algorithms.

RADAR: Point-based methods return an estimated point as a localization result. A primary example of a point-based method is the RADAR scheme [16]. In RADAR, during the off line phase, a mobile transmitter with known position broadcasts beacons periodically, and the RSS readings are measured at a set of landmarks. Collecting together the averaged RSS readings from each of the landmarks for a set of known locations provides a radio map. At runtime, localization is performed by measuring a transmitter's RSS at each landmark, and the vector of RSS values is compared to the radio map. The record in the radio map whose signal strength vector is closest in the Euclidean sense to the observed RSS vector is declared to correspond to the location of the transmitter. In this work, instead of using the averaged RSS in the traditional approach, we use the RSS centroids obtained from the K-means clustering algorithm as the observed RSS vector for localizing a MAC address.

Area Based Probability (ABP): Area-based algorithms return a most likely area in which the true location resides. One major advantage of area-based methods compared to point-based methods is that they return a region, which has an increased chance of capturing the transmitter's true location. ABP returns an area, a set of tiles on the floor, bounded by a probability that the transmitter is within the returned area [15]. ABP assumes the distribution of RSS for each landmark follows a Gaussian distribution. The Gaussian random variable from each landmark is independent. ABP then computes the probability of the transmitter being at each tile L_i on the floor using Bayes' rule:

$$P(L_i|\mathbf{s}) = \frac{P(\mathbf{s}|L_i) \times P(L_i)}{P(\mathbf{s})}. \tag{8.4}$$

Given that the transmitter must reside at exactly one tile satisfying $\sum_{i=1}^{L} P(L_i|\mathbf{s}) = 1$, ABP normalizes the probability and returns the most likely tiles up to its confidence α.

Both RADAR and ABP are employed in our experiments to localize the positions of the attackers.

8.5.2   Experimental Evaluation

In order to evaluate the effectiveness of our localization system in finding the locations of the attackers, we are interested in the following performance metrics:

Localization Error CDF: We obtain the cumulative distribution function (CDF) of the location estimation error from all the localization attempts, including both the original nodes and the spoofing nodes. We then compare the error CDF of all the original nodes to that of all the possible spoofing nodes on the floor. For area-based algorithms, we also report CDFs of the minimum and maximum error. For a given localization attempt, these are points in the returned area that are closest to and furthest from the true location.

Relationship between the true and estimated distances: The relationship between the true distance of the spoofing node to the original node $||P_{org} - P_{spoof}||$ and the distance of the location estimate of the spoofing node to that of the original node $||L_{org} - L_{spoof}||$ evaluates how accurate our attack localizer can report the positions of both the original node and the attackers.

We first present the statistical characterization of the location estimation errors. Figure 8.7 presents the localization error CDF of the original nodes and the spoofing nodes for both RADAR and ABP in the 802.11 network as well as the 802.15.4 network. For the area-based algorithm, the median tile error $ABP - med$ is presented, as well as the minimum and maximum tile errors, $ABP - min$ and $ABP - max$. We found that the location estimation errors from using the RSS centroids in signal space are about the same as using averaged RSS as the input for localization algorithms [15]. Comparing to the 802.11 network, the localization performance in the 802.15.4 network is qualitatively better for both RADAR and ABP algorithms. This is because the landmark placement in the 802.15.4 network is closer to that predicted by the optimal and error minimizing placement algorithm as described in [73].

More importantly, we observed that the localization performance of the original nodes is qualitatively the same as that of the spoofing nodes. This is very encouraging as the similar performance is strong evidence that using the centroids obtained from the K-means cluster analysis is effective in both identifying the spoofing attacks as well as localizing the attackers.

The challenge in localizing the positions of the attackers arises because the system does not know the positions of either the original MAC address or the node with the masquerading MAC. Thus, we would like to examine how accurate the localization system can estimate the distance between P_{org} and P_{spoof}. Figure 8.8 displays the relationship between $||L_{org} - L_{spoof}||$ and $||P_{org} - P_{spoof}||$ across different localization algorithms and networks. The blue dots represent the cases of the detected spoofing attacks. While the red crosses indicate the spoofing attack has not been detected by the K-means spoofing detector. Comparing with Figure 8.4, i.e. the detection rate as a function of the distance between P_{org} and P_{spoof}, the results of the undetected spoofing attack cases represented by the red crosses are in line with the results in Figure 8.4, the spoofing attacks are 100% detected when $||P_{org} - P_{spoof}||$ equals to or is greater than about 50 feet.

Further, the relationship between $||L_{org} - L_{spoof}||$ and $||P_{org} - P_{spoof}||$ is along the 45 degree straight line. This means that $||L_{org} - L_{spoof}||$ is directly proportional to $||P_{org} - P_{spoof}||$ and indicates that our localization system is highly effective for localizing the attackers. At a fixed distance value of $||P_{org} - P_{spoof}||$, the values of $||L_{org} - L_{spoof}||$ fluctuate around the true

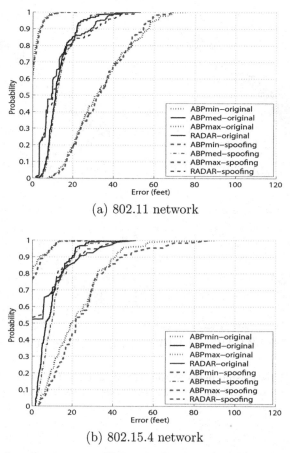

(a) 802.11 network

(b) 802.15.4 network

FIGURE 8.7. Localization error CDF across localization algorithms and networks.

distance value. The fluctuation reflects the localization errors of both P_{org} and P_{spoof}. The larger the $||P_{org} - P_{spoof}||$ is, the smaller the fluctuation of $||L_{org} - L_{spoof}||$ becomes, at about 10 feet maximum. This means that if the attacker is farther away from the original node, it is extremely likely that the K-means spoofing detector can detect it. In addition, our attack localizer can find the attacker's position and estimate the distance from the original node to the attacker at about 10 to 20 feet maximum error.

8.6 Discussion

So far we have conducted K-means cluster analysis in signal space. Our real-time localization system also inspired us to explore packet-level localization at the server, which means localization is performed for each packet received at the landmarks. The server utilizes each RSS reading vector for

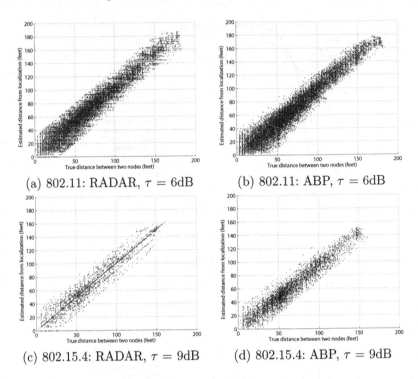

(a) 802.11: RADAR, $\tau = 6$dB   (b) 802.11: ABP, $\tau = 6$dB

(c) 802.15.4: RADAR, $\tau = 9$dB   (d) 802.15.4: ABP, $\tau = 9$dB

FIGURE 8.8. Relationship between the true distance and the estimated distance for the original node and the spoofing node across localization algorithms and networks.

localization. Over a certain time period (for example, 60 seconds), for a MAC address there will be a cluster of location estimates in physical space. Intuitively, we think that, during a spoofing attack there will be distinctive location clusters around the original node and the spoofing nodes in physical space. Our intuition was that the clustering results from the per-packet localization would allow the detection and localization of attackers in one step.

However, we found that the performance of clustering packet-level localization results for spoofing detection is not as effective as deriving the centroids in signal space. The relationship between $||P_{org} - P_{spoof}||$ and $||L_{org} - L_{spoof}||$ is shown in Figure 8.9. Although it also has a trend along the 45 degree line, it shows more uncertainties along the line. Therefore, we believe that given a set of RSS reading samples for a MAC address, working with the signal strength directly preserves the basic properties of the radio signal, and this in turn is more closely correlated with the physical location, and thus working with the RSS values directly better reveals the presence of the spoofing attacks.

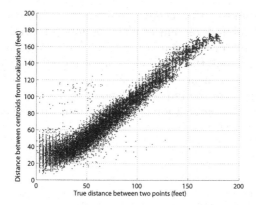

FIGURE 8.9. Packet-level localization: relationship between the true distance and the estimated distance for the original node and the spoofing node when using RADAR in the 802.11 network.

8.7 Conclusion

In this chapter, we proposed a method for detecting spoofing attacks as well as localizing the adversaries in wireless and sensor networks. In contrast to traditional identity-oriented authentication methods, our RSS based approach does not add additional overhead to the wireless devices and sensor nodes. We formulated the spoofing detection problem as a classical statistical significance testing problem. We then utilized the K-means cluster analysis to derive the test statistic. Further, we have built a real-time localization system and integrated our K-means spoofing detector into the system to locate the positions of the attackers and as a result to eliminate the adversaries from the network.

We studied the effectiveness and generality of our spoofing detector and attacker localizer in both an 802.11 (WiFi) network and an 802.15.4 (ZigBee) network in a real office building environment. The performance of the K-means spoofing detector is evaluated in terms of detection rates and receiver operating characteristic curves. Our spoofing detector has achieved high detection rates, over 95% and low false positive rates, below 5%. When locating the positions of the attackers, we have utilized both the point-based and area-based algorithms in our real-time localization system. We found that the performance of the system when localizing the adversaries using the results of K-means cluster analysis are about the same as localizing under normal conditions. Usually the distance between the spoofing node and the original node can be estimated with median error of 10 feet. Our method is generic across different localization algorithms and networks. Therefore, our experimental results provide strong evidence of the effectiveness of our approach in detecting the spoofing attacks and localizing the positions of the adversaries.

During the course of the security analysis for localization systems, we found that the landmark placement plays an important role on localization performance. While most research has focused on improving the localization algorithm, we took the viewpoint that it is perhaps just as important to improve the deployment of the localization system. In the next chapter, we will investigate the impact of landmark placement on localization performance using a combination of analytic and experimental analysis.

Part III

Defending Against Radio Interference

Part III

Defending Against Radio Interference

9

A Brief Survey of Jamming and Defense Strategies

9.1 Introduction

The shared nature of the wireless medium, combined with the commodity nature of wireless technologies and an increasingly sophisticated user-base, allows wireless networks to be easily monitored and broadcast on. At the most basic level, adversaries may easily observe communications between wireless devices, and just as easily launch simple denial of service attacks against wireless networks by injecting false messages. Traditionally, denial of service is concerned with filling user-domain and kernel-domain buffers [113]. However, in the wireless domain, the availability of the wireless network may be subverted quite easily through radio interference.

In this book, we examine the problem of ensuring the availability of wireless networks in the presence of radio interference. Such interference may be either incidental or intentional– corresponding to the accidental scenario where many transmitters share the same "radio" environment (e.g. the same 802.11 channel may be used by many users in an apartment building), or may involve one or more intentional adversaries seeking to disrupt the continuity of network services by actively transmitting or occupying the wireless channel. A general characteristic of such Radio Frequency (RF) interference scenarios is that the wireless medium is blocked and other wireless devices are prevented from communicating and/or receiving.

In this chapter, we will briefly overview the problem of radio interference by providing examples of how easily RF-interference can disrupt wireless communications in various wireless networks. We will then provide a high-level overview of a general strategy to overcome interference, which consists of a multi-phased approach involving interference detection (see Chapter 11) and then the use of an *evasion* strategy. Although, in this chapter, we will outline two different evasion strategies, channel surfing and spatial retreats, we will focus our attention only on a more detailed analysis of channel surfing in Chapter 12.

We begin in subsection 9.2 by presenting case studies on the RF-interference problem. We introduce the three different wireless network scenarios that we will study in this section in subsection 9.3. Following the setup of the problem, we present *channel surfing*, a defense against MAC/PHY-layer denial of service attacks in subsection 9.4. Channel surfing involves valid participants changing the channel they are communicating on when a denial of service attack occurs. In subsection 9.5, we examine *spatial retreats*, which involves legitimate network devices moving away from the adversary.

9.2   Interference Case Studies

We now look at three different case studies that examine the performance of a wireless network in the presence of interference. The first case study demonstrates an adversarial case of interference involving a jammer that either bypasses the MAC-layer or uses a waveform generator to disrupt communications, the second case study examines the impact of cross-channel interference, while the third involves examining the effect of interference that arises due to natural congestion in carrier sense multiple access based (CSMA-based) networks (e.g. 802.11). Our case studies show that both malicious jammers and MAC compliant network nodes can cause interference within wireless communications.

In most forms of wireless medium access control, there are rules governing who can transmit at which time. For example, one popular class of medium access control protocols for wireless devices are those based on carrier sense multiple access (CSMA). CSMA is employed in Berkeley motes as well as in both infrastructure and infrastructureless (ad hoc) 802.11 networks. In this section, we examine the impact that various interferers have on networks employing CSMA. The metrics we used to quantify the impact of the interferers are, (1) carrier sensing time, i.e. the amount of time a device spent in sensing the channel before transmission; (2) throughput, i.e. the amount of data bits that have been delivered from a network node to another; and (3) packet delivery ratio (PDR), i,e. the ratio of the number of data packets received by the destination node over the number of data packets sent by the source node.

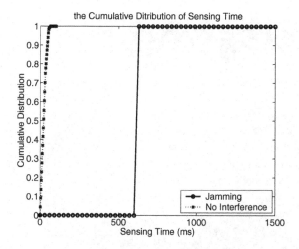

FIGURE 9.1. The cumulative distribution for the carrier sensing times measured using MICA2 Motes.

9.2.1 Non-MAC-compliant Interferer Case Study

In this section, we study the impact that a Non-MAC-compliant interferer has on communication. In particular, we examine the impact the interferer has on carrier sensing time as well as packet delivery ratio (PDR).

During normal operation of CSMA, when A tries to transmit a packet, it will continually sense the channel until it detects the channel is idle, after which it will wait an extra amount of time (known as the propagation delay) in order to guarantee the channel is clear. Then, if RTS/CTS is used it will send the RTS packet, or otherwise will send the data packet. Suppose the adversary X is continually blasting on a channel and that A attempts to transmit a packet. Then, since X has control of the channel, A will not pass carrier-sensing, and A may time-out or hang in the carrier-sensing phase. As a result, no packets from A can be sent out.

To validate our analysis carrier sensing time and PDR we performed a set of experiment using MICA2 motes. Each mote had a ChipCon CC1000 RF transceiver and used TinyOS 1.1.1 as the operating system, which used a fixed threshold for determining idleness. To build the jammer X, we disabled the back off operations to bypass the MAC protocol.

Carrier Sensing Time: To validate the effect of jammers on the carrier sensing time in a real wireless network, we performed an experiment using two Motes, X and A. Here, Mote A corresponds to a network node trying to send out a 33-byte packet every 100ms, and which measures the sensing time while doing so. Mote X, the jammer, continuously sends out random bits without a time gap in between transmissions. Additionally, we measured the sensing time without any interference, i.e. X does not send any bits.

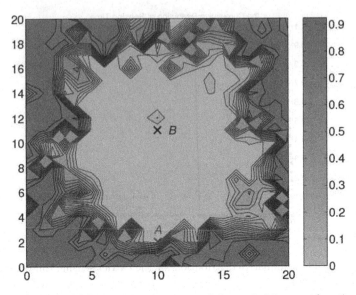

FIGURE 9.2. The PDR contours as measured for varied jammer locations on a 20X20 square feet grid with the sender's and the receiver's locations fixed. The cross marked as B represents the receiver, and the cross marked as A is the sender. The color goes from light gray (0: no packets arrived at the receiver correctly) to dark gray (all packets arrived correctly).

Fig. 9.1 depicts the cumulative distribution of the sensing time for those two scenarios. Fig. 9.1 shows that most of the carrier sensing time in the non-interference scenario is less than $100ms$, while the cumulative distribution of the jamming scenario jumps at the point where the sensing time equals to 640ms. This is caused by a timeout we added to the TinyOS. In our experiment, if the device does not start to send the packet within 640ms after the packet was passed to the MAC-layer from application layer, a timeout will occur, the packet will be discarded, and its sensing time will be counted as 640ms.

PDR: In this set of experiments, we have a sender A, receiver B and jammer X. Rather than constantly transmitting random bits, the jammer X here listens to the channel, and upon detection of preambles of a data packet, it immediately blasts on the channel by sending a random 2-byte jamming packet, causing the data packet garbled when it arrives at the receiver. We measured the PDR at the receiver B by calculating the ratio of the number of packets that have been correctly received with respect to the total number of packets that are expected to send. The number of expected packets can be obtained through tracking the sequence number of each packet. If no packets are received, the PDR is defined to be 0.

The interference level caused by the jammer X is governed by a number of factors. The most obvious one is the relative location of the jammer and the receiver. We have studied the degradation of the PDR caused by the jammer X as a function of the jammer's location in a 20x20 ft. square grid. We have evaluated a total of 439 points in space by varying the jammer's location while keeping the sender's and the receiver's locations fixed.

We illustrated the degradation of the PDR by means of a color map in Figure 9.2. The resulting PDR contours, i.e. the lines that connect points on the grid with the same PDR levels, encompass the sender and the receiver which are located at $(10, 2)$ and $(10, 11)$ respectively. The obstruction of the PDR is roughly determined by the separation distance between the jammer and the receiver with a steep threshold occurring between 7 and 9 feet away from the receiver. When the jammer X is closer than 7 feet to the receiver, almost all the communication is corrupted. Due to the type of the jammer, the PDR is also sensitive to the separation between the sender and the jammer. In theory, the contours of the PDR between the sender and the receiver are circles centered at the receiver. However, due to the radio irregularity in an indoor environment which is not accounted for in theoretical model, the PDR contours are irregular. There is also a special case of greater than zero PDR when the jammer is in the immediate proximity of the receiver, appearing inside the diamond-shaped contours around $(10, 12)$, north to the receiver.

9.2.2   Cross-channel Interference Case Study

In this discussion we examine the issue of channel selection and the impact it can have on communications. Unlike the interferers studied in the previous case study, where the jammers do not follow MAC rules and can completely take the channel, the interferers discussed in this sub-section can be regular network nodes that follow a proper MAC protocol.

In order to demonstrate how channel selection can interfere with network communications, we conducted a set of experiments using Berkeley motes. In these experiments, two motes act as the communicator and receiver, denoted by A and B. A continuously sends out 31-byte packets to B, resulting in a throughput of 3.6Kbps. We then placed interferers or jammers in different locations. In the first set of experiments, we used motes as interferers. We tried three interferer scenarios, which are illustrated in Figures 9.3. These interferers also continuously send out packets of the same size. The interferers completely followed the default MAC protocol of Berkeley motes. All the motes transmit at the same power level. The default frequency of the motes was 916.7MHz. The results for this set of experiments are summarized in Figure 9.4 (a). When all the motes transmit at the default frequency, the measured throughput between A and B significantly drops compared to the scenario with no interferers. Due to the fact that all devices follow the MAC protocol, the throughput did not become zero. We then incremented the transmission/reception frequency of the communicator and receiver by 50KHz. As the frequency gap between

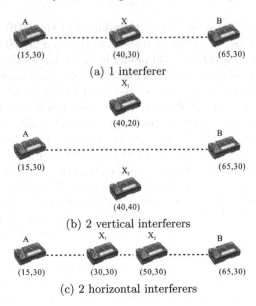

(a) 1 interferer

(b) 2 vertical interferers

(c) 2 horizontal interferers

FIGURE 9.3. The locations of the communicators (A and B) and interferers (X, X_1 and X_2). The unit of distances is centimeter.

the communicators and the interferers increases, but before it reaches a threshold, the measured throughput worsens. This is because the transmissions still interfere with each other, yet the MAC protocol is not able to coordinate transmissions across different frequencies, resulting in a much higher collision rate and a lower throughput. Finally, when the communicators increase their frequency to (or above) 917.5MHz, they are no longer interfered with.

The lessons we learned here is that when there are nodes working on multiple channels, it is very important for each node to choose an interference-free channel, i.e. orthogonal channel, to avoid cross-channel interference. Although the specifications for the radio employed in Berkeley motes state that a channel separation of 150kHz is recommended in order to prevent cross-channel interference. From our experiment, we have found that 800kHz is a safer value for channel separation in order to maintain *effective network-layer orthogonality*, resulting in 32 orthogonal channels on MICA2 platform.

To understand how the jammer affects the network communication when the interfering signal is not centered on the channel that the network is operating on, we used a waveform generator as the jammer interfering with the two communicators. The position setup is the same as depicted in Figure 9.3(a), where X represents the location of waveform generator. The waveform generator continuously emitted a narrow AM signal at 916.7MHz frequency and with an amplitude of -10dBm. Unlike the interferers in the

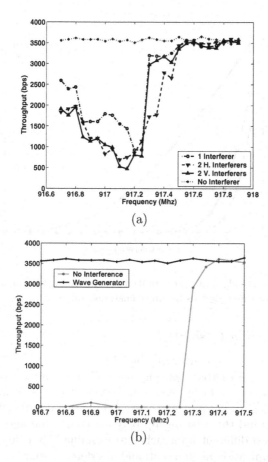

FIGURE 9.4. Experimental results indicating throughput versus channel assignment for (a) MAC-compliant Mote, (b) Waveform generator continuously emitting an AM signal.

above experiment, the jammer does not follow MAC rules and can completely take the channel so that the communicators do not have a chance to transmit. The results are shown in Figure 9.4(b). Before the communicators move out of the spectral interference range, the measured throughput is 0. As soon as the communicators move to (or above) 917.4MHz, they can transmit without any interference. The reason for a narrower gap in this scenario compared to the mote interferer cases is that the spectral width of the waveform generator's signal is narrower than the spectrum of a Berkeley motes' signal.

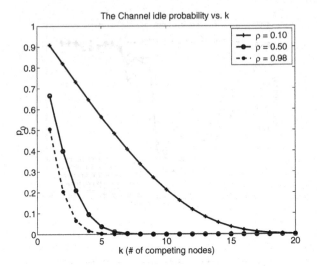

FIGURE 9.5. The steady-state probability that the channel is idle as the number of competing nodes increases using finite customer queue.

9.2.3 Congestion Case Study

When the adversary X is continually blasting on a channel, if A attempts to transmit a packet, A's carrier-sensing time will be infinity or some time-out value. This time-out, however, could have occurred for legitimate reasons, such as congestion. Congestion can cause unintentional radio interference as well. To understand the relationship between congestion and interference, we employed two different approaches to examine how congestion effects network communication as unintentional interference sources. We start by looking at a theoretical channel occupancy model, and then present our empirically study.

The theoretical model: For the theoretical channel occupancy model, we will employ a simplified model for CSMA as well as some simplifying assumptions about the underlying network traffic. We assume that there are a finite set of radio sources that each acts as an independent Poisson source with a packet generation rate of λ packets/sec. Further, we assume that transmission of packets is completed at an average rate of μ according to an exponential distribution[1]. λ represents the traffic load. For example, λ is larger in the congested scenario than a light-weighted traffic scenario. μ depends on the size of a packet and channel capacity.

Under these assumptions, we may model the channel as a two-state Markov model, where state 0 is the Channel-Idle state and state 1 is

---

[1]We note that the validity of our model depends on λ/μ, and that the approximations made become increasingly accurate as $\lambda/\mu \to 0$, c.f. [114].

the Channel-Occupied state. The two-state model for the channel is simply an $M/M/1/k/k$ queue, where we have finite number of customers (k customers) and a single server. In our case, the single server is the channel, and the customers are the network nodes that potentially compete with each other for the channel. We may use queuing theory to derive $p_0 = P(\text{state-0})$, the steady-state probability that the channel is idle at an arbitrary time [115]. Let $\rho = \lambda/\mu$, then at steady state $p_0 = 1/\sum_{i=0}^{k} \frac{k!\rho^i}{(k-i)!}$. For a given ρ, p_0 is a function of k, the number of competing nodes. We plot $p_0(k)$ for three different levels of ρ in Figure 9.5. In particular, $\rho = 0.1$ represents a light weighted traffic load scenario, where each node tries to occupy one tenth of the channel capacity; $\rho = 0.98$ corresponds to a heavily loaded traffic scenario, where every node tries to saturate the channel. The plots are inline with the intuition, i.e. as the number of competing nodes increases, the channel idle probability decreases. The higher the packet generation rate per node, the faster the channel gets saturated. Therefore, in a congested scenario, a node has to wait longer time before it can access the channel and start to transmit.

Now, let's look at the carrier sensing time. Suppose that A senses the channel at an arbitrary time t. We are interested in the amount of time D that A must, under CSMA, wait before it starts transmission. A will sense the channel until it is idle. Under popular implementations of CSMA, when the channel becomes idle, node A will continue sensing the channel for an additional τ time before transmitting in order to ensure distant transmissions have arrived at A[2]. If the channel becomes busy during that time, then A declares the channel busy and continues sensing until it experiences an idle period that lasts τ units of time.

In order to capture the distribution for D, we will introduce a few auxiliary variables. Let N denote the number of times we visit state 0 before witnessing an idle period of at least τ duration. Also, let us define the random variable $T_{0,k}$ to be the duration in state 0 for the k-th visit to state 0, and similarly $T_{1,k}$ to be the duration in state 1 for the k-th visit to state 1. Observe that, if we are at state 0, then to exit CSMA requires $T_{0,k} \geq \tau$, and hence we must introduce $q = P(T_{0,k} \geq \tau) = e^{-\lambda\tau}$. We may now look at the chain of events, as depicted in Figure 9.6. If the channel was idle when we started, then we enter at the left-most state 0, whereas if the channel was occupied when we entered, then we enter at the left-most state 1.

The probability density function for D can be shown to be

$$f_D(d) = p_0 \sum_{k=1}^{\infty} P(N = k) f_{0,k}(d) \tag{9.1}$$

$$+ (1 - p_0) \left(f_{T_{1,1}} + \sum_{k=1}^{\infty} P(N = k) f_{0,k}(d) \right) \tag{9.2}$$

---

[2]Typically, the propagation delay τ is on the order of 50μseconds.

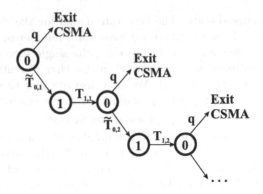

FIGURE 9.6. The events leading to exiting CSMA, and the corresponding probability parameters used in the derivation.

$$= (1 - p_0)f_{T_{1,1}} + \sum_{k=1}^{\infty} P(N = k)f_{0,k}(d) \qquad (9.3)$$

Here N is geometric with $P(N = k) = (1 - q)^{k-1}q$. The distributions $f_{0,k}(d)$ are the pdfs describing the duration contributed by exiting during the k-th visit to state 0 when we start at state 0. For example, if we exit after the first visit to state 0, then the only contribution comes from the propagation delay τ, that is, $f_{0,1}(d) = \delta(\tau)$, the point mass at $d = \tau$. As another example, let us look at the time, Y, contributed when exiting at the second visit to state 0. Since we did not exit during the first visit to state 0, Y is composed of time contributed by $T_{0,1}$ that is strictly less than τ. We will call this conditional random variable $\tilde{T}_{0,1}$. Next, Y also consists of $T_{1,1}$, and finally the τ needed to exit CSMA. Hence $Y = \tilde{T}_{0,1} + T_{1,1} + \tau$, and thus

$$f_{0,2}(d) = f_{\tilde{T}_0} * f_{T_1} * \delta(\tau) = f_{0,1} * f_{\tilde{T}_0} * f_{T_1} \qquad (9.4)$$

where $*$ denotes convolution and arises due to independence between the random variables involved, and f_{T_j} is the distribution for occupancy time for any state j. Similar recursive representations can be derived for $f_{0,k} = f_{0,k-1} * f_{\tilde{T}_0} * f_{T_1}$.

From this theoretical pdf, we may calculate the mean of carrier sensing time $E[D] = \int_0^{\infty} f_D(x)x\,dx$. Now let's examine the trend of $E[D]$. When k the number of competing nodes increases and λ the aggregated traffic load increase, the probability q that a node can exit CSMA at a kth stage decreases. Thus, the shape of $P(N = k) = (1 - q)^{k-1}q$ over k becomes flatter with longer tail, as show in Figure 9.7. The flatter shape of $P(N = k)$ over n indicates that the $E[D]$ becomes larger as the traffic load increase. Although the regular network nodes is CSMA compliant, the increasing density of the nodes plus the increasing traffic load together will lead to higher carrier sensing time, and thus higher interference.

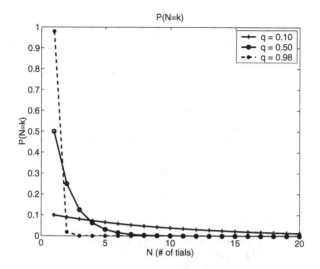

FIGURE 9.7. The probability that a node will exit CSMA at the kth' trial.

The experimental study: A second approach involves each network device collecting statistics regarding the amount of time D that a device must wait before it can start transmission during normal, or even somewhat congested, network conditions. After collecting enough statistics, we calculate the distribution $f_D(d)$ which describes the amount of time spent in sensing before the channel becomes idle during various network traffic conditions, including light-weighted and congested scenarios.

In order to measure the distribution $f_D(d)$, we carried out several experimental studies using the 802.11 extensions to the ns-2 simulator. We modified ns-2 by disabling the MAC layer retransmission, so that we could focus our investigation on the channel sensing behavior. In our experiments we have two nodes, A and B. Once every 19 msecs, node A senses the channel by trying to send out a beacon to node B. We obtained the MAC-layer delay time D by calculating the difference between the time when beacon packets reach MAC-layer and the time when the MAC successfully senses the channel as idle and sends out RTS. In order to capture the statistical behavior of the delay time, we calculated the cumulative distribution of the delay time for several scenarios involving different levels of background traffic loads. As shown in Figure 9.8 (a), we introduced several streams (from sender S_i to receiver R_i) that are within the radio range of A and B in order to increase the interfering traffic. Each stream's traffic was chosen to represent an MPEG-4 video stream suitable for a wireless video application. We used traffic statistics corresponding to the movie Star Wars IV [116], where each sender transmitted packets with the packet size governed by an exponential distribution with a mean size of 268 bytes, and the packet inter-arrival times following an exponential distribution with

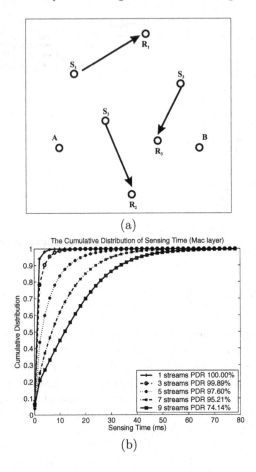

FIGURE 9.8. The MAC-layer sensing time experiment: (a) basic underlying experimental setup, (b) cumulative distributions of D for different traffic scenarios and the corresponding packet delivery ratio.

mean 40msecs, resulting in each stream having an average traffic rate of 53.6Kbps. The corresponding cumulative distributions of D are shown in Figure 9.8 (b). These observations can be explained as follows. When there are only a few streams, there are few nodes competing for the channel, and node A can get the channel quickly with high probability. As the number of streams increases, the competition for the channel becomes more intense, thus taking longer for A to acquire the channel. This observation agrees with the result we get from our theoretical channel occupancy model.

From this figure, we can observe that the cumulative distribution of carrier sensing time reaches its saturation point faster when the number of streams is small. For instance, in the 1 stream scenario, 90% of the time, a node waits less than 5ms before it starts to transmit, while in 9 stream scenario, only 20% the carrier sensing times are less than 5ms. Thus, the interference condition is more severe when the number of streams is bigger.

Additionally, we can observe that as the number of streams increases the average packet delivery ratio decreases. Packet delivery ratio is another important parameter to quantify the network interference condition. The lower the packet delivery ratio is, the more severe the interference is. For instance, when the number of streams is 9, the average packet delivery ratio is 74.1% and corresponds to a poor quality of service for each application.

9.3   Defense: Detection and Evasion

In order to cope with interference, whether intentional or accidental, we have proposed a two-phase strategy involving the diagnosis of the radio interference/jamming, followed by a suitable defense strategy. We highlight the challenges associated with detecting jamming, and refer the reader to Chapter 11 for more discussion on detection. Coping with the problem of overcoming interference is very challenging, as well. At one level, if the interferer is malicious and relatively powerful, then the only viable strategy for defense is to employ sophisticated physical layer methods with large anti-jam margins. On the other hand, for interference scenarios that arise from accidental interference or only modest levels of jamming, there are other defense methods that don't involve costly physical layer methods. Although it might seem that there is nothing that can be done since assuming that an adversary can continually blast on a channel grants the adversary considerable power, we nonetheless take inspiration for our work from the famous Chinese war strategy book, *Thirty-Six Stratagems*:

> He who can't defeat his enemy should retreat.

Translating this philosophy into the wireless domain, we may propose that wireless devices employ spectral and spatial evasion strategies in order to protect against interference.

Although there are many different scenarios where jamming may take place, we will focus on three basic classes of wireless networks, as depicted in Figure 9.9:

1. **Two-Party Radio Communication:** The two-party scenario is the baseline case, in which A and B communicate with each other on a specific channel. As long as interferer X is close enough to either A or B, its transmission will interfere with the transmission and reception of packets by A and B.

2. **Infrastructured Wireless Networks:** Infrastructured wireless networks, such as cellular networks or wireless local area networks (WLANs), consist of two main types of devices: access points and mobile devices.

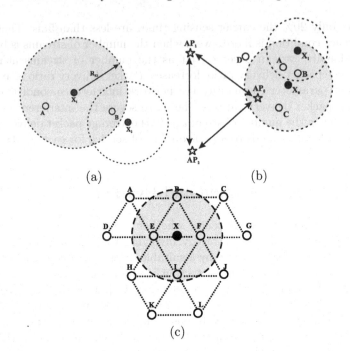

FIGURE 9.9. The three wireless communication scenarios studied in this paper:
(a) two-party radio communication, (b) infrastructured wireless networks, and
(c) ad hoc wireless networks. The adversary is depicted by X.

Access points are connected to each other via a separate, wired infrastructure. Mobile devices communicate via the access point in order
to communicate with each other or the Internet. The presence of an
interferer, such as X_0 or X_1, might make it impossible for nodes to
communicate with their access point.

3. **Mobile Ad Hoc Wireless Networks:** Ad hoc networks involve
   wireless devices that establish opportunistic connections with each
   other in order to form a communication network. Typically, ad hoc
   networks employ multi-hop routing protocols in order to deliver data
   from one network node to another. The presence of an interferer may
   bring down whole regions of the network.

It should be noted that in all three cases, the interferer could have limited
interference range, thereby affecting only one of two communicating participants. For example, in Figure 9.9 (a), interferer X_2 is located to the right
of B and blocks B from being able to acquire the communication channel
while A might not be able to detect X_2.

9.4 Channel Surfing Overview

The first escape strategy that we present is channel surfing. Typically, when radio devices communicate they operate on a single channel. When an interferer comes in range and blocks the use of a specific channel, it is natural to migrate to another channel. The idea of channel surfing is motivated by a common physical layer technique known as frequency hopping. We assume throughout this section that the interferer/jammer blasts on a single channel at a time, and that the interferer/jammer cannot pretend to be a valid member of the network (i.e. the jammer does not hold any authentication keys used by the network devices). In this section, we examine the channel surfing strategies for three basic classes of wireless networks listed in previous section. We will explore the case of an ad hoc network in more detail in Chapter 12.

9.4.1 Two-Party Radio Communication

Consider the radio scenario depicted in Figure 9.9(a). In this scenario, jammer X_1 or X_2 has disrupted communication between A and B. We desire both A and B to change to a new channel in order to avoid X's interference. Using some detection techniques discussed in Chapter 11, once A and B have detected jamming, they will change to a new channel, and resume communication in the new channel.

To facilitate channel surfing, both parties have to agree on the channel adaptation sequence, since A and B cannot negotiate with each other while they are within the jamming range. Additionally, we emphasize the importance of using orthogonal channels, since if A and B evade to a new channel that is not orthogonal to the original one, A or B will still be interfered with by X's signal. For many wireless networks, it is necessary to determine the number of orthogonal channels experimentally. For example, the specifications for the radio employed in Berkeley motes state that a channel separation of 150kHz is recommended in order to prevent cross-channel interference. As noted earlier, we have found through experiments that 800kHz is a safer value for channel separation in order to maintain *effective network-layer orthogonality*.

Another issue that naturally arises in employing such a channel changing strategy is whether or not one should continually change channels regardless of whether the adversary is blocking the current channel. Although physical layer frequency hopping employs a strategy of constantly changing the underlying frequency, there are reasons why this might not be desirable at the link-layer. In particular, although changing the frequency of the carrier wave is easy to accomplish in the case of frequency hopping spread spectrum, changing channels at the link-layer is more involved as it requires synchronization between both parties, which necessitates additional time cost.

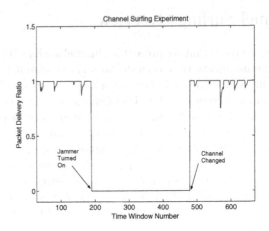

FIGURE 9.10. Packet delivery measurements from the Mote channel surfing prototype.

Prototype: We built a proof-of-concept prototype system using two Berkeley motes A and B. The application running on these two motes involved A sending out a packet to B every 200 msecs. Each packet contained a sequence number starting at 1. We partitioned the time axis into windows, and B kept track of how many packets it received in each window (n_{recv}). It can also determine how many packets A has attempted to send in each window by looking at the sequence number of the last message it receives (n_{send}). In order to capture the quality-of-service of the application, we employed the *packet delivery ratio*, $r = n_{recv}/n_{send}$.

In the experiment, we used a waveform generator as the jammer X. As soon as the jammer is turned on, A cannot access the channel, and so no packets can be sent out. As a result, the packet delivery ratio becomes 0. In order for the application to survive the jamming, both A and B should incorporate a jamming detection and defense strategy. For the discussion here, our prototype detection algorithm works as follows. At the application level, each mote sets a jamming check timer (30 seconds). Each time the timer expires, it attempts to send out a beacon by making a `SendMsg.Send()` call. The send call will return SUCCESS if the channel monitor component identifies an idle period so that the message send can start. (Later on, a notification will be sent to the application after the MAC-layer ACK is received from the receiver.) If the channel monitor component cannot sense the channel as idle after a long time (e.g., by using empirical threshold), the send call will return FAILURE. After it returns FAILURE continuously for several time, the mote can conclude that it is under a jamming attack. Following the detection of jamming, the mote will change its frequency to an orthogonal channel (e.g, from 916.7MHz to 917.6MHz). In order to avoid

---

Algorithm: Infrastructured Channel Surfing

if $DETECT\_JAM(Self) == TRUE$ **then**
  | Change_Channel()
else
  | **if** $AM\_AP == True$ **then**
    | Calc_ChildrenLastCalledHome()
    | Calc_NegligentChildren()
    | RESPONSES = Probe_NegligentChildren()
    | **if** $Any(RESPONSES) == NULL$ **then**
      | Broadcast_ChangeChannelCommand()
      | Change_Channel()
    | **end**
  | **else**
    | ListenForBeacons()
    | **if** $TimeToLastBeacon() > BIG$ **then**
      | Change_Channel()
    | **end**
  | **end**
end

Algorithm 2: Channel surfing for wireless infrastructured networks. This algorithm runs on each network device.

collision, A and B should not send beacons at the same time. The code is included below:

```
task void checkJamming(){
    sent = call SendMsg.send(TOS_BCAST_ADDR,
                             sizeof(uint16_t),
                             &beacon_packet);
    if(!sent){
        if(failures++ < thresh)
            post checkJamming();
        else post changeChan();
    }else{
        failures = 0;
    }
}
```

After both A and B change their frequencies, they can resume their application behavior, and the packet delivery ratio will go up again. We present measurements for the prototype channel surfing experiment in Figure 9.10. The channel surfing strategies works as expected.

9.4.2 Infrastructured Network

Now consider an infrastructured wireless network as depicted in Figure 9.9(b). Here, we have an access point AP_0, which has four wireless devices A, B, C, and D connected to it. There are two main scenarios for a radio interference/jamming attacks against the access point, corresponding to jammer X_0 and X_1. In the first scenario, jammer X_0 interferes with AP_0, A, B and C, but does not interfere with node D since it is outside of X_0's radio range. In the second scenario, jammer X_1 interferes with A and B, but not with AP_0 or any of the other nodes. The main difference between these two scenarios lies in the fact that one has the access point blocked by the jammer, while the other does not.

Algorithm: Basic Ad Hoc Channel Surfing

```
if DETECT_JAM(Self)==TRUE then
 |   Change_Channel()
else
 |   RESPONSES = Random_ProbeNeighbors()
 |   if Any(RESPONSES == NULL) then
 |    |   STRANDED = Test_NextChannel()
 |    |   if Any(STRANDED)==TRUE then
 |    |    |   Broadcast_ChangeChannelCommand()
 |    |    |   Change_Channel()
 |    |   end
 |   end
end
```

Algorithm 3: Basic channel surfing for wireless ad hoc networks.

We need a strategy for changing channels whereby *all* nodes connected to the access point will change channels with the access point. We do not want to have scenarios where some devices are on the old channel while some are on the new channel. Algorithm 2 presents the sequence of events needed for a network device to determine whether to change channels.

Periodically, the algorithm checks to see whether the device has been blocked from communicating by a jammer. This can be done using the methods described in Section 11. If jamming has been detected, then the device will change channels. Otherwise, the device checks to see whether it is an access point. Access points will examine their list of children to see which devices have not communicated recently. Those devices whose last communication with the AP was greater than some threshold amount of time will be probed by the AP to ascertain whether they were remaining silent for a long time. If any device does not respond to their probe, the AP will conclude that the device has disappeared due to radio interference/jamming.

The rationale behind this is that, during normal operation of an infrastructured wireless network, if a network node wishes to leave the network it will perform a disassociation request, allowing the AP to free up any resources allocated to managing that network device. Further, when a network node moves to another access point, the device will perform reassociation with the new access point. The new access point will relay this information to the old access point, and acquire any data that might be buffered at the old access point. In both cases, the access point will know when a user legitimately leaves its domain.

If the AP concludes that the device disappeared due to jamming, then it will broadcast an emergency change channel packet that is signed by the AP's private key. This packet can be authenticated by each of the AP's children that are not blocked by the adversary. Following the issuance of the change channel command, the AP will change its channel and commence beaconing on the new channel in hopes to elicit associations from its children.

Algorithm: Dual-Radio Ad Hoc Channel Surfing

if $DETECT\_JAM(Self)==TRUE$ then

   | Change_Channel()
   | InformNeighbors()
   | EstablishNewLinks()

end

Algorithm 4: Dual-radio channel surfing for wireless ad hoc networks.

If the device is not an access point, then it will check to see if it has not heard its access point's beacons in a long time. It may even probe the access point. Either way, the child device will decide that it needs to catch up with its parent and change channels. In Algorithm 2, the default condition is to remain on the same channel. Performing an channel surfing procedure using Algorithm 2 allows the node A, B, C, and D in Figure 9.9(b) to survive all two jamming situations, regardless whether the AP is jammed or not. As a final comment, we note that in our overview discussion above, we have implicitly assumed that devices do not run out of battery or hardware failures. Thus, the issue of a client device disappearing will be solely due to interference.

9.4.3 Ad Hoc Network

When a jammer starts to interfere with an ad hoc network, he severs many of the links between network devices and can possibly cause network partitioning. Channel surfing can counteract such network faults by having the network or regions of the network switch to a new channel and re-establish network connectivity. In order to use channel surfing to address jamming for ad hoc networks, we assume that each network device keeps a neighbor list. However, since we are operating in ad hoc mode, we do not assume that if a device moves that it will inform its neighbors of its intent to relocate. Also, during unhindered network operation, we assume that no network partitions arise due to network mobility.

Algorithm 3 presents an outline of the channel surfing operations that run on each device. First, devices check to see if they have been blocked by a jammer. If so, they change channels and monitor the new channel to assist in reforming the ad hoc network on the new channel.

If a device has not been blocked by a jammer, then there is a chance that its neighbors have been blocked. Therefore, at random times a node will probe its neighbors to see that they are still nearby. There are several reasons that a neighbor node might not be present: it might have moved to a different part of the network, or it might have been blocked by a jammer. The device checks to see if a jammer is present by testing to see if any devices are stranded on the next channel. If the test returns positive, then the device returns to the original channel and broadcasts a signed change channel command to its neighbors, which is flooded through the rest of the

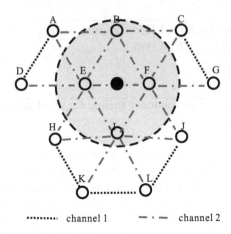

.... channel 1   — · —   channel 2

FIGURE 9.11. Channel surfing for an ad hoc network consisting of dual radio devices.

network. It will then change channels and assist in reforming the ad hoc network on the new channel. Other devices will authenticate the command, and switch channels. If the test returned negative, then the device will assume that its absent neighbor has merely migrated to a different portion of the network and remove it from the neighbor list.

One unfortunate drawback of channel surfing for ad hoc networks is that it requires the use of flooding messages to promptly initiate a channel change across the entire ad hoc network. A simpler and more efficient alternative channel surfing strategy is possible if the network devices employ a dual-radio interface, that is they are capable of operating on two channels simultaneously.

Algorithm 4 describes the operations that run on each network device when a dual-radio interface is available. The default operation of the ad hoc network is for devices to employ one radio channel for communication, yet monitor both channels. When a device detects that it has been blocked, it will switch to the next channel. Once on the new channel, the device will contact its neighbors via the new channel to inform them of its new channel policy, warn of possible radio interference/jamming, and establish new links in order to maintain network connectivity. In addition to the usual routing information, each network device must maintain an additional channel assignment field for each of its neighbors. The end result is that a network will consist of some links on the old channel and some links on the new channel, as depicted in Figure 9.11.

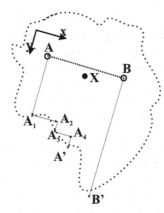

FIGURE 9.12. The spatial retreat strategy for a two-party communication scenario. The region depicted by the dotted line is the interference range of the adversary.

9.5   Spatial Retreats

The second escape strategy that we propose is *spatial retreats*. The rationale behind this strategy is that when mobile nodes are interfered with, they should simply move to a safe location. Spatial retreat is often a desirable defense strategy to employ in those wireless networks that involve mobile participants, such as users with cell phones or WLAN-enabled laptops. The key to the success of this strategy is to decide where the participants should move and how should they coordinate their movements. In this section, the discussion on spatial retreats primarily focuses on a simple interference scenario in which the jammer is stationary. This interference model might arise in cases where the jammer is unknowingly or unintentionally jamming the communication. More powerful adversarial models where the adversary is mobile and can stalk the communicating devices has been investigated and we refer the reader to [117].

9.5.1   Two-Party Radio Communication

Let us again start by examining the two-party communication scenario. We present an example jamming scenario in Figure 9.12, where the jammer X interferes with both A and B so that these two nodes cannot communicate with each other. In a spatial retreat, as soon as the communicating parties (i.e., A and B) detect the jamming scenario, they try to move away from the jammer. It is a daunting task, however, to decide on a retreat plan as both parties must agree on the direction of the retreat and how far to retreat. This task is complicated by the fact that A and B cannot communicate with each other while they are within the jammer's broadcast radio range. Further, even after they leave the adversary's radio range, they may not remain within each other's radio range due to the lack of synchronization between them and the irregularity of the interference region.

Considering the above factors, any functional retreat plan must satisfy the following two conditions: (1) it must ensure that both parties leave the adversary's interference range; and (2) it must ensure that the two parties stay within each other's radio range. In order to accomplish these two requirements, we propose a three-stage protocol:

1. *Establish Local Coordinates:* We assume the two parties know each other's initial position prior to the introduction of the adversary. This assumption is reasonable since it is becoming increasingly popular to incorporate positioning capabilities in mobile devices [118]. Using both parties' positions, we can decide on a local coordinate system (for example, we may define the x axis of our local coordinate system to be aligned with the segment $\overline{AB}$, as shown in Figure 9.12, and determine the y axis accordingly).

2. *Exit the Interference Region:* After the local coordinates are established, both parties move along the y-axis. While they move, they periodically check the interference level, e.g. the ambient signal strength or energy level emitted by the interferer. As soon as a node detects that it is out of the interference range, it stops moving. We would like to emphasize that, in practice, the two parties will stop asynchronously (as shown in Figure 9.12) because the radio range of the adversary is irregular in shape and the two parties cannot talk to each other before they move out of the range. In the example, A stops at location A_1, while B stops at B'.

3. *Move Into Radio Range:* There is a possibility, after exiting the adversary's radio range, that the two parties will be outside of each other's radio range, as shown in Figure 9.12. In this scenario, the two parties must move closer to each other so that they can resume their communication. If we let both parties move around, then they may not be able to find each other. Rather than giving both nodes the freedom to move in the third phase, we propose that one entity act as the Master, who will remain stationary, while the other entity acts as the Slave, who will move in search of the Master. In our figure, B is the Master and stays at B' while A moves to find B. Since every node knows the other node's initial location, A can move along the x-axis to approach B. One issue that comes up is that A may enter the interference range while searching for B. As soon as A detects that it is in the interference range, it must stop moving along the x-axis and return to moving along the y-axis to exit the interference range. While moving along the x-axis, A will not move beyond B's x-position, and if A's x-coordinate ever equals B's, A will move towards B directly.

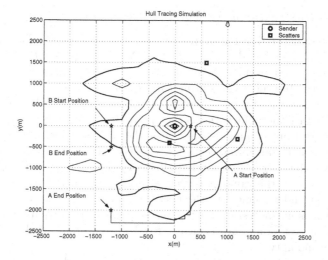

FIGURE 9.13. Simulated hull tracing scenario in which B is the Master and A is the Slave. Upon escaping the radio region of the adversary, A seeks to get within 1000 meters of B.

This protocol achieves the two necessary conditions, and can easily be modified to handle scenarios where only one node is blocked by the interferer.

We studied the behavior of the proposed spatial retreat strategy by conducting a simulated radio communication scenario involving an adversary emitting a jamming signal in the 916.7 Mhz unlicensed band. The two entities A and B were initially located at $(300, 0)$ and $(-1200, 0)$ meters respectively. In order to capture a realistic non-isotropic radio pattern for the interferer, we placed three scatterers at $(600, 1500)$, $(1200, -300)$, and $(-100, -400)$ meters. The radio environment was simulated through ray tracing [119]. The scatterers were assumed to introduce random phase shifting in the transmitted signal. The hull tracing algorithm presented above was employed, with entity B acting as the Master while entity A acted as the Slave. Both A and B were assumed to know each other's initial position, and that they could measure the energy emitted by the interferer. We present the results of the simulation in Figure 9.13. In this figure, we have presented contours of equal energy for different (x, y) locations relative to the interferer. The paths taken by entities A and B are depicted. As both A and B escape the radio region of the adversary, A moves too far from B and cannot maintain radio connectivity. Therefore, A performs the *Move Into Radio Range* portion of the procedure.

9.5.2 Infrastructured Network

In the infrastructured scenario, there are several access points $AP_0, AP_1, \cdots,$ AP_N that are connected via a backbone. Wireless devices, R_j, connect to access points and perform communication between themselves (or with devices on the Internet) via routing through the APs.

As noted earlier, during jamming in the infrastructured network, the interferer can either block the access point from the receivers, or block the receivers from communicating with the access point, or do both. A spatial retreat for an infrastructured wireless network must be a strategy that allows the user R_j to survive all three situations. The basic idea of spatial retreat in this context is that a mobile device will move to a new access point and reconnect to the network under its new access point. We note that it is not necessary for the access point to participate in a spatial retreat as access points are typically fixed infrastructure and not usually capable of mobility.

All three situations described above can be detected by an appropriate jamming detection strategy, as discussed in Chapter 11. In order to perform a spatial retreat for an infrastructured wireless network, we assume that each mobile device has an *Emergency Access Point List* assigned to it, and that the mobile device knows how to move in order to reach its Emergency Access Point. The Emergency Access Point can be assigned to each device by its current Access Point prior to the jamming/ interference.

When a device R_j detects that a jamming/interference occurs (either it cannot communicate, or it cannot receive beacons from its access point), it will begin to move towards its Emergency Access Point. While moving, it will occasionally pause and attempt to re-establish communication with its home access point. This is done in order to avoid any unnecessary handovers to other access points, and arises in scenarios where the adversary only blocks the user and not the access point. However, if R_j is not able to re-establish communication with its original AP, it will continue to move towards its Emergency Access Point. When R_j receives beacons from the new access point, it will initiate access point handoff. The purpose of access point handoff is to perform mutual authentication and establish authorization to use the new access point's services. There are many variations of authenticated handoff that can be employed, such as [120, 121].

We note that one problem that might result from spatial retreats in the infrastructured network is that all the mobile devices under an AP might move to the same Emergency AP. In order to prevent the resulting congestion at other APs, it is wise for the current AP to assign the Emergency Access Point lists in such a way as to divide the load across all of its neighboring APs.

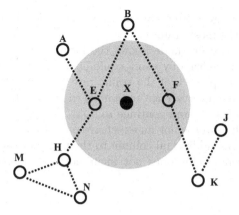

FIGURE 9.14. Scenarios for spatial retreat strategies in an ad hoc network setting. The adversary is marked by X.

9.5.3 Ad Hoc Network

It is much harder to design a spatial retreat strategy for ad hoc network scenarios because each node is not only involved in the communication it initiates but is also involved in forwarding packets. For ad hoc networks, it is critical to maintain network connectivity and if a node must leave its original position as a response to jamming attacks, it should move to a new location that minimizes degradation to network connectivity. For this preliminary study, we assume that only those nodes who are interfered with by the jamming attack need to escape, while other nodes should stay where they are. A globally optimized topology reformation strategy is beyond the scope of this paper [122].

Figure 9.14 illustrates a spatial retreat scenario in an ad hoc network setting, which attempts to minimize the network connectivity degradation. In Figure 9.14, node E originally connects to nodes A, B, and H, thus participating in flows $\overrightarrow{AEB}$, $\overrightarrow{AEH}$, and $\overrightarrow{HEB}$. After adversary X starts jamming the channel, E decides to move away. As shown in the figure, it is impossible for E to find a new location where it can avoid X but still maintain connection to A, B, and H. It has two choices: (1) move closer to A and B, or (2) move closer to A and H. (It cannot maintain $\overrightarrow{HEB}$ any more.) It compares these two options, and chooses the one which leads to a smaller loss in local network behavior by trying to maintain the local flow with the most value. Suppose the local flow $\overrightarrow{AEB}$ had a much higher traffic rate than the local flow $\overrightarrow{AEH}$. In this scenario, E decides to move to a new location between A and B in order to maintain the high-valued flow $\overrightarrow{AEB}$. In the other example shown in the same figure, F connected B and K before jamming occurs. No matter where F moves to, it cannot connect both B and K. Network partitioning cannot be avoided in this case, and F should move to a location away from the interferer where it will be able serve as the endpoint for the most traffic.

We assumed that every node knows the location of its neighbors. This assumption can be realized by using equipment such as GPS. In addition, every node must keep track of the traffic rate of each stream it connects. Following a jamming/interference, each node will escape to a location where it can avoid the jammer, and continue to serve as much traffic as possible. Tracking the traffic rates is not an expensive operation, and can be accomplished by adding an additional column to the neighbor table for recording traffic measurements. This does not incur noticeable energy or memory overhead.

9.6   Conclusion

In this chapter, we have given an overview on the problem of jamming/radio interference. Our case studies for RF-interference show that jamming can be caused both by Non-MAC-compliant wireless devices and by MAC-compliant nodes. Non-MAC-compliant radio devices could jam network by either continuously sending packets ignoring the MAC-layer protocols, or just emitting jamming signals. The network nodes, which are compliant with the underlining MAC protocol but have lots of data packets to send, or have failed to choose a proper orthogonal channel, could interfere with other nodes in the network, as well.

We have discussed strategies to cope with non-MAC-compliant interferers. In particular, we have presented two different strategies that may be employed to mitigate the effects of interference. The rationale behind both strategies is that legitimate wireless users should avoid the interference as much as possible because it is very hard to combat the adversary. The first strategy involves changing the transmission frequency to a range where there is no interference from the adversary. The second strategy involves wireless users moving to a new location where there is no interference. We examined both strategies for three general classes of wireless networks: a generic two-party radio communication scenario, infrastructured wireless networks, and multi-hop ad hoc networks.

In the next chapter we examine the impact of jamming on wireless communication. We will then examine the issues of detecting interference in Chapter 11, and provide a more detailed analysis of channel surfing in ad hoc networks in Chapter 12.

10

Jamming Attacks and Radio Interference

10.1 Introduction

Jamming attacks/radio interference can severely interfere with the normal operation of wireless networks and, consequently, mechanisms are needed that can cope with jamming attacks/radio interference. The entity (either malicious or an unintentional wireless device) that launches such radio interference is referred to as the *jammer* or *interferer*. In the following three chapters, we examine radio interference from both sides of the issue: first, we study the problem of conducting radio interference on wireless networks in this chapter, and second we examine the critical issue of diagnosing the presence of jamming and resuming communication in the presence of jamming in Chapter 11 and Chapter 12, respectively.

The first stage towards defending a wireless network is to understand what is the impact that a jammer could have on wireless networks. Thus, in this chapter, we analyze the network performance degradation caused by radio interference using two methodologies, i.e. theoretical analysis, and real system implementation. In section 10.2, we first model the channel capacity under radio interference, and the geographical impact the radio interferers have on the networks according to the physical layer metrics signal to noise ratio. Then in section 10.3, we study the effectiveness of radio interference in real system implementation. In particular, we introduce four typical radio interference models, though by no means all-inclusive, which represent a broad range of radio interference strategies covering both intentional and unintentional radio interference in section 10.3.2. Finally, we implemented

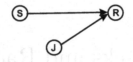

FIGURE 10.1. Two-party radio communication scenario.

four jamming models and present the experimental results of their impact on communication in section 10.3.3.

10.2   Theoretical Analysis on the Effectiveness of Jamming

We start by examining the theoretical performance degradation caused by radio interference. In particular, we study the channel capacity in the presence of interference, and the geometric impact of the interferers based on Physical layer considerations.

There are many different strategies an interferer can use to jam wireless communications. For instance, the jammer either proactively jam the channel or reactively adjust its jamming strategy based on the network traffic J observes. We will describe the jamming strategies in Section 10.2.2. In this section, we focus our discussion on a proactive jammer, where the jammer continuously send out an interference signal.

10.2.1   Jamming Impact on Channel Capacity

The Shannon's capacity theorem tells us that the capacity of an additive white Gaussian noise channel is given by $C = Blog_2(1 + S/N)$, where B is the transmission bandwidth and S/N is the signal to noise ratio. There are several questions that arise regarding the effect of interference on the channel capacity.

Let's start by considering a simple network consisting of a source S, a receiver node R, and a jammer J that is interfering (either intentionally or accidentally) with legitimate wireless communications between S and R, as depicted in Figure 10.1. We assume the channel between node S and node R, and between jammer J and node R are both additive white Gaussian noise (AWGN) channels.

For simplicity, we shall abuse notation and let S be the signal sent by the sender S and J the one blasted by the jammer J. We assume that our signals are transmitted over a channel with bandwidth B. Without loss of generality, we assume the channel response for both the sender and the jammer is frequency non-selective (i.e. constant). This allows us to absorb the role of the $S \rightarrow R$ and $J \rightarrow R$ channel amplitude responses into the transmit power of S and J respectively, so that the received signal at node

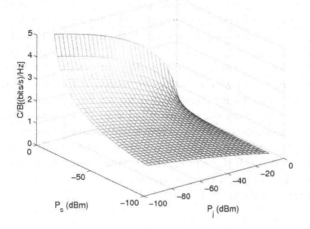

FIGURE 10.2. Normalized channel capacity as a function of P_s and P_j.

R is:

$$R = S + J + N, \tag{10.1}$$

where N is the Gaussian noise with variance σ^2, the signal S has power $P_s = 2B\sigma_s^2$ and the signal J has power $P_j = 2B\sigma_j^2$. The capacity of the communication channel between node S and node R is formally defined as

$$C = \max_{p(s)} I(S; R), \tag{10.2}$$

where $I(S; R)$ is the average mutual information that can be inferred about the transmitted signal S by observing the received signal R. The capacity of the channel, which is the maximum bit rate from $S \to R$, is the maximum value of $I(S; R)$ over all input symbol probability distributions. It is well-known that $I(S; R)$ is maximum when S is a Gaussian random variable [123], which is parameterized by the variance (power) term σ_s^2. Therefore, we shall assume that the sender's transmitted signal is a zero-mean Gaussian signal. Similarly, we also assume that the jammer is employs an optimal interference strategy, so that J's signal is a zero-mean Gaussian random variable with power σ_j^2. Straight-forward calculations give that the channel capacity as

$$\begin{aligned} C &= \max_{p(s)} I(S; R) \\ &= B \log_2(1 + \frac{2\sigma_s^2}{2\sigma_j^2 + 2\sigma^2}). \end{aligned} \tag{10.3}$$

In the case where the noise power is much smaller than the jamming power, the channel capacity is approximately

$$C = B \log_2(1 + \frac{P_s}{P_j}) \tag{10.4}$$

where P_s is the average transmission power of signal S, P_j is the average transmission power of jamming signal J. We plot the normalized capacity C/B versus the signal power P_s and the jamming power P_j in Figure 10.2. For a given jamming power and bandwidth B, as the signal power P_s increases, the channel capacity increases. We note that the signal power P_s and the jamming power P_j are the one measured at the receiver end. Given that the sender retains its transmission power level the same, the closer the receiver is away from the sender, the larger P_s is, which suggests that moving the sender closer to the receiver while keeping the jammer where is was will increase the channel capacity. This naturally leads to the fact that the relative locations of the sender, the receiver, and the jammer affect the channel capacity. Thus, in the next subsection, we will study geographical issues associated with interference.

10.2.2 Non-isotropic Model for Jamming

We now provide a geographical interpretation of the impact a jammer has on wireless communications. Typically, most recent papers on jamming and wireless networks have modeled the impact of the jammer as an isotropic effect. This has caused many authors to depict the effective jamming range as a circular region that is centered at the jammer's location. In reality, though, such an interpretation of jamming is overly simplistic and does not provide a complete depiction of the complex relationships between the transmission power of the source and the jammer, and the geometry of the deployment.

The circular jamming model does not capture the fact that the success reception of a packet is primarily determined by S/J at the receiver R. This ratio, S/J, depends on multiple factors, which include the transmission power of the source and jammer, as well as the distances between the receiver R, the source S, and the jammer J, i.e. d_{SR} and d_{JR}.

For a given pair of the source S and the jammer J, changing the location of the receiver R results in changing d_{SR} and d_{JR}, which in turn changes ratios S/J. To better understand the effectiveness of the jammer, we now derive the contours of constant S/J. Let $\gamma_{S/J}$ denote the signal-to-jamming ratio at each contour.

To begin, let us use the standard quadratic propagation loss model (i.e. a free space model). In this case, the received power is

$$P_R = \frac{P_T G_T G_R}{4\pi(d/d_0)^2} \tag{10.5}$$

where, P_T is the transmission power of the transmitter; G_T is the antenna gain of that transmitter in the direction of the receiver; G_R is the antenna gain of the receiver in the direction of the transmitter. d is the distance between the transmitter and the receiver, while d_0 is a reference distance (typically chosen so that $d_0 = 1$). Assume that the jammer uses the same

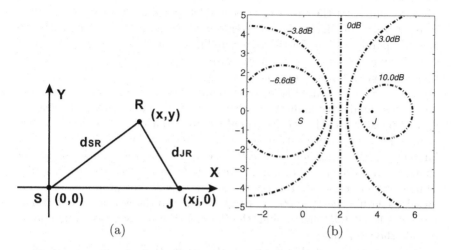

FIGURE 10.3. Coordinate system for constant S/J contours, (a) the positions of the sender S, the receiver R, and the jammer J; (b) constant S/J contour plot, where the distance between S and J is 4 units. The contour labels are the ratio of P_{ST}/P_{JT}, and the contour lines correspond to the edge where $\gamma_{S/J} = 0dB$.

type of device as the sender, e.g. both use omni-directional antennas. Then, this signal propagation model can be applied to the signal sent by the sender and the jammer as well, and the antenna gains G_T of the sender and the jammer are the same in all directions.

The signal-to-interference ratio at the receiver thus becomes

$$\gamma_{S/J} = \frac{P_{SR}}{P_{JR}} = \frac{P_{ST}d_{JR}^2}{P_{JT}d_{SR}^2}, \tag{10.6}$$

where P_{ST} is the transmission power of the sender S, P_{JT} is the transmission power of J. Using the coordinate system shown in Figure 10.3, the coordinate of the sender S $(0,0)$, the jammer J is located at $(x_j, 0)$, and the receiver R is arbitrarily placed at (x, y). Noting that $d_{SR}^2 = (x_j - x)^2 + y^2$, and $d_{JR}^2 = x^2 + y^2$, we may substitute to get the contours of constant $\gamma_{S/J}$

$$(x - \frac{x_j}{1 - \beta})^2 + y^2 = \frac{\beta x_j^2}{(1 - \beta)^2}, \tag{10.7}$$

where $\beta = \frac{\gamma_{S/J}}{P_{ST}/P_{JT}}$. For a given P_{ST} and P_{JT}, the constant contours of $\gamma_{S/J}$ are circles centered at $(\frac{x_j}{1-\beta}, 0)$ with radius $\frac{\sqrt{\beta}x_j}{|1-\beta|}$.

We provide a depiction of equation (10.7) in Figure 10.3 (b). In this figure, we provide several contours corresponding to different transmit signal-to-jamming ratios, i.e. $P_{ST}/P_{JT} \in \{-6.6, -3.8, 0, 3.0, 10.0\}$ dB. Each of these contours map out loci of constant received signal-to-interference ratio corresponding $\gamma_{S/R} = 0dB$. We now discuss the interpretation of this figure.

First, we note that the case $P_{ST}/P_{JT} = 0$dB splits the figure into two separate categories of curves: those centered near the source and those centered near the jammer. First, for those centered near the source, we may interpret these regions as areas where, if we were to place a receiver within one of these contours, it would be able to successfully decode transmissions from the source if the required signal-to-jammer ratio is $\gamma_{S/R} = 0dB$. Hence, a receiver located within the $P_{ST}/P_{JT} = -6.6$dB curve, would be able to decode packets (i.e. the receive signal-to-interference level would be better than 0dB) from a source that is transmitting at a level 6.6dB below the jammer. On the other hand, the circles centered near the jammer imply the contrary– a receiver within one of these circles would receive packets at a signal-to-interference level *worse* than $\gamma_{S/R} = 0dB$. For example, for those locations within the $P_{ST}/P_{JT} = 10$dB curve, even though the source transmits at a level 10dB higher than the jammer, the receiver is still unable to decode the transmission.

The contours depicted in Figure 10.3(b) shows that the effective jamming range cannot be simply modeled as a cycle centered at the jammer within which the receivers will be blocked. The effective jamming range depends on the network geographical topology. We note that, in a realistic deployment, the contours will not be so regular, due to the non-uniformity of the underlying multipath propagation environment. Thus, it is crucial to study the practical issues associated with the real system implementation.

10.3   System Study on Jamming/Interference Models and their Effectiveness

The first stage to defending a wireless network is to understand what types of radio interference are feasible, and how effective they are in a real system. Therefore, in this section, we introduce radio interference that may be implemented using wireless network devices. We first define the characteristics of a jammer's behavior, and then enumerate metrics that can be used to measure the effectiveness of jamming. These metrics are closely related to the ability of a radio device to either send or receive packets. We then introduce four typical jammer attack models, though by no means all-inclusive, which represent a broad range of interference strategies that cover both intentional jamming attacks and unintentional radio interference, and will serve as the basis for our discussion throughout the remainder of the thesis. Throughout this thesis work, we will use the Berkeley MICA2 Mote platform for conducting our experiments with jammers. The observed characteristics of the jammers and the detection schemes presented later should hold for different wireless platforms, such as 802.11.

10.3.1   Jamming Characteristics and Metrics

Although several studies [122, 124–126] have targeted jamming-style attacks, the definition of this type of attack remains unclear. A common assumption is that a jammer continuously emits RF signals to fill a wireless channel, so that legitimate traffic will be completely blocked [122,126]. We believe, however, that a broader range of behaviors can be adopted by a jammer. For example, a jammer may remain quiet when there is no activity on the channel, and start interference as soon as it detects a transmission. The common characteristic for all jamming attacks is that their communications are not compliant with MAC protocols. Therefore, *we define a jammer to be an entity who is trying to interfere with the physical transmission and reception of wireless communications either purposefully or accidentally.*

The outcome of a jammer is interference with legitimate wireless communications. A jammer can achieve this result by either preventing a real traffic source from sending out a packet, or by preventing the reception of legitimate packets. Let us assume that A and B denote two legitimate wireless participants, and let us denote X to be the jammer. A legitimate participant may be unable to send out packets for many reasons. To name just a couple, X can continuously emit a signal on the channel so that A will never sense the channel as idle, or X can keep sending out regular data packets and force A to receive junk packets all the time. On the other hand, however, even if A successfully sends out packets to B, it is possible for X to blast a radio transmission to corrupt the message that B receives. We thus define the following two metrics to measure the effectiveness of a jammer:

- **Packet Send Ratio (PSR):** The ratio of packets that are successfully sent out by a legitimate traffic source compared to the number of packets it intends to send out at the MAC layer. Suppose A has a packet to send. Many wireless networks employ some form of carrier-sensing multiple access control before transmission may be performed. For example, in the MAC protocol employed by Mica2, the channel must be sensed as being in an idle state for at least some random amount of time before A can send out a packet. Further, different MAC protocols have different definitions on an idle channel. Some simply compare the signal strength measured with a fixed threshold, while others may adapt the threshold based on the noise level on the channel. A radio interference attack may cause the channel to be sensed as busy, causing A's transmission to be delayed. If too many packets are buffered in the MAC layer, the newly arrived packets will be dropped. It is also possible that a packet stays in the MAC layer for too long, resulting in a timeout and packets being discarded. If A intends to send out n messages, but only m of them go through, the PSR is $\frac{m}{n}$. The PSR can be easily measured by a wireless device by keeping track of the number of packets it intends to send and the number of packets that are successfully sent out.

- **Packet Delivery Ratio (PDR):** The ratio of packets that are successfully delivered to a destination compared to the number of packets that have been sent out by the sender. Even after the packet is sent out by A, B may not be able to decode it correctly, due to the interference introduced by X. Such a scenario is an unsuccessful delivery. The PDR may be measured at the receiver B by calculating the ratio of the number of packets that pass the CRC check with respect to the number of packets (or preambles) received. PDR may also be calculated at the sender A by having B send back an acknowledge packet. In either case, if no packets are received, the PDR is defined to be 0.

10.3.2   Jamming Attack/Radio Interference Models

There are many different interference strategies that a jammer can perform in order to interfere with other wireless communications. As a consequence of their different interference intention or attack philosophies, these various interference models will have different levels of effectiveness, and may also require different detection strategies. While it is impractical to cover all the possible interference models that might exist, in this study, we discuss a wide range of radio interference and somewhat malicious jamming attacks that have proven to be effective in disrupting wireless communication. Specifically, we have designed and built the following jammers:

- **Constant Jammer:** The constant jammer continually emits a radio signal, and it can be either a malicious jamming attack where adversary purposely interfere with network communication, or unintentional radio interference where the interferer is always emitting RF signal. We have implemented a constant jammer using two types of devices. The first type of device we used is a waveform generator which continuously sends a radio signal. The second type of device we used is a normal wireless device. In this paper, we will focus on the second type, which we built on the MICA2 Mote platform. Our constant jammer continuously sends out random bits to the channel without following any MAC-layer etiquette. Specifically, the constant jammer does not wait for the channel to become idle before transmitting. If the underlying MAC protocol determines whether a channel is idle or not by comparing the signal strength measurement with a fixed threshold, which is usually lower than the signal strength generated by the constant jammer, a constant jammer can effectively prevent legitimate traffic sources from getting hold of channel and sending packets.

- **Deceptive Jammer:** We use deceptive jammer to model a malicious jammer. Instead of sending out random bits, the deceptive jammer constantly injects regular packets to the channel without any gap between subsequent packet transmissions. As a result, a normal communicator will be deceived into believing there is a legitimate packet and will be duped to remain in the receive state. For example, in TinyOS, if a preamble is detected, a node remains in the receive mode, regardless of whether that node has a packet to send or not. Hence, even if a node has packets to send, it cannot switch to the send state because a constant stream of incoming packets will be detected. Further, we also observe that it is adequate for the jammer to only send a continuous stream of preamble bits (0xAA in TinyOS) rather than entire packets.

- **Random Jammer:** Instead of continuously sending out a radio signal, a random jammer alternates between sleeping and jamming. Specifically, after jamming for t_j units of time, it turns off its radio, and enters a "sleeping" mode. It will resume jamming after sleeping for t_s time. t_j and t_s can be either random or fixed values. During its jamming phase, it can either behave like a constant jammer or a deceptive jammer. Throughout this paper, our random jammer will operate as a constant jammer during jamming. The distinction between this model and the previous two models lies in the fact that this model tries to take energy conservation into consideration, which is especially important for those jammers that do not have unlimited power supply. By adjusting the distribution governing the values of t_j and t_s, we can achieve various levels of tradeoff between energy efficiency and jamming effectiveness. Random jammer can either represent a malicious jammer or unintentional interferer that interfere with network communication from time to time.

- **Reactive Jammer:** Finally, we have a malicious jammer called reactive jammer. The three models discussed above are active jammers in the sense that they try to block the channel irrespective of the traffic pattern on the channel. Active jammers are usually effective because they keep the channel busy all the time. As we shall see in the following section, these methods are relatively easy to detect. An alternative approach to jamming wireless communication is to employ a reactive strategy. For the reactive jammer, we take the viewpoint that it is not necessary to jam the channel when nobody is communicating. Instead, the jammer stays quiet when the channel is idle, but starts transmitting a radio signal as soon as it senses activity on the channel. As a result, a reactive jammer targets the reception of a message. We would like to point out that a reactive jammer does not necessarily conserve energy because the jammer's radio must contin-

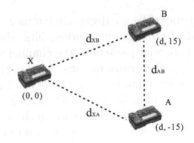

FIGURE 10.4. Placement of the Motes during jammer effectiveness experiments.

uously be on in order to sense the channel. The primary advantage for a reactive jammer, however, is that it may be harder to detect.

10.3.3 Experimental Results

We have implemented the above four jammer models using Berkeley Motes that employ a ChipCon CC1000 RF transceiver and use TinyOS as the operating system. We disabled channel sensing and back off operations to bypass the MAC protocol, so that the jammer can blast on the channel irrespective of other activities that are taking place. The level of interference a jammer causes is governed by several factors, such as the distance between the jammer and a normal wireless node, the relative transmission power of the jammer and normal nodes, and the MAC protocol employed by normal nodes. The closer a jammer is to a node, or the higher transmit power it employs, the greater the impact it will have on network operation. The MAC protocols employed by the network also play a role. Usually, MAC protocols decide the channel is idle if the measured signal strength value is lower than a threshold. Many MAC protocols, such as the one in TinyOS release 1.1.1, uses a fixed threshold value. Some MAC protocols, however, such as BMAC [127], adapt the threshold value based on the measured signal strength values, i.e. they choose the minimum signal strength among the most recent n readings when channel is idle as the current threshold value. Consequently, if a constant jammer transmits at a constant power, and both the jammer and the nodes are static, these adaptive MAC protocols will consider the channel as idle since they will regard the energy emitted by the jammer as ambient noise. In addition to these network configuration parameters, the impact of a jammer is also affected by jammer-specific parameters, such as the sleep interval for a random jammer. In order to understand the interactions of these parameters and quantify the impact of a jammer in different scenarios, we conducted a set of experiments involving three parties: A, B, and X, where A and B are normal wireless nodes with A being the sender, B the receiver, and X a jammer using one of our four models. The transmission power levels employed by A, B, X are all $-4dBm$.

TABLE 10.1. The resulting PSR and PDR for different jammer models under various scenarios.

d_{XA} (inch)		Constant Jammer			
		BMAC		1.1.1 MAC	
		PSR (%)	PDR (%)	PSR (%)	PDR (%)
38.6		74.37	0.43	1.00	1.94
54.0		77.17	0.53	1.02	2.91
72.0		99.57	93.57	0.92	3.26
d_{XA} (inch)		Deceptive Jammer			
		BMAC		1.1.1 MAC	
		PSR (%)	PDR (%)	PSR (%)	PDR (%)
38.6		0.00	0.00	0.00	0.00
54.0		0.00	0.00	0.00	0.00
72.0		0.00	0.00	0.00	0.00
d_{XA} (inch)		Random Jammer			
		BMAC		1.1.1 MAC	
		PSR (%)	PDR (%)	PSR (%)	PDR (%)
$t_j = U[0,31]$ $t_s = U[0,31]$	38.6	79.45	0.26	70.19	16.77
	44.0	80.15	17.48	70.30	21.95
	54.0	80.43	99.00	76.98	99.75
$t_j = U[0,31]$ $t_s = U[1,8]$	38.6	60.47	0.06	56.49	0.00
	44.0	60.72	47.41	56.00	0.41
	54.0	61.77	96.75	100.0	99.64
d_{XA} (inch)		Reactive Jammer			
		BMAC		1.1.1 MAC	
		PSR (%)	PDR (%)	PSR (%)	PDR (%)
m = 7bytes	38.6	99.00	0.00	100.0	0.00
	54.0	100.0	99.24	100.0	99.87
	72.0	100.0	99.35	100.0	99.97
m = 33bytes	38.6	99.00	0.00	100.0	0.00
	44.0	99.00	58.05	100.0	87.26
	54.0	99.25	98.00	100.0	99.53

These three nodes are carefully placed so that X has the same impact on both A and B. In particular, we set d_{XA}, the distance between X and A, equal to d_{XB}, the distance between X and B, and we fixed the distance between the sender A and the receiver B at $d_{AB} = 30$ inches, as depicted in Fig. 10.4.

The resulting PSR and PDR for each jammer model are summarized in Table 10.1. As the Table 10.1 shows, if A employs 1.1.1 MAC, a constant jammer that is reasonably close to A can completely block A, from sending out packets, resulting in a very low PSR. However, if A employs BMAC, which adapts the threshold based on the surrounding signal strength, A can still manage to send out a large portion of the packets, i.e, with PSR being 74.37% even when X is only 38.6 inches away from A. The reason why A cannot send out all of the packets is that the signal strength produced by X varies with time. The corresponding PDR in both cases, however, is poor because most of the packets are corrupted by the constant jammer, especially when the constant jammer is close to the sender.

However, the same trend cannot be observed for a deceptive jammer. Since a deceptive jammer continuously sends out packets with valid preamble, both A and B are forced to constantly stay in the reception mode no matter which MAC protocol they use. Hence, A and B cannot send out any packets at all and the PSR are 0% all the time. PDR in this case is defined as 0.

For the random jammer, in addition to studying the impact of network configuration parameters, such as the distance between the jammer and the nodes, and the MAC protocol on the effectiveness of the jammer, we also look at jammer-specific parameters, such as the on-off periods. Specifically, we studied two random jammers. For the first random jammer, the duration of the jamming period t_j is a uniform random number between 0 and 31 spibus interrupts in TinyOS [128], denoted by $t_j = U[0,31]$, and the duration of the sleeping period t_s is a uniform random number between 0 and 31 as well, denoted by $t_s = U[0,31]$. For the second random jammer, $t_j = U[0,31]$, and $t_s = U[1,8]$. On average, the second jammer sleeps less, and switches to the jamming mode more often. Thus, the PSR measured in the second jammer scenario is less than the PSR in the first jammer scenario. Additionally, since the random jammer alternates between jamming and sleeping, BMAC, which always chooses the minimum signal strength value among the recent readings, cannot increase the threshold quickly enough to consider the channel idle. Thus, BMAC considers the channel as busy when the random jammer is jamming, resulting in a lower PSR.

A reactive jammer starts interference as soon as it hears a transmission on the channel. Consequently, the effectiveness of a reactive jammer is also dependent on size of legitimate network packets as well as the size of packet the jammer emits. In Table 10.1, we explore the behavior of the reactive jammer for network packets of size $m = 7$ and $m = 33$ bytes, where the jammer emitted a 20 byte jamming packet. First, we observe that in all cases the sender is able to reliably send out its packets. Ideally, if m is short, one would infer that there may not be enough time for a reactive jammer to corrupt a network packet in transmission. However, as we see in Table 10.1, for different network packet sizes, although there is a difference in the resulting PDR, the difference is in fact negligible. Hence, even for short packets of a few bytes in length, a jammer employing the reactive strategy is able to effectively disrupt network communication.

10.4 Conclusion

The shared nature of the wireless medium will allow adversaries to pose non-cryptographic security threats by conducting radio interference attacks. Therefore, understanding the nature of jamming attacks and radio interference is critical to assuring the operation of wireless networks. Thus, in this chapter, we have analyzed the network performance degradation caused by radio interference from two aspects, i.e. theoretical analysis, and real system implementation.

First, we have studied channel capacity in radio interference scenarios theoretically. The channel capacity is the average mutual information that can be inferred from the received signal. Theoretical analysis shows that the channel capacity is given by $C = B \log_2(1 + \frac{P_s}{P_j})$, where B is the transmission bandwidth, P_s is the average transmission power of signal S, and P_j is the average transmission power of jamming signal J. Starting from the fact that the channel capacity is a function of $\frac{P_s}{P_j}$, we have derived the contour of $\frac{P_s}{P_j}$ which represents the geographical impact that a jammer has on wireless communications. Our study shows that the effective jamming range cannot be simply modeled as a circle centered at the jammer. Instead, the effective jamming range depends on the geographical location of the source, the jammer and the receiver.

Theoretical study gives an insight on the impact that an interferer can have on the network. It, however, does not capture many practical issues. In the second part of this chapter, we have studied the effectiveness of radio interference in a real system implementation. In particular, we introduce four typical radio interference models, though by no means all-inclusive, which represent a broad range of radio interference strategies, and cover both intentional and unintentional radio interference. We implemented four jamming models and the experiment results show that they are effective in disturbing the network communications.

11

Detecting Jamming Attacks and Radio Interference

11.1 Introduction

Detecting jamming attacks is important because it is the first step towards building a secure and dependable wireless network. *Detecting radio interference attacks is challenging as it involves discriminating between legitimate and adversarial causes of poor connectivity.* Specifically, we need to differentiate a jamming scenario from various network conditions: congestions that occur when the aggregated traffic load exceeds the network capacity so that the packet send ratio and delivery ratio are affected; the interrupt of the communication due to failures at the sender side; and other similar condition.

In this chapter, we address the problem of detecting jamming attacks and radio interference. In particular, we focused on detecting the four jamming models presented in Chapter 10, which cover a broad range of malicious jamming attacks and unintentional radio interference. There are several measurements that naturally lend themselves to detecting jamming, such as signal strength, carrier sensing time, and packet delivery ratio. In Section 11.2, we explore these measurements in detail and present scenarios where they may not be effective in detecting a jamming attack, and in fact could cause false detections. For each of these measurements, we develop statistics upon which to make decisions. Since statistics built upon individual measurements may lead to false conclusions, in Section 11.3 we develop two improved detection strategies. These two detection strategies are both built upon the fundamental assumption that communicating parties should have some basis for knowing what their characteristics should be if they are not jammed, and consequently can use this as a basis for differentiating jammed scenarios from mere poor link conditions.

11.2   Basic Statistics for Detecting Jamming Attacks and Radio Interference

11.2.1   Signal Strength

One seemingly natural measurement that can be employed to detect jamming is signal strength, or ambient energy. The rationale behind using this measurement is that the signal strength distribution may be affected by the presence of a jammer. In practice, since most commodity radio devices do not provide signal strength or noise level measurements that are calibrated (even across devices from the same manufacturer), it is necessary for each device to employ its own empirically gathered statistics in order to make its decisions. Each device should sample the noise levels many times during a given time interval. By gathering enough noise level measurements during a time period prior to jamming, network devices can build a statistical model describing normal energy levels in the network.

We now explore two basic strategies that employ signal strength measurements for detecting a jamming attack and radio interference. The first approach uses either the average signal value or the total signal energy over a window of N signal strength measurements. This is a simple approach that extracts a single statistic for basing a hypothesis test upon. Since a single statistic loses most of the shape characteristics of the time series, a second strategy would seek to capture the shape of the time series by representing its spectral behavior. The second strategy that we discuss uses N samples to extract spectral characteristics of the signal strength for the basis of discrimination. In the discussion below, we assume that we have measured the channel's received energy levels $s(t)$ at different times and collected N of these samples to form a window of samples $\{s(k), s(k-1), \cdots, s(k-N+1)\}$.

Basic Average and Energy Detection

We can extract two basic statistics from signal strength readings, namely, the average signal strength and the energy for detection. In both cases, the statistical hypothesis testing problem is binary and essentially involves deciding between signal absent and signal present hypotheses.

The use of the signal average arises naturally when the jammer emits a constant amplitude signal. In this case, the detection statistic is $T(k) = (\sum_{j=k-N+1}^{k} s(j))/N$. The use of the signal energy arises when the jammer emits a powerful noise-like signal, such as a white Gaussian process. Here, the detection statistic is $T(k) = (\sum_{j=k-N+1}^{k} s(j)^2)/N$. In either case, the detection decision is made by comparing $T(k)$ to a threshold γ that is

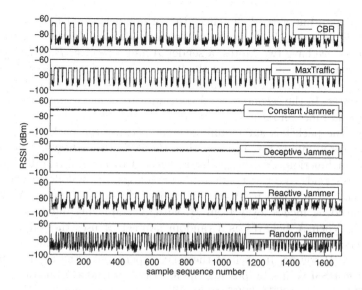

FIGURE 11.1. RSSI readings as a function of time in different scenarios. RSSI values were sampled every 1msec.

suitably chosen by considering tradeoffs between probability of detection and false alarm, such as through application of the Neyman-Pearson theorem [91, 129].

Signal Strength Spectral Discrimination

The average signal strength or the signal energy over a window of N samples does not reflect the fact that there may be many different received signal sample paths that could have led to the same mean or energy value. For example, a signal that has half of its ADC values as 50 and half as 150 would be considered the same as a signal whose samples are all 100 if we use the average signal strength as our decision statistic.

In order to have more robustness to false decisions and enhance the ability to classify scenarios, it is natural to use spectral discrimination techniques to classify the signal. One possible spectral discrimination mechanism is to employ higher order crossings (HOC). We refer the reader to the treatise on HOC [130] for explicit definition of HOC statistics. We have chosen to study higher order crossings since the calculation of these statistics only involves differences between samples, and is thus simple and practical to implement on resource-constrained wireless devices, such as sensor nodes. More complicated spectral techniques that involve the estimation of power spectral densities are possible and yield comparable performance but require more computational complexity.

Effectiveness Analysis:

In order to understand the effect that a jammer would have on the received signal strength, we performed six experiments. In the first two experiments, we have two Motes, a sender A and a receiver B, which are 30 inches apart from each other. In the first case, A transmits 20 packets per second, corresponding to a traffic rate of 5.28kbps, which we refer to as a CBR source. In the second case, A transmits at its maximum rate; as soon as the send function returns to the application level asynchronously, either because the packet is successfully sent or because the packet is dropped (the packet pumping rate is larger than the radio throughput), it posts the next send function. Such a sender is referred to as a MaxTraffic source, and corresponds to a raw traffic rate of 6.46kbps. In the following four experiments, in addition to A and B, we introduced the jammer X, which was placed 54 inches away from B, with X employing our four jammer models. When X behaves as a random jammer, it uses the following parameters: $t_j =$ U[0,31] and $t_s =$ U[0,31]. In these four jamming scenarios, A is a CBR source. In each of these six experiments, the receiver B obtains the RSSI values by posting the RSSIADC.getData() function on the port TOS_ADC_CC_RSSI_PORT every millisecond. The reported RSSI values in Fig. 11.1, in dBm, are converted from the raw values following the analog-to-digital conversion of the received voltage levels [131]. We present time series data for each of the six scenarios in Fig. 11.1. From these results, we observed that the average values for the constant jammer and the MaxTraffic source scenario, are roughly the same. Further, the constant jammer and deceptive jammer have roughly the same average values, with the slight difference in the plot resulting from experimental setup. Additionally, the signal strength average from a normal CBR source does not differ much from that measured for the reactive jammer scenario. Similar statements can be made for using the signal energy. These results suggest the following important observation: we may not be able to use simple statistics, such as average signal strength or energy, to discriminate jamming scenarios from normal traffic scenarios because it is not straightforward to devise a threshold that can separate these two scenarios.

There is a practical issue that arises from the locations the nodes and jammers relative to each other. Nodes that are very close to each other will naturally lead to high signal strength measurements, while nodes separated by more distance will yield lower signal strength measurements.

From the time series in Fig. 11.1, we observe that there are some differences in the shapes underlying the time series for these scenarios. For example, the measured signal strength for the constant jammer and the deceptive jammer exhibit a much lower variation (the time series curve is almost flat) compared to the signal strengths for MaxTraffic source.

We next examined the issue of whether spectral discrimination techniques would be able to distinguish between normal and jammed scenarios. We calculated the first two higher order crossings for the time series, D_1 and D_2, using a window of 240 samples. We plot D_1 versus D_2 in Fig. 11.2. From

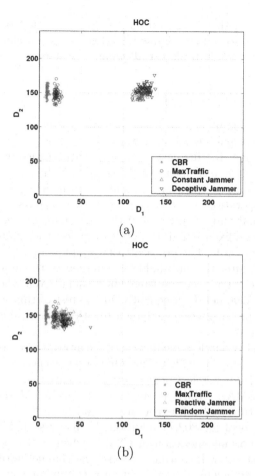

FIGURE 11.2. Plot of the first two higher order crossings, D_1 vs. D_2, for different jammer and communication scenarios.

the Fig. 11.2 (a), we observe that the points gather in two clusters, one cluster corresponding to the constant and deceptive jammers, while the other cluster corresponding to normal CBR and MaxTraffic sources. Hence, using HOC, we can distinguish normal traffic scenarios from the constant and deceptive jammer. However, examining Fig. 11.2 (b) we see that we cannot distinguish the reactive or random jammer from normal traffic scenarios. The reason for this is that a reactive jammer or random jammer causes the channel state to alternate between busy and idle in much the same way as normal traffic behaves. In particular, because the reactive jammer does not change the underlying busy and idle periods for a normal traffic scenario, it is particularly difficult to distinguish between signal readings for a reactive jammer and signal readings from the underlying traffic.

Hence, based on these observations, we conclude that employing HOC (or even other spectral methods), will work for some jammer scenarios, but are not powerful enough to detect all jammer scenarios.

11.2.2 Carrier Sensing Time

As discussed in Chapter 10.3, a jammer can prevent a legitimate source from sending out packets because the channel might appear constantly busy to the source. In this case, it is very natural for one to keep track of the amount of time it spends waiting for the channel to become idle, i.e. the carrier sensing time, and compare it with the sensing time during normal traffic operations to determine whether it is jammed. We would like to emphasize that this is only true if the legitimate wireless node's MAC protocol employs a fixed signal strength threshold to determine whether the channel is busy or idle. For protocols that employ an adaptive threshold, such as BMAC, after the threshold has adapted to the ambient energy of the jammer, the carrier sensing time will be small even when a jammer is blasting on the channel. Consequently, in the rest of this section, we only focus on MAC protocols that employ a fixed threshold, such as the MAC in TinyOS 1.1.1.

In most forms of wireless medium access control, there are rules governing who can transmit at which time. One popular class of medium access control protocols for wireless devices are those based on carrier sense multiple access (CSMA). CSMA is employed in MICA2 Motes as well as in both infrastructure and infrastructureless (ad hoc) 802.11 networks. The MAC-layer protocol for 802.11 additionally involves an RTS/CTS handshake. During normal operation of CSMA, when A (the sender) tries to transmit a packet, it will continually sense the channel until it detects the channel is idle, after which it will wait an extra amount of time (known as the propagation delay) in order to guarantee the channel is clear. Then, if RTS/CTS is used it will send the RTS packet, or otherwise will send the data packet. Suppose we assume that the jammer X continuously emits radio signal on a channel and that A attempts to transmit a packet. Then, since the channel is occupied by X, A will either time-out the channel sensing operation (if a time-out mechanism is available in the MAC protocol) or be stuck in the channel sensing mode.

Unfortunately, a large carrier sensing time could have occurred in non-jammed scenarios as well, such as congestion. It is therefore important to have some mechanism to distinguish between normal and abnormal failures to access the channel. In order to do so, a thresholding mechanism based on the sensing time can be used to identify jamming: Each time A wishes to transmit, it will monitor the time spent sensing the channel, and if that time is above a threshold (or if it is consistently above the threshold), it will declare that a jamming is occurring. The threshold may be determined theoretically based on a simple channel occupancy model, or empirically. The problem with theoretically calculating the threshold is that it is extremely difficult to build a complete mathematical model that captures a realistic MAC protocol. A well-known $M/M/1/1$ queuing model may be used to describe the MAC protocol [114, 115, 126], but doesn't capture the notion of collisions, or retransmissions. Therefore, we focus on the second approach to determining the threshold, which involves each network device collecting statistics regarding the amount of time D that a device must wait before it can start transmission during normal, or even somewhat congested, network conditions. With a distribution $f_D(d)$ describing carrier sensing times during acceptable network conditions, we may classify any new measured sensing time as either normal or anomalous by employing significance testing [91]. In this case, our null hypothesis is that the measured delay D corresponds to the distribution $f_D(d)$. If we reject the null hypothesis, then we conclude the network is experiencing a jamming attack. Since it is undesirable to falsely conclude the presence of jamming when the network is merely experiencing a glitch, we need to use a conservative threshold to reduce the probability of a false positive.

Effectiveness Analysis: In order to quantify the validity of detecting jamming at the MAC-layer using carrier sensing time in a real wireless network, we performed an experiment using two Motes, X and A. Here, Mote A corresponds to a network node trying to send out a 33-byte packet every 100msecs, and which measures the sensing time while doing so. Mote A employed the MAC protocol from TinyOS release 1.1.1, which used a fixed threshold for determining idleness. Mote X cycles through the four different types of jammers, as well as the MaxTraffic source. Additionally, we measured the sensing time when there is no background traffic, i.e. X does not send any traffic.

Fig. 11.3 depicts the cumulative distribution of the sensing time for the six different scenarios. Fig. 11.3 (a) shows that the cumulative distribution of the constant jammer and the deceptive jammer jumps at the point where the sensing time equals to 640msces. This is caused by a timeout we added to the TinyOS. In our experiment, if the device does not start to send the packet within 640msecs after the packet was passed to the MAC-layer from application layer, a timeout will occur, the packet will be discarded, and its sensing time will be counted as 640msecs.

The drawback of carrier sensing time is that it exhibits significant missed detections in the presence of other types of jammers. As Fig. 11.3 (b) shows, most of the sensing time in other jammer scenarios is smaller than the sensing time in a congested scenario. The reactive jammer will exhibit normal carrier sensing times because the jammer will not attempt to jam until another node has successfully started transmission. As a result, the transmitting node A will observe normal carrier sensing times. In particular, in our experiment the reactive jammer produces sensing time cumulative distributions that overlap completely with the case of no background traffic.

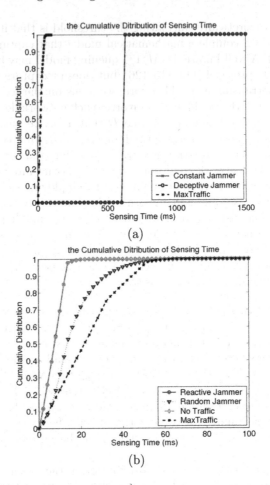

FIGURE 11.3. The cumulative distribution for the carrier sensing times measured using MICA2 Motes.

We note that, if the MAC protocol employs an adaptive threshold for determining channel idleness, instead of the fixed threshold in our experiment, then the cumulative distribution of the sensing time for the constant jammer would have shifted to the left, while there would have been no difference for the deceptive jammer since node A would still have been locked in a received state. The reactive jammer would have exhibited the same characteristics. Similar to the constant jammer, the random jammer also shifts the cumulative distribution to the left. We verified these observations through identical experiments to the ones described above where we used BMAC instead of the MAC protocol from TinyOS release 1.1.1.

In summary, both signal strength and carrier sensing time, under certain circumstances, can only detect the constant jammer and deceptive jammer.

Neither of these two statistics is effective in detecting the random jammer or the reactive jammer.

11.2.3 Packet Delivery Ratio

A jammer may not only prevent a wireless node from sending out packets, but may also corrupt a packet in transmission. Consequently, we next evaluate the feasibility of using packet delivery ratio (PDR) as the means of detecting the presence of jamming. The packet delivery ratio can be measured in the following two ways: either by the sender, or by the receiver. At the sender side, the PDR can be calculated by keeping track of how many acknowledgements it receives from the receiver. At the receiver side, the PDR can be calculated using the ratio of the number of packets that pass the CRC check with respect to the number of packets (or preambles) received. Unlike signal strength and carrier sensing time, PDR must be measured during a specified window of time where a baseline amount of traffic is expected. If no packet is received over that time window, then the PDR within that window is zero.

Since a jamming attack will degrade the channel quality surrounding a node, the detection of a radio interference attack essentially boils down to determining whether the communication node can send or receive packets in the way it should have had the jammer not been present. More formally, let us use π_0 to denote the PDR between a sender and a receiver, who are within radio range of each other, assuming that the network only contains these two nodes and that they are static. As shown in Table 10.1, any one of the four jammers, if placed within a reasonable distance from the receiver, can cause the corresponding PDR to become close to 0. In the cases shown in Table 10.1, π_0 is 100%. From these results, we can conclude that a jammer can cause the PDR to drop significantly. We would like to point out that a non-aggressive jammer, which only marginally affects the PDR, does not cause noticeable damage to the network quality and does not need to be detected or defended against.

Next, we need to investigate how much PDR degradation can be caused by non-jamming, normal network dynamics, such as congestion, failures at the sender side, etc. In order to study the impact of congestion on PDR, we introduced 3 MaxTraffic sources, resulting in a raw offered traffic rate of 19.38kbps[1], to model a rather highly congested scenario. Even under such a congestion level, the PDR measured by the receiver is still around 78%. As a result, a simple thresholding mechanism based on the PDR value can be used to differentiate a jamming attack, regardless of the jamming model, from a congested network condition.

---

[1] At 100% duty cycle, the MICA2 radio's maximum bandwidth capacity is 12.364kbps, though the effective maximum throughput is typically much less than that.

Though PDR is quite effective in discriminating jamming from congestion, it is not as effective for other network dynamics, such as a sender battery failure, or the sender moving out of the receiver's communication range, because these dynamics can result in sudden PDR drop in much the same way as a jammer does. Specifically, if the sender's battery drains out, it stops sending packets, and the corresponding PDR is 0%.

Consequently, compared to signal strength and carrier sensing time, PDR is a powerful statistic in that it can be used to differentiate a jamming attack from a congested network scenario, for different jammer models. However, it still cannot differentiate the jamming attack from other network dynamics that can disrupt the communication between the sender and the receiver.

11.3   Jamming Detection with Consistency Checks

In the previous section we saw that no single measurement is capable of detecting all kinds of jamming attacks. Since the purpose of a jammer is to influence the channel quality between a node and its neighbors, it is not reasonable, or needed, to try to detect a jammer if that jammer does not effectively interfere with the receipt/send of packets at a node. While a node losing its sending ability is a clear sign that it is being jammed, a weak reception capability (i.e. a low PDR) can be caused by several factors besides jamming, such as a low link quality due to the relatively large distance between the sender and the receiver.

We observed in the previous section that PDR is a powerful measurement that is capable of discriminating between jammed and congested scenarios, yet is unable to identify whether an observed low PDR is due to natural causes of poor link quality. In order to compensate for this drawback, and enhance the likelihood of detection, we will examine two strategies that build upon PDR to achieve enhanced jammer detection. We augment the use of PDR by applying signal strength measurements to conduct consistency checking to determine whether low PDRs are due to natural causes or due to radio interference. Later, in Section 11.3.2, we discuss a complementary technique that uses location information to augment PDR measurements for jamming detection.

Throughout this section, we assume that a node is only responsible for detecting whether it is jammed, and is not responsible for detecting the jammed condition of its neighbors. This follows from the fact that a wireless node is the best source of information regarding its local radio environment and is a less reliable predictor of the radio condition at distant locations. We assume that each node maintains a neighbor list, obtained from the routing layer, which will assist in making more reliable detection decisions. Additionally, we assume that the deployment of the network is sufficiently dense to guarantee that each node has several neighbors. All legitimate nodes in the network will participate in the detection protocol by transmitting a

Algorithm:PDRSS_Detect_Jam

$\{PDR(N) : N \in Neighbor\_list\}$ = Measure_PDR()
$MaxPDR = \max\{PDR(N) : N \in Neighbor\_list\}$
if $MaxPDR < PDRThresh$ **then**
    SS = Sample_Signal_Strength()
    $CCheck$ = SS_ConsistencyCheck($MaxPDR$, SS)
    if $CCheck == False$ **then**
       | post NodeIsJammed()
    end
end

Algorithm 5: Jamming detection algorithm that checks the consistency of PDR measurements with observed signal strength readings.

baseline amount of traffic, e.g. by sending heartbeat beacons. This allows each node to reliably estimate PDR over a window of time, and conclude that the PDR is 0 if no packets are observed during that time period.

11.3.1 Signal Strength Consistency Checks

The packet delivery ratio serves as our starting point for building the enhanced detector. Rather than rely on a single PDR measurement to make a decision, we employ measurements of the PDR between a node and each of its neighbors. In order to combat false detections due to legitimate causes of link degradation, we use the signal strength as a consistency check. Specifically, we check to see whether a low PDR value is consistent with the signal strength that is measured. In a normal scenario, where there is no interference or software faults, a high signal strength corresponds to a high PDR. However, if the signal strength is low, which means the strength of the wireless signal is comparable to that of the ambient background noise, the PDR will be also low. On the other hand, a low PDR does not necessarily imply a low signal strength. It is the relationship between signal strength and PDR that allows us to differentiate between the following two cases, which were not possible to separate using just the packet delivery ratio. First, from the point of view of a specific wireless node, it may be that all of its neighbors have died (perhaps from consuming battery resources or device faults) or it may be that all of a node's neighbors have moved beyond a reliable radio range. A second case would be the case that the wireless node is jammed. The key observation here is that in the first case, the signal strength is low, which is consistent with a low PDR measurement. While in the jammed case, the signal strength should be high, which contradicts the fact that the PDR is low. Table 11.1 summarizes typical network scenarios that can cause low PDR values and how the signal strength measurements can help further isolate the cause of the low PDR values.

Based on these observations we propose the detection protocol shown in Algorithm 5. In the **PDRSS_Detect_Jam** algorithm, a wireless node will declare that it is not jammed if at least one of its neighbors has a high PDR value. However, if the PDRs of all the neighbors are low, then the node may

TABLE 11.1. A combination of PDR and signal strength improves jamming detection accuracy.

Observed PDR	Observed signal strength	Typical scenarios
PDR = 0 (no preamble is received)	low signal strength	non-jammed: neighbor failure, neighbor absence, neighbors being blocked, etc.
PDR = 0 (no preamble is received)	high signal strength	node jammed
PDR low (packets are corrupted)	low signal strength	non-jammed: neighbor being faraway
PDR low (packets are corrupted)	high signal strength	node jammed

or may not be jammed and we need to further differentiate the possibilities by measuring the ambient signal strength. Rather than continually sample the ambient signal levels, which may use precious energy and processor cycles, the function `Sample_Signal_Strength()` instead reactively measures the signal strength values for a window of time after the PDR values fall below a threshold (the threshold we have identified in Section 11.2.3), and returns the maximum value of the signal strengths during the sampling window[2], which is denoted as SS. We note that the duration of the sampling window should be carefully tuned based upon the traffic rate, the jamming model, the measuring accuracy, and the desired detection confidence level.

The function `SS_ConsistencyCheck()` takes as input the maximum PDR value of all the neighbors, denoted as $MaxPDR$, and the signal strength reading SS. A consistency check is performed to see whether the low PDR values are consistent with the signal strength measurements. If the signal strength SS is too large to have produced the observed $MaxPDR$ value, then `SS_ConsistencyCheck()` returns False, else it returns True.

The consistency check may be conducted empirically as follows. During deployment, or during a guaranteed time of non-interfered network operation, a table (PDR, SS) of packet delivery ratios and signal strength values are measured. We may divide the data into PDR bins and calculate the mean and variance for the data within each bin. Or, we may conduct a simple regression to build a relationship between PDR and SS. The output of the binning or the regression is a relationship from which we may calculate an upper bound for the maximum SS that would have produced a particular PDR value in a non-jammed scenario. Using this bound, we may partition the (PDR, SS) plane into a benign-region and a jammed-region.

We conducted an experiment using MICA2 Motes to validate Algorithm 5. We gathered (PDR, SS) values for a source transmitting to a receiver node at a power level of roughly -5dBm. The PDR values were calculated using a window of 200 packets, while the SS values were sampled every 1msec for 200msecs in order to provide sufficient resolution to capture the

---

[2]In order to prevent spurious readings and have improved stability, in practice we use the average of the top three signal strength readings.

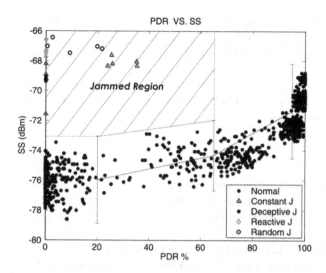

FIGURE 11.4. The (PDR, SS) measurements, indicating the relationship between PDR and signal strength. Also presented are the (PDR, SS) values measured for different jammers. The data was binned into three PDR regions, $(0, 40)$, $(40, 90)$ and $(90, 100)$, and the corresponding 99% confidence intervals are presented. The shaded region is the jammed-region, and corresponds to (PDR, SS) values that are above the 99% signal strength confidence intervals and whose PDR values are less than 65%.

jammer behavior during a reactive jammer attack. The packets were 33 byte long and transmitted at a rate of 20 packets per second. The source receiver separation was varied in order to produce a full spectrum of normal (PDR, SS) values, as depicted in Fig. 11.4. Using these values, we found the 99% SS confidence bars values for $(0, 40)$ $(40, 90)$ and $(90, 100)$ PDR regions. We depict these confidence bars, and define the corresponding jammed-region to be the region of (PDR, SS) that is above the 99% signal strength confidence intervals and whose PDR values are less than 65%. The jammed-region is shaded and appears in the upper-left corner of Fig. 11.4. We then performed experiments where we introduced the different jammers. The reactive jammer that we used sent out a 20-byte long interference packet as soon as it detects activities on the channel, while the random jammer had $t_j = U[0,31]$ and $t_s = U[0,31]$. We varied the source-receiver configurations as well as the location of the jammer, and measured the resulting PDR and SS values. As can be seen in Fig. 11.4, the (PDR, SS) values for all jammers distinctively fall within the jammed-region.

It is to be noted that the jammer in this experiment had a transmission power level of roughly -4dBm, which is stronger than that of the source. In fact, in order for the jammer to be more effective, it needs to operate at a relatively higher power level. However, a jammer using higher power will further decrease the PDR value and increase the SS measurement, thus pushing the resulting (PDR, SS) pair further towards the upper left corner, making it more distinct the benign-region. On the other hand, a jammer that operates on a lower power level is not as effective in interfering with the network operations. As a result, the combination of PDR and signal strength is quite powerful in discriminating a jammed scenario from various network conditions.

11.3.2 Location Consistency Checks

We now discuss a second consistency checking algorithm for detecting the presence of a radio interference attack. Whereas PDRSS_Detect_Jam employs signal strength to validate PDR measurements, the LOC_Detect_Jam algorithm employs location information. In addition to the assumptions listed earlier, for LOC_Detect_Jam we also assume that all legitimate neighbor nodes transmit with a fixed power level, such as the default settings when the sensor or ad hoc network was originally deployed. While this assumption holds for many real network settings, we would like to point out that scenarios where nodes have varying transmission powers can be addressed by easy extensions to our algorithm.

In PDRSS_Detect_Jam, the sampling granularity and the window length for measuring signal strength are two parameters that must be carefully set based upon the assumed jammer models as well as the underlying network traffic conditions. As noted earlier, it may not be practical to sample the signal strength with a fine granularity over a long window of time, and for this reason PDRSS_Detect_Jam employs a reactive consistency checking strategy in the sense that signal strength measurements are performed after PDR measurements fall below a threshold.

Instead of employing a reactive consistency check, the LOC_Detect_Jam algorithm uses a proactive consistency check. Rather than a node reacting to conduct measurements, the location consistency checking scheme involves information that is already made available to the wireless node prior to determining that PDR values are suspicious. As a consequence of this, the granularity and window length at the detector is no longer an issue. We note, in our specification of LOC_Detect_Jam that, although we require each node to transmit a location advertisement message, the issue of window length and granularity of signal strength sampling has been translated from a complicated issue involving assumptions regarding the jamming model into an issue regarding a node's mobility. As shall be seen, the analogous notion of position message frequency may be simply addressed using knowledge of node mobility and an assumption regarding the nominal packet delivery ratio of the network.

The LOC_Detect_Jam protocol requires the support of a localization infrastructure, such as GPS [6], or other localization techniques [16, 24, 77], which provides location information to wireless devices. We assume that this

---

Algorithm:LOC_Detect_Jam

$\{PDR(N) : N \in Neighbors\}$ = Measure_PDR()
$(n, MaxPDR)$ = $(\arg\max, \max)\{PDR(N) : N \in Neighbors\}$
if $MaxPDR < PDRThresh$ **then**
     $P_0 = (x_0, y_0)$ = GetMyLoc()
     $P_n = (x_n, y_n)$ = LookUpLoc(n)
     d = dist(P_0, P_n)
     $CCheck$ = LOC_ConsistencyCheck($MaxPDR$, d)
     if $CCheck == False$ **then**
         | post NodeIsJammed()
     end
end

Algorithm 6: Jamming detection algorithm that checks the consistency of PDR measurements with location information.

localization infrastructure is not able to be attacked or exploited by potential adversaries. Recently, countermeasures have been proposed to protect localization services from being exploited by adversaries including mechanisms proposed in Chapter 5, and methods presented in [35, 54]. In the LOC_Detect_Jam protocol, we again use PDR as the metric indicating link quality. A node will decide its jamming status by checking its PDR and deciding whether the observed PDR is consistent with what it should see given the location of its neighbor nodes. Conceptually, neighbor nodes that are close to a particular node should have high PDR values, and if we observe that all nearby neighbors have low PDR values, then we conclude that the node is jammed.

In our protocol, we let every node periodically advertise its current location and further let each node keep track of both the PDR and the location of its neighbors. Due to node mobility, it is necessary that the location advertisements occur with sufficient frequency to be able to reliably capture the migration of neighbors from regions of high PDR near node A to regions of lower PDR further from node A. If a jammer suddenly comes into the network near node A, then the location information that node A has will correspond to the location of the neighbors prior to the start of the interference. Analogous to PDRSS_Detect_Jam, if node A finds that the PDR values of all of its neighbors are below the threshold $PDRThresh$, then node A will perform a consistency check by using the position P_n of the neighbor who had the maximum PDR. The distance between P_n and P_0 (i.e. the location of node A) is calculated, and together $MaxPDR$ and d are used as input into LOC_ConsistencyCheck() to conduct a location-based consistency check.

LOC_ConsistencyCheck() operates similarly to SS_ConsistencyCheck(). During deployment, a table of (PDR, d) values are gathered to represent the profile of normal radio operation for node A. As in SS_ConsistencyCheck(), we may define a jammed-region and a benign-region using either a binning procedure or regression to obtain lower bounds on the PDR that should be observed for a given distance under benign radio conditions using measured data. If the point $(MaxPDR, d)$ falls in the jammed-region, then the node declares it is jammed.

We note that, just as in the operation of PDRSS_Detect_Jam, the assumption that every legitimate node transmits a minimal baseline amount of traffic with which to estimate PDR is paramount to the operation of the LOC_Detect_Jam protocol. This baseline amount of traffic may coincide with the transmission of location advertisements in order to reduce the overhead of the protocol. The baseline traffic assumption allows us to declare the PDR to be 0 when no packets are received from a neighbor node within a given time period. This assumption is particularly important for handling scenarios where every neighbor node is jammed, as it allows LOC_Detect_Jam to pass into the location-based consistency check, which will allow the algorithm to declare that the node is jammed since its neighbors should have delivered at least a minimal amount of packets. Finally, we note that we have disregarded the extremely unlikely event that all neighboring devices have faulted or depleted their power resources.

We conducted an experiment to validate Algorithm 6. The setup of the experiment was the same as the experiment used to validate Algorithm 5. We gathered (PDR, d) values for normal operation as well as for scenarios involving the different jammers, as depicted in Fig. 11.5. As can be seen in Fig. 11.5, the (PDR, d) values for the jammer scenarios, where the source-receiver separation was small, are distinctly separated from normal operation values, and hence fall in the jammed-region. Again, we would like to point out that, for a reasonably dense network where every node has one or more neighbors that are close to itself, a jammer's presence can be easily identified, as shown in Fig. 11.5. If a node, on the other hand, does not have a nearby neighbor, then the PDR of that node, even without the jammer, is rather poor (Fig. 11.5). For these nodes, the effect of a jammer will not be noticeable anyway.

We now address the frequency of node position advertisement. There are two factors that affect the frequency: first, nodes may move towards or away from each other, and second, position messages may be missed, especially for neighbors farther away from node A. We may address the first factor by setting a requirement that a node announces its location whenever it has moved a distance δ from its previous position. By using the device's velocity v, we find that a device should update its position at least every $\tau = \delta/v$ seconds. To address the second issue, we assume that each device seeks a guarantee of η that its position announcement will arrive to neighbors who are sufficiently close to have at least a nominal packet delivery ratio of q. Assuming independence of successive transmissions of position announcement messages, the cumulative distribution for the amount of transmissions T before the first successful delivery is

$$F_T(T) = 1 - (1 - q)^T, \quad \text{for } T \in \{1, 2, 3, \cdots\} \tag{11.1}$$

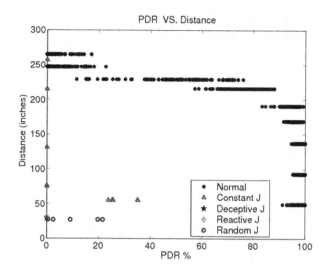

FIGURE 11.5. The (PDR, d) measurements, indicating the relationship between PDR and distance between source and receiver. Also presented are the (PDR, d) values measured for the different jammer models.

From the cumulative distribution, we may find the amount of transmissions, $\tilde{T}$, needed to have a guarantee of η that the position announcement will have been heard. Combining the two factors, a node should announce its position every $\tau/\tilde{T}$ seconds. The frequent announcement of position information guarantees that nodes will have knowledge of their neighbor's position.

11.4 Conclusion

The shared nature of the wireless medium will allow adversaries to pose non-cryptographic security threats by conducting radio interference attacks. In this chapter, we have studied the issue of detecting the presence of jamming attacks, and examined the ability of different measurement statistics to classify the presence of a jammer. We showed that by using signal strength, carrier sensing time, or the packet delivery ratio individually, one is not able to definitively conclude the presence of a jammer. Therefore, to improve detection, we introduced the notion of consistency checking, where the packet delivery ratio is used to classify a radio link as having poor utility, and then a consistency check is performed to classify whether poor link quality is due to jamming. We introduced two enhanced detection algorithms: one employing signal strength as a consistency check, and one employing location information as a consistency check. We evaluated the effectiveness of each scheme through empirical experiments and showed that each of the four jammer models we introduced in Chapter 10 can be reliably classified using our consistency checking schemes.

12

Channel Surfing: Defending Wireless Networks against Radio Interference:

12.1 Introduction

In this chapter, we examine the ability of a wireless network to cope with radio interference. We propose the use of *channel surfing*, whereby the sensor nodes adapt their frequency allocations as needed to avoid interference. The challenging research question here is how to establish network connectivity between multiple frequency zones. The inherent diversity in network configuration, interference model, and platform setup suggests that no single solution is sufficient. We examine four strategies to restore network connectivity across multiple channels, each having unique characteristics and advantages. Although channel surfing may be applied to more general wireless networks (e.g. 802.11), in order to validate our strategies, we focused our study on a sensor network platform, and have implemented our methods on a 30-node Mica2 sensor network testbed. During the process of implementation, we have overcome a number of challenges and have demonstrated that all four strategies can maintain network operations in the presence of jamming/interference.

We begin the chapter in Section 12.2 by providing an overview of the sensor network and interference model used in our studies. We next give an overview of channel surfing in Section 12.3 and Section 12.4, where we detail a set of increasingly sophisticated channel surfing protocols. In Section 12.5 and Section 12.6, we describe our validation effort on our sensor testbed.

12.2   System Models

The objective behind this chapter is to examine networking issues associated with adjusting channel assignments in a sensor network in order to avoid radio interference. In this section we outline the basic sensor communication and jamming model that we use throughout this chapter.

12.2.1   Our Sensor Communication Paradigm

There are many choices for sensor platforms and data dissemination models available to the sensor network designer. The broad range of choices implies that there are many different directions that one can take in order to tackle the problem of radio interference. Early on in our studies we found that it was impractical to devise a generic approach that worked across all varieties of sensor networks, and instead found that it was necessary to tailor the design of our solutions to a specific communication paradigm.

Channel surfing requires that sensor radios change their *channel* allocations. In order to commence with channel surfing, the radio devices employed must have a notion of a channel. Most sensor platforms have a natural form of channelization that is accomplished by changing the carrier frequency. For example, in our validation efforts, we use the 916.7MHz Mica2 platform and separate the channels by 800KHz. In the algorithm discussion, we assume that channelization exists.

Another important factor in the sensor communication paradigm is the choice of the data dissemination method and the associated routing protocol. From a sink's viewpoint, there are two main data dissemination paradigms: few-to-one data dissemination (whereby a sink is connected to one source node or a small number of source nodes), as in Directed Diffusion [132] and SPIN [133]; and many-to-one (whereby a sink node is connected to a large portion of a network), as in Zebranet [134] and TAG [135]. In this chapter, we have chosen to focus on the many-to-one model, and we assume that there is only one sink that collects data. For the many-to-one model, our studies focus on tree-based routing schemes, a popular family of routing protocols whereby the network establishes and maintains a forwarding tree [135].

We briefly touch upon the salient features of tree-based routing needed for our discussion. In these schemes, a routing tree is formed with the sink node serving as the root of the tree. A node selects its routing *parent* as its best radio neighbor in the direction of the tree's root. In such a tree-based routing structure, a node usually has a parent (except the sink) and one or more children (except the leaf nodes), wherein it receives data packets from its children, and sends packets to its parent. A node's parent and children are considered its *neighbors*. Besides the parent and children, there may be other nodes that are within a node's communication range. These nodes, are *not* neighbors of the node. In this chapter, we use the term neighbors to refer to the topology-based relationship rather than a physical location-based relationship.

12.2.2   Our Interference Model

We now turn our attention to describing the interference model. When considering issues of radio interference and jamming, it must be emphasized that there is a very broad range of capabilities that one might assume is available to the interferer, ranging from whether the interferer is incidental or intentional, powerful or resource-constrained, narrowband or broadband, or static or adaptive.

It is immediately apparent that if the jammer is a high-powered, non-coherent, broadband source of interference (e.g. capable of occupying all channels simultaneously), then there is no hope for building a resilient sensor network short of choosing a different PHY-layer transceiver with a powerful anti-jam margin [123, 136]. Further, it should also be noted that an aggressive adversary may jam a single channel (e.g. by reprogramming another sensor) at a time and rapidly switch between channels to effectively disrupt network services across all channels. Both cases represent powerful, aggressive broadband jamming adversaries. Instead of considering powerful interference models for which the only viable defense might be powerful physical layer techniques, in this work, we consider a non-intentional or a relatively benign adversary in which the interferer blocks one (or even a few) of the channels at a time. This model has been considered elsewhere in the literature [124, 126, 137, 138], and it can be accomplished by employing a narrowband RF source (e.g. such as a waveform generator) or by another sensor node that has been reprogrammed to jam a particular channel. Further, even if the interferer hops to different channels, we assume that the interferer stays on one channel for a brief period of time before switching to another channel. Related to these assumptions, we note that we consider traditional security threats [139, 140](such as authentication, communication confidentiality, and coping with node compromise) to be orthogonal to the issues discussed in this paper.

A jammer, whether incidental or intentional, can significantly affect the network's communication. As we shall point out in the next section, our channel surfing algorithms operate on-demand– when interference is detected, channel adaptation is used. Hence, as the starting point for coping with jamming, it is necessary to detect jamming. Since spectrum utilization is a local phenomena, detecting the presence of a jammer must be done by each individual sensor node– that is, a node can only detect whether it is jammed, not whether other nodes are jammed. Several jamming detection approaches have been proposed, ranging from measuring simple properties (e.g. ambient signal strength and packet delivery ratio [122, 126]), to more complicated consistency checks. In this chapter, we utilize our detection scheme presented in chapter 11, which involves a consistency checking process on each node to ensure reliable identification of jamming attacks. We note that even if an interferer only degrades the network link quality partially (e.g. by blocking half of the packets), the classification of such a situation is the responsibility of the detection algorithm and the policy the network operator has used to define "interference". The implication of this fact is that in the channel surfing algorithms we describe, we would like for the jamming/interference detection process to be a separate module that is utilized by the channel surfing algorithm. As long as a node is declared "jammed", channel surfing will kick in.

12.3   Channel Surfing Overview

Typically, when radio devices communicate they operate on a single channel. Here, the concept of channel may be a single operating frequency or more generally may be any division of the operating spectrum. For the sake of discussion, we shall assume that the notion of a channel is associated with a single carrier frequency. Channel surfing is motivated by frequency hopping, a physical layer technique used in spread spectrum communication. As noted earlier, we assume that the adversary blasts on only one channel (or at most a few channels) at a time.

The basic idea behind channel surfing is that the link layer channel assignments should be changed in order to avoid interference. When thinking of how to achieve this, a natural idea is to directly apply the philosophy of constantly changing the channels (as is done in frequency hopping spread spectrum). However, employing such a strategy at the link layer, can be detrimental and costly to providing network services. First, if one rapidly changes the channel assignment, then it is necessary to have a fine granularity of synchronization across the entire network. Even if channels are changed less rapidly, for example on the order of a hundred milliseconds, constantly changing the channel incurs switching overhead. For example, routes that exist on one channel may not be guaranteed to exist on other channels, and frequently changing channels may cause the network to become unstable, thereby necessitating frequent route maintenance and discovery each time a channel is changed. Further, such penalties are incurred even if there is no interference presence.

Based on these arguments, a better strategy would be to only adapt channel allocations when needed, i.e. when interference is detected. In order to achieve an interference-resistant sensor network, we propose a collection of distributed on-demand channel surfing algorithms. As the starting point for these algorithms, each node runs a jamming detection process to determine whether it is jammed. In channel surfing, those nodes that detect themselves as *jammed* nodes should immediately switch to another orthogonal channel and wait for opportunities to reconnect to the rest of the network. After the jammed nodes lose connectivity, their neighbors, which we refer

Algorithm: Channel Surfing Framework

```
while (1) do
    if (NeighborsLost() == TRUE) then
        working_channel = next_channel;
        for (i = 0; i < m; i + +) do
            SendInquiryPacket();
            if (ReceiveReplyWithinT() == TRUE) then
            |   break;
            end
        end
        if (i==m) then
        |   working_channel = original_channel;
        else
        |   Use a Channel Surfing Strategy;
        end
    end
end
```

Algorithm 7: The channel surfing framework.

to as *boundary* nodes, will follow Algorithm 7 to discover the disappearance of their jammed neighbor nodes (e.g. via a drop in link quality) and temporally switch to the new channel to search for them. If the lost neighbors are found on the new channel, the boundary nodes will participate in rebuilding the connectivity of the entire network.

Before we move onto our specific algorithms, we first discuss a few challenging issues in the above channel surfing framework. The first challenge concerns the potential boundary nodes. The basic scheme specifies that a boundary node should switch to a new channel after its neighbors are jammed and have escaped to the new channel. However, if a node immediately probes the next channel whenever it experiences a poor link quality with any of its neighbors, the system will enter a non-stable state because wireless sensor networks inherently experience frequent link quality degradations or even topological changes [141, 142]. Fortunately, after carefully studying the working of the underlying system, we found that it is possible for boundary nodes to correctly differentiate jammed neighbors from those neighbors that just had a poor link with the boundary node. This can be explained as follows. For tree-based routing, a node has precisely two types of neighbors/links (one link with its parent, and possibly several links with its children). Thus there are two possibilities for broken links and we can address these each separately. First, if a node witnesses degradation in the link with its parent, then the underlying routing protocol will first attempt to find another, suitable parent node. If the node finds a replacement parent, then it will announce its parent selection in a routing update. However, only if there is no suitable replacement parent, the node will probe the next channel to find its parent. In this way, the node does not need to switch channels if it can still maintain its normal network operations. Next, we cover the case of a node losing one of its children. If the lost child node was not jammed, but just connected to a new parent, the node *should* hear its former child's routing announcement. If it does not witness the child's routing announcement within a specified window of time, then it will probe the next channel looking for its lost child. Thus, a node will become a boundary node either because it is the parent of jammed nodes, or because it is the jammed nodes' child that cannot find a new parent.

After detecting the loss of a neighbor, the boundary node should not switch to the new channel too quickly. If it switches too soon, it may arrive at the new channel before the jammed nodes. To understand this, consider a scenario where the jammer starts interference at time t_0. At that time, the jammed nodes will not be able to send out packets. However, since it takes less time for a node to detect the absence of a neighbor than it does for a node to decide it is jammed, the boundary nodes will detect the absence of a jammed node at time $t_0 + \delta_1$, while the jammed node will declare itself jammed at $t_0 + \delta_2$, where $\delta_2 > \delta_1$. If the boundary nodes switch to the new channel immediately after $t_0 + \delta_1$, they will not find the jammed nodes there. Rather than have the boundary node wait on the next channel, which would prevent it from conducting its primary objective of relaying messages to the sink, or having the node constantly flip-flop between channels looking for its children, we should make the boundary nodes wait for at least an additional $\delta_2 - \delta_1$ amount of time before switching to the next channel. The values of δ_1 and δ_2 are characteristic of the particular routing protocol, and the jamming detection scheme.

In addition to the switch timing for a boundary node, it is important to have a discovery protocol by which a boundary node can find its neighbors on the new channel. After a node switches to the new channel searching for its neighbors, it should send out an "inquiry" message, such as "Is my neighbor X here?" If it receives a reply from X, it will start working on repairing the connectivity between X and the sink. Otherwise, it waits for time δ to send another message. If the node does not hear from X after a few trials, it assumes the child is not jammed, returns to the original channel and resumes its original operation in the network. In total, the time spent probing the next channel should be less than δ_2 to avoid cascading channel probing.

Finally, it is desirable to choose the next channel so that the adversary cannot predict what channel the nodes will surf to. We may choose to chain the channel selections using a keyed pseudo-random generator [143]. If the n-th channel assignment is $C(n)$, then we take $C(n+1) = E_K(C(n))$, where K is a key shared by all nodes in the network that is used exclusively for channel assignment. If ever $C(n + 1) = C(n)$, then the channel assignment proceeds to $C(n+2)$ and so on until a different channel is selected. Finally, if the jammer can block several channels, then after a jammed node escapes to a new channel, it should first detect whether the channel is jammed before it starts working on that channel. In practice, one typically has to check at most a few iterations in order to find a new channel that is not jammed.

12.4   Channel Surfing Strategies

After the boundary nodes discover that their neighbors are jammed and have escaped to another channel, they will attempt to reconnect the jammed nodes with the rest of the network. In this chapter, we propose two different classes of techniques that the boundary nodes can use to repair network connectivity: (1) *coordinated channel switching*, in which the boundary nodes participate in transitioning the entire network to the new channel, thereby reestablishing the network on the new channel; and (2) *spectral multiplexing*, where the boundary nodes multiplex between the old channel and the new channel, serving as a "bridge" that connects nodes operating on different channels. In this section, we discuss these two strategies, outline their challenges, and highlight their advantages and disadvantages. Further practical issues are discussed in Section 12.5.

12.4.1   Coordinated Channel Switching

The idea behind coordinated channel switching is rather simple: the entire network must coordinate its evasion of the interference by switching to the next channel and resuming network operation there. These strategies are characterized by a transition phase during which an increasing amount of the nodes switch to the next channel. Following the transition, the entire network resumes stable operation on the next channel. Within the family of coordinated channel switching protocols, the strategies are characterized according to the coordination strategy governing the transition. In this section, we examine two different strategies: first, a technique whereby each node autonomously follows its neighbors to the next channel; second, a strategy whereby nodes are accelerated through the transition phase through the broadcasting of channel changing commands by the boundary nodes.

Autonomous Channel Switching

In the autonomous channel switching scheme, a node will autonomously switch to the new channel if it detects that some of its neighbors have moved to the new channel. In particular, each node is responsible for determining the next channel $C(n+1)$ it should switch to by using $C(n+1) = E_K(C(n))$. The scheme begins with the jammed nodes detecting they are jammed. After a set amount of time, the boundary nodes will notice that they have not received messages from their jammed neighbors. The boundary nodes will then probe the new channel, searching for their lost neighbors. If the boundary node finds a neighbor residing on the new channel, it will stay in the new channel, extending the zone of nodes in the new channel. This process repeats until the entire network reconnects on the new channel. Overall, a network with n hops from the jammer to the boundary of the network, will take n rounds to complete the channel switching.

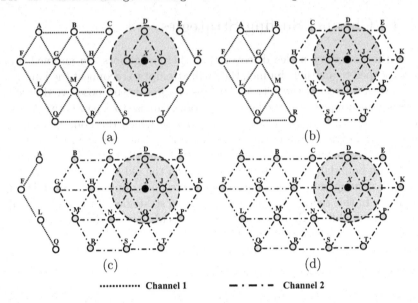

FIGURE 12.1. A walk-through of autonomous channel switching. The shaded circle illustrates the jammed area, with the jammer X at the center. The entire network switches to channel 2 within four rounds, each round shown in one plot.

Algorithm Walk-through: In order to illustrate the autonomous algorithm, let us walk through the example depicted in Figures 12.1(a)-(d). Let channel 1 be the old channel and channel 2 be the new channel. Here, the jammer X affects nodes $\{D, I, J, O\}$. Upon detecting they are jammed, these four nodes switch to channel 2, as shown in Figure 12.1(a). The dashed-dot lines indicate links that exist in channel 2, while the dotted lines correspond to links in the first channel. The boundary nodes $\{C, E, H, K, N, P, S, T\}$ will notice that their jammed neighbors are no longer on channel 1, and will probe channel 2. When they discover their neighbors on the new channel, they will remain on channel 2 and form a network on channel 2, extending the size of the channel 2 subnetwork. In the next round, the channel 1 neighbors of $\{C, E, H, K, N, P, S, T\}$ will notice that some of their neighbors are missing and will probe channel 2. Upon finding their neighbors in channel 2, they will remain, and the channel 2 subnetwork will grow. This process repeats until the entire network has moved to the new channel. Including the jammed nodes, there are four hops from the jammer to the boundary of the network. Thus, after the jammed nodes evade to the new channel, we need three more rounds for the rest of the network to convene on channel 2, as shown in Figure 12.1(b), (c), and (d).

Algorithm Challenges: This algorithm relies primarily upon the ability of each node to detect the absence of its neighbors, and to differentiate between whether a node is missing due to fluctuations in link connectivity or from jamming. Hence, when implementing this algorithm, the issues identified in Section 12.3 become very essential to the performance and operation of this scheme.

Discussion: One of the main advantages of this scheme is the simplicity of its description. Further, this scheme introduces minimal additional communication overhead. The nodes in the network merely detect the absence of their neighbors, and probe subsequent channels. Unfortunately, this scheme requires a long latency for the entire network to switch channels. If there are n hops from the jammer to the boundary of the network, then it will take at least $n\delta_2$ time for the network to stabilize (recall, from the discussion in Section 12.3 that we require boundary nodes to wait at least an additional $\delta_2 - \delta_1$ time).

To differentiate between jamming and natural causes of poor link quality, it is necessary for δ_2 to be large. Thus, since the protocol latency is linear in the amount of hops in the network (or roughly square root in the amount of nodes in the network), the issue of scaling becomes a dominant factor for this protocol. As the number of sensors in the network increases, the latency associated with the transition phase can become prohibitive and a large fraction of the sensor data will not be delivered.

Broadcast-Assist Channel Switching

Switching channels autonomously incurs significant latency because (i) a node can only switch after its neighbors have done so, and (ii) it must wait for some time even after its neighbors have switched. Broadcast-assist channel switching addresses both problems. Instead of requiring that every node detect whether its neighbors have switched, a boundary node that has found a jammed neighbor residing on the next channel will facilitate a more rapid phase transition by broadcasting a command. In particular, a boundary node switches back to the old channel, broadcasts a *channel switch command*, and returns to the new channel. Once a node receives this notice, it rebroadcasts the command and switches to the channel specified. Overall, broadcast-assist channel switching facilitates parallel channel switching and, just as parallel execution is usually faster than sequential execution, it can significantly reduce the network switching latency.

Algorithm Walk-through: We now examine the broadcast-assist channel switch algorithm using the same example as used to describe the autonomous scheme. Nodes $\{D, I, J, O\}$ are jammed by the jammer X and consequently switch to the new channel, e.g. channel 2, as shown in Figure 12.2(a). As a result, nodes $\{D, I, J, O\}$ form the channel 2 subnetwork, and the rest of nodes form the channel 1 subnetwork. After time δ_2, the boundary nodes, $\{C, H, N, S, T, P, K, E\}$ will notice that their jammed neighbors are no longer on channel 1, and will probe them on channel 2 (Figure 12.2(b)). After finding the jammed nodes on the new channel, the

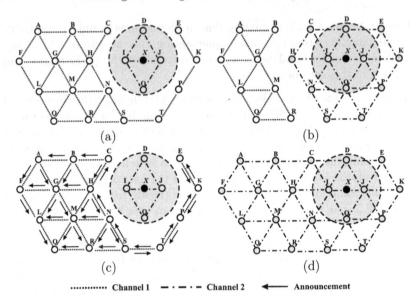

FIGURE 12.2. A walk-through of broadcast-assist channel switching. The shaded area depicts the jammed area, with the jammer X at the center.

boundary nodes return to the original channel temporarily and broadcast a switching notice $(n_{id}, C(n + 1))$, where n_{id} is the ID of the sender node, and $C(n + 1)$ is the new channel to switch to. The switching notice is sent to the rest of the network through the channel 1 subnetwork, as shown in Figure 12.2(c). The boundary nodes join the jammed nodes on the new channel after broadcasting the notice. Shortly thereafter, the rest of the nodes will switch to the new channel after receiving the switching notice, and reestablish the network on channel 2, as shown in Figure 12.2(d).

Algorithm Challenges: The major challenge facing this scheme is the fact that unreliable/variable links can cause some nodes to miss a channel switch notice. However, the channel switching command is typically broadcasted independently by multiple boundary nodes. Thus a node is very likely to receive at least one notice, and be able to switch to the new channel. Even should a case arise where a node does not receive a switch notice, it will still autonomously move to the next channel as it will detect that it cannot receive messages from neighbors that have already switched to the new channel.

Discussion: Compared to autonomous switching, the broadcast-assist channel switching incurs much less switching latency. After jamming is detected, the boundary nodes will switch their channel within time δ_2, and then broadcast the switching notice. Suppose the network has n hops between the jammer and the boundary of the network, and that μ is the delay for one-hop transmission ($\mu \ll \delta_2$). Then the overall latency is roughly

$\delta_2 + (n - 1)\mu$, which is much less than the $n\delta_2$ latency required by autonomous channel switching.

Additionally, the success of performing a broadcast-assist channel switching doesn't depend on the likelihood that each individual node can detect the loss of its neighbors but, rather, as long as one of the boundary nodes finds its lost neighbors in the new channel and informs the rest of network, the network will resume its connectivity in the new channel in spite of the radio interference. Finally, we note that the broadcasted channel switch command should be authenticated [133, 144–147] to prevent malicious message injection by an adversary. Thus, in practice, the channel switch notice has the format of $E_{K_A}(n_{id}, C(n + 1), nonce)$, where K_A is a network-wide authentication key that is distinct from the key used to generate $C(n + 1)$, and *nonce* is included to prevent replay attacks.

12.4.2 Spectral Multiplexing

Performing a coordinated channel switch requires the entire network to reestablish the routing tree as the link connectivity will not be the same on the new channel. The global nature of coordinated channel switching can be a source for significant network cost, and a natural alternative is to employ a local response where only jammed nodes switch channels, while non-jammed nodes remain on the original channel. To guarantee the communication between these two frequency zones, boundary nodes have to work on both channels by repeatedly switching back and forth between two channels to *relay* packets, a process we call *spectral multiplexing*.

Initially, for spectral multiplexing, jammed nodes that were originally on channel $C(n)$ will switch to $C(n + 1) = E_K(C(n))$. However, for spectral multiplexing, the primary challenge lies in the fact that the boundary nodes must carefully decide when they should stay on which channel, and for how long, so that they can minimize the number of packets that cannot be delivered due to the frequency mismatch between the sender and the receiver. If the boundary nodes are configured with dual radios, this scheduling is unnecessary. However, commercial sensor platforms including Berkeley motes only have one radio interface and, as a result, a node can work on only one channel at a time. It is therefore crucial to make sure that the sender and the receiver are able to work on the same channel when they want to exchange messages. Towards this end, boundary nodes must transmit an announcement to its neighbors that they are operating in a "dual-mode" on two channels. Further, these boundary nodes must employ a synchronization mechanism to coordinate the spectral schedules of the sender and the receiver when one party needs to work in "dual-mode". The overall

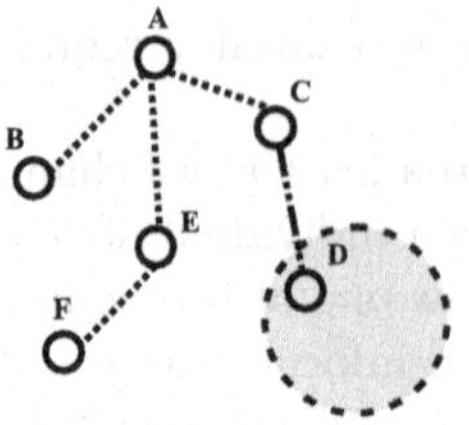

slot 1	A, B, C, E, F
slot 2	C, D

(a) the network topology   (b) the global schedule

FIGURE 12.3. Illustration of the synchronous spectral multiplexing algorithm.

scheduling objective is to ensure that dual-mode nodes are present on the right channel when the neighbors on that channel are ready to transmit[1].

In general, there are two ways of coordinating schedules from different entities: one is to have all the entities adopt synchronous schedules, and the other is to operate in an asynchronous fashion. We thus propose the corresponding methods to coordinate the frequency schedules for neighboring nodes: (1) Synchronous Multiplexing, in which all the nodes share the same schedule by dividing the global time axis into different slots and assigning one slot to a channel; and (2) Asynchronous Multiplexing, in which a node operates on a local schedule, and the boundary nodes make local decisions about when to switch channel.

Synchronous Spectral Multiplexing

In synchronous spectral multiplexing, the entire network is governed by one global clock. The global time axis is divided into slots, and multiple slots form a round. The number of slots in a round is determined by the number of channels the network has to operate on at any specific time. (In this chapter, we limit our discussions on situations where the network works on 2 channels simultaneously, and the discussion can be easily extended to cover situations with more than 2 channels.) Each slot is assigned to a single channel, and during that time slot, the network may only use the corresponding channel– regardless of whether they are jammed, boundary nodes, or not. At the end of a time slot, the entire network utilizes the next channel and, again, the nodes that are not using the next channel do not transmit, nor must they switch channels unless they are dual-mode boundary nodes. By following this global schedule, we can avoid frequency mismatch between a pair of communicators.

Algorithm Walk-through: Figure 12.3(a) presents an example network scenario in which D is jammed, and switches to channel 2. Its parent node, C, thus becomes a boundary node, and has to multiplex between two chan-

---

[1]The need for scheduling transmissions has also been considered in the context of duty cycling, as in S-MAC [148], in order to preserve energy.

nels. The rest of the nodes continue to work on channel 1. The global schedule for this case is shown in Figure 12.3(b), which has two slots for each round, with slot 1 allocated to channel 1, and slot 2 to channel 2. Following this schedule, during slot 1, nodes $\{A, B, E, F\}$ work as normal. Node C sends out packets to its parent A, but does not receive any packets from D. At the end of slot 1, these nodes stop their activities on channel 1, and node C switches to channel 2. During slot 2, the only transmitting node is D, and C buffers all the packets it receives from D. At the end of slot 2, D ends its transmissions and C switches to channel 1. These two slots keep alternating in this fashion until the radio interference ends, or for the lifetime of the network.

Algorithm Challenges: There are several challenging issues associated with this scheme: (1) How to synchronize the schedules of every node, (2) how to start multiplexing, and (3) how to determine the slot duration?

Synchronization: One natural synchronization approach is to have the entire network work under a global synchronized clock, wherein each node maintains a unique and global timescale. Many protocols can be used to establish a global timescale across the entire network, e.g. TPSN (Timing-sync Protocol for Sensor Networks) [149]. A closer investigation, however, reveals that perfect synchronization across the entire network is not only inefficient, but also unnecessary. Instead, since communication takes place locally amongst neighboring nodes, we can focus on achieving a fine synchrony within any local region. Additionally, instead of employing traditional parewise synchronization, we let the root initiate the synchronization process by broadcasting SYNC packets to its children.

Nodes use a Timer to start/end a slot; at the beginning of a slot, a node sets its Timer to the duration of a slot, and when the timer expires, a node switches to the next slot. Without periodic synchronization, the timers on different nodes may drift significantly. To avoid this, $SYNC$ packets are sent periodically at an interval much larger than the slot duration. Upon receiving a $SYNC$ packet, a node will immediately terminate the current slot and start a new slot by resetting the Timer to the slot duration. This simple protocol can effectively minimize the synchronization error between a pair of neighbors. The resulting synchronization error, $\Delta\tau$, only includes delays involved in sending, propagating, and receiving $SYNC$ packets.

Building a "global" synchronization from a local synchronization protocol requires a starting reference point that initiates the synchronization process. In a tree-based forwarding structure, it is natural to choose the root of the tree as the reference point. Therefore, in the first round the root first sends out $SYNC$ packets to its children, whose depth is 1, and whose clock will be synchronized within $\Delta\tau$ of the root's clock. Similarly, in the $(i + 1)^{th}$ stage, the nodes with depth i send $SYNC$ packets to their children. Finally, after every node receives a $SYNC$ packet, for a tree with depth of n, some leaf nodes will be $n\Delta\tau$ behind the root. This synchronization procedure is

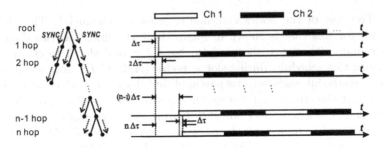

FIGURE 12.4. Illustration of the synchronization mechanism.

presented in Figure 12.4. After synchronizing, each node will start a new slot upon the expiration of its timer. However, in order to compensate for the synchronization error between its time and its neighbor's time, we require that a node wait for a short, random time period prior to transmission.

Further complicating synchronization is the fact that it must be maintained when nodes work on different channels (coping with drift while multiplexing). For example, in the case where the parent node is on channel 1, while the child is on channel 2, the *SYNC* packet from the parent will be lost. To address this complication, the *SYNC* packet should specify which channel the current slot is associated with. In order to guarantee that nodes on both channels are synchronized, the node should send these *SYNC* packets in rapid succession across both channels. This mechanism ensures the *SYNC* packets will reach the destination.

Initiation: As soon as a boundary node discovers that jammed nodes have evaded to the new channel, it will send a message to the entire network on the original channel that contains a list of the channels it will be working on. After a node receives this message, it will compare its own channel with the channel list included in the message, and append its channel if it is not already in the list. In this way, when the message reaches the root of the routing tree, it will contain all the channels the network has to operate on, and the root will create a slotted channel schedule based on this list, and broadcast the schedule down the tree, along with the clock synchronization packets.

Slot Duration: Slot duration is an important parameter in the synchronous spectral multiplexing algorithm. At first glance, it seems intuitive that a shorter slot duration is more desirable because, if a node stays on one channel short enough, the required buffer space will be smaller and, more importantly, less latency would be incurred. However, we found that a smaller slot can be problematic as well, mainly due to the overhead associated with switching channels. Before a node switches to a new channel, it has to complete receiving all the packets that are in transmission. In order to guarantee this, after the Timer expires, we let each node wait for a small amount of time for all the possible transmissions to complete. In our implementation, this translates into the parent node waiting a little longer because the receiving side is usually the parent node in a tree-based routing. Another problem with short slot durations is that proportionately the synchronization errors will be relatively large with respect to the duration of a slot, thereby affecting this scheme's efficiency. Finally, we note that there is a radio startup cost associated with switching channels(e.g. 250msec for the CC1000 radio chip in the Mica2 mote). Overall, a good slot duration should be determined based on several factors: the available buffer space, the traffic rate, the required message latency, and the synchronization error.

In this study, we choose to adopt the largest slot durations that can satisfy the available buffer space constraints, and we consider our underlying sensing application model to have a periodic traffic pattern. Further, we have empirically witnessed that boundary nodes are more likely parent nodes of the jammed nodes (children of jammed nodes typically look for alternate parents on the original channel), and thus boundary nodes will merely receive on channel 2, but will both receive and send on channel 1. Specifically, based on the traffic rate from each channel and the buffer size, each boundary node calculates the longest stay time it can have on each channel (usually a node should stay on each channel for the same amount of time). In order to understand the calculation, let us look at an example. Suppose a boundary node A has a buffer that can support 10 slots, and it has 1 child on channel 1 that produces 20 packets per second, and 2 children on channel 2 each of which produces 10 packets per second. A can at most stay on each channel for 250 msecs. By spending 250 msecs on each channel, it will receive 10 packets in a round (5 from each channel), which will fill up its buffer. After each boundary node independently calculates a slot duration, the sink will collect all the information, chooses the smallest one as the global slot duration and announces this.

Discussion: Synchronous multiplexing adopts a deterministic global schedule that governs the channel assignment of every node in the network. The deterministic nature of this algorithm guarantees that it can work well even under complex scenarios where multiple nodes need to work on multiple channels and these nodes are neighbors of each other. However, in order to achieve this, every node in the network must pay the extra overhead needed to maintain synchrony between nodes.

Asynchronous Multiplexing

In the asynchronous multiplexing algorithm, a node is only aware of its neighbors' channel information, but not the channel information of a remote node. The simplest spectral scheduling method is to have a boundary node flip its radio frequency between two channels in a round-robin fashion. However, a completely random round-robin multiplexing strategy ignores the schedules of the communicating parties, and would thus fare poorly. For example, suppose a jammed node, working on the new channel, sends

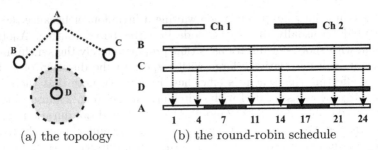

(a) the topology   (b) the round-robin schedule

FIGURE 12.5. Illustration of the round-robin asynchronous spectral multiplexing algorithm.

packets at times 10, 20, 30, 40, and 50. If the corresponding boundary node stays on the new channel during time windows [1, 6], [13, 18], [25, 30], [37, 42], [49, 54], then it will miss the packets sent at 10, 20, and 30. The resulting packet loss ratio for the jammed node is now as high as 60%. The above example illustrates the limitation of a random round-robin scheme and highlights the need for some level of coordination between the boundary node and its neighbors for the asynchronous multiplexing scheme.

Algorithm Walk-through: Figure 12.5 illustrates the idea behind the asynchronous multiplexing scheme. In this example, the boundary node A has to receive packets from three nodes, B, C, and D, with the first two working on channel 1 and the last one working on channel 2. Suppose all three nodes send packets every 10 seconds, starting at time 1, 4, and 7 respectively. In this case, starting from time 0, A decides to stay on one channel for 5 seconds and then switches to the next channel for 5 seconds. In this way, A can receive every packet from its neighbors.

Algorithm Challenges: The challenges associated with this schemes include synchronization and slot duration.

Synchronization: To coordinate the schedules of a boundary node and its children, we have adopted a simple protocol that involves the boundary node announcing its schedule (the duration it will stay in the each channel) by notifying its children just after it switches to a new channel. In the example in Figure 12.5(a), A notifies nodes B and C of its schedule as soon as it switches to channel 1, so that they can start transmissions when A enters channel 1, and stop transmissions after A leaves channel 1. Similarly, it also notifies D whenever it is on channel 2. We note that a child node must buffer both its own packets as well as packets coming from its own children while waiting for the dual-mode parent to return to the channel it is working on. To counteract the possibility that the notifications could be lost, a child should start to send its buffered packets immediately after it hears from its parent.

Slot Duration: Determining the slot duration in the asynchronous spectral multiplexing is easier than in the case of the synchronous spectral multiplexing, because in the former situation, not only can different boundary nodes employ different schedules, but they can also stay for a different amount time on each channel. For example, considering the boundary node usually is the parent of the jammed nodes, the boundary node should stay longer on the channel 1 than channel 2, as it has to forward all the packets received in both channels to its parent via channel 1.

Due to the nature of asynchronous spectral multiplexing, nodes can determine their slot durations in a more flexible fashion. Suppose a boundary node decides to stay on channel 1 for t_1 time and channel 2 for t_2 time ($t_1 \geq t_2$), where t_1 and t_2 are chosen according to the traffic volume on each channel and its buffer size. For example, it can choose to have $\frac{t_1}{r_1} + \frac{t_2}{r_2} = B$ where r_1 and r_2 are the traffic rates on the two channels respectively, and B is the buffer size. After setting this baseline schedule, the boundary node can adapt its switching rate as a response to varying network conditions (e.g. topology change, traffic rate change, etc).

Discussion: Compared to synchronous multiplexing, asynchronous multiplexing does not maintain a global schedule, and thus incurs less synchronization overhead. The advantage of asynchronous multiplexing, however, is more pronounced when the jammed region is small and regular. For larger jammed areas, we will have more boundary nodes that work on multiple channels. In this case, the overhead gap between synchronous and asynchronous techniques lessens. A final advantage of the asynchronous method its ability to adapt to local traffic and buffer conditions.

12.5   Sensor Testbed and Metrics

We now focus our discussion on our experimental validation efforts. We have built a 30-node mote testbed, and have conducted numerous experiments with the testbed to evaluate the effectiveness of the four channel surfing strategies in providing interference-resistance.

12.5.1   Testbed Configuration

We have built our sensor network testbed using 30 Mica2 sensor motes. These devices each have a $902 - 928$ MHz Chipcon CC1000 radio. We used 916.7MHz as the original channel and separated our channels by 800KHz, effectively giving us 32 channels. The operating system running on each mote was TinyOS version 1.1.7 [150]. We attached one of the motes to a MIB510CA programming board in order to act as the network sink. In order to conduct experiments that exhibit repeatable characteristics, we chose an indoor laboratory area where we could fix the deployment across the exper-

FIGURE 12.6. Our Mica2 testbed consists of 30 motes that are placed on the floor. Nodes are roughly separated from each other by 2.5 feet. The sink is located at the bottom of the figure, with a programming board attached to it.

iments, as illustrated in Figure 12.6. However, due to our space limitations (there are walls just beyond the boundary of the picture), we were forced to reduce the radio range of each mote: in the depicted configuration, we have a node separation of roughly 2.5 feet. We tuned the transmission power of each mote down to -5dBm in order to restrict the radio range of each sensor node and to increase the network hop count.

The primary objective of this thesis is building a jamming-resistant sensor network, and we have focused our efforts on exploring how a networked system can maintain connectivity in the presence of jamming/interference, regardless of the type of the application data. As a result, we have not attached any specific sensors to the motes. Rather, we modified the Surge application, which comes with TinyOS, to convey experimental statistics. By default, Surge uses a tree-based routing algorithm (which we shall call $SP(t)$) for a single network sink, as detailed in the `MultiHopRouter.nc` file in the TinyOS 1.1.7 release. Since our focus was on networking issues associated with jamming resistance, we did not employ message acknowledgements or retransmissions. In addition to this basic communication model, we note that each sensor message contains a sequencing field (in the routing header) that can be used to estimate performance statistics, such as link quality. Finally, we note that our packet size was 32 bytes, and that a node can buffer at most 24 packets (across all channels).

12.5.2 Implementation of a Sensor Network

Before we could develop the proposed channel surfing strategies on the testbed, we had to modify the existing TinyOS code to address a number of implementation-related issues. In this section, we list a few of the more relevant issues.

The first challenge is devising a reliable link quality estimation technique as this directly affects the efficiency of the routing process. A good estimation technique must be both accurate and resilient to fluctuations. For example, our routing protocol maintains the tree-based topology based on the measured link quality: a node chooses the neighbor that has the best link quality as its parent. Similarly, if a node observes low link quality from the current parent, it will attempt to choose a new parent. On the other hand, due to the low-power and low-fidelity nature of the Mica2 radio, the resulting wireless link quality usually fluctuates greatly, which makes measuring link quality a challenging task.

The existing estimation method utilizes the windowed mean with exponentially weighted moving average estimator, where link quality is defined as:

$$\theta_{new} = \frac{N_{recv}(t)}{\max(N_{exp}(t), N_{recv}(t))} \tag{12.1}$$

$$\theta = \theta_{old} \times \alpha + \theta_{new} \times (1 - \alpha). \tag{12.2}$$

Here θ_{new} is the currently estimated value, $N_{recv}(t)$ is the number of received packets within time t, and $N_{exp}(t)$ is the number of expected packets within time t. In Equation 12.2, the value of α governs how much the past estimation affects the current estimation. The default value for α is set to 0.25 in TinyOS 1.1.7, but we found that this value is too low and the estimated link quality was overly variable. Therefore, the routing topology changes frequently and was hard to stabilize. Through our experiments, we found that $\alpha = 0.75$ yields much better estimates, which incidentally agrees with the value recommended in [142].

After adopting suitable parameters for estimating link quality, we faced the challenge of choosing a parent node. A node chooses its parent based on two criteria: the hop count from the sink and the estimated link quality. Ideally, the parent node should have the smallest hop count and best link quality. However, in reality, nodes that satisfy both criteria may not exist, and we have to prioritize them. In our implementation, we give more importance to link quality than to hop count. For example, we specify that a parent node must have a link quality better than 75% (compared to 10% in the original $SP(t)$). At the same time, we avoid frequent parent switching by adopting the rule that a node can only choose a new parent when the resulting link quality is at least 20% better than the quality of the current link.

In the TinyOS 1.1.7 release of Surge, the application attempts to push packets to the network queue at a fixed rate. However, if the send operation of the previous Surge packet is incomplete, i.e. the SendDone has not been received, then the latest packet is discarded. This posed a problem for us as it did not guarantee that a fixed amount of traffic packets would reach the network buffers. In order to guarantee a nominal data rate, we modified Surge to push data regardless of whether the SendDone callback was received.

12.5.3   Building a Jamming-Resistant Network

After preparing the underlying sensor network testbed, we next implemented the channel surfing framework and the four strategies. In the implementation of these strategies, we had to address the following issues.

First, we modified the buffering mechanism to address the fact that buffered packets may need to be sent on different channels. As a result, for each buffered packet, we associated it with an identifier indicating the channel on which it is to be transmitted.

The second issue we addressed was related to the replacement policy of the mote neighbor table. Each Mica2 mote maintains a neighbor table recording the link quality between each node in its radio range (e.g. *radio neighbors*) and itself. Though a node may have many radio neighbors, especially in a dense deployment, the motes in our testbed only have 16 entries in their neighbor table. As a result, an appropriate buffer replacement policy must be employed to sort a node's radio neighbors. The default replacement policy in TinyOS sorts the radio neighbors based on the link quality between each radio neighbor and the considered node. This policy, however, must be modified to implement channel surfing strategies. To understand this, suppose node A's child, B, is jammed. A is supposed to find the link quality between B and itself has degraded, and then probe the next channel to search for B. However, since B is jammed, the estimated link quality between A and B may be so low that B is evicted from A's neighbor table. In this case, A loses awareness of the existence of B, and will not look for it on the new channel. In order to address this problem, we modified the replacement policy so that it always sorts a node's topological neighbors ahead of non-topological radio neighbors. Hence, we can ensure that a jammed node stays in its former parent's neighbor table until the former parent finds out that it is jammed.

The third issue is that it is necessary to revisit the problem of link quality estimation, and make some further modifications. Estimating the link quality involves comparing the number of packets a node receives with the number of packets it expects by using the sequence numbers of the packets. However, when we operate on multiple channels, this estimation mechanism may cause problems because a dual-mode node will have roughly a 50% link quality, and thus nodes on both channels will not choose the dual-mode nodes as their parents, hindering the realization of spectral multiplexing. We address this by assigning independent sequence numbers on different channels, so that the estimated link quality for a dual-mode node on both channels is sufficiently high.

12.5.4   Performance Metrics for Channel Surfing

The overall impact of channel surfing strategies on a network can be examined in two aspects: the additional overhead it may introduce when the network is in its normal state (i.e. no jamming or radio interference), and the benefit it may have when the network experiences radio interference. To minimize the protocol overhead, we designed the channel surfing strategies so that the protocols do not rely on extra beacons or messages. Instead, we reuse the beacons and the link estimation methods employed in the routing layer. Thus, if the network is in a normal state, it will have comparable performance regardless of whether it employs channel surfing or not. Meanwhile, if the network undergos some form of radio interference, channel surfing can immediately take effect to repair network connection. Finally, a gray area in which channel surfing may impose unnecessary overhead is caused by possible false positives, when channel surfing reacts to neighbor losses due to non-interference reasons. Nonetheless, as pointed out earlier in Section 12.3, we minimize the likelihood of having false positives by only checking a lost neighbor in the new channel under very few circumstances and by adopting a large time threshold. As a result, we believe the proposed channel surfing strategies are both effective in interfered situations and non-intrusive in normal situations. To demonstrate these points, we choose to use the following two performance metrics:

- *Network Recovery:* The main objective of channel surfing is to repair normal network functionalities in the presence of jamming or radio interference. Therefore, we measure their merit by comparing network characteristics in the following three scenarios: (1) under normal conditions, (2) under jamming but without channel surfing, and (3) after applying channel surfing strategies. Specifically, we focus on two network characteristics: network performance (number of packets delivered to the sink), and the latency required to recover network performance.

- *Protocol overhead:* Another class of metrics are related to the overhead introduced by the channel surfing strategies. One obvious concern is that nodes may unnecessarily switch channels. As a result, we have measured the number of channel switches every node experiences throughout the duration of an experiment.

We recorded the packet delivery statistics at the sink by using the sequencing field to maintain a running record of the packets that the sink receives from each sensor. Additionally, as noted earlier, we utilized the sensor messages themselves as a means to measure experimental statistics, such as the operating channel for each sensor node.

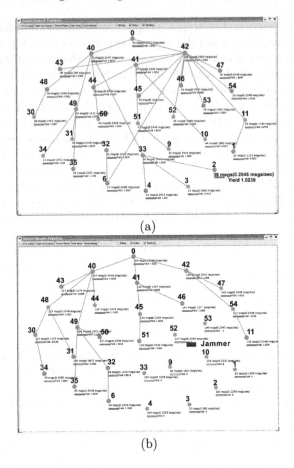

(a)

(b)

FIGURE 12.7. The network topology for jamming experiments. (a) Prior to introducing the jammer, (b) shortly after the jammer is introduced.

12.6   Experimental Results

12.6.1   The Impact of Jamming/Interference

We deployed the testbed as illustrated in Figure 12.6, and the resulting tree-shaped routing topology (Figure 12.7 (a)) was captured using the Surge Network Viewer. In the routing tree, the root node, node 0, corresponds to the network sink. We then conducted an experiment to study the network behavior under normal indoor conditions as well as the impact that a single jammer can have on the network. In this experiment, we first let the network run for more than 20 minutes prior to introducing a jammer. For our jammer, we used the same device as legitimate nodes, thus the jammer can only jam one channel at a time. In particular, we used the constant jammer of [137], which is a mote that bypasses the MAC-layer to continually

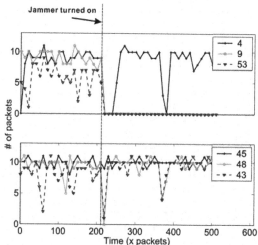

FIGURE 12.8. Packet delivery time series for a 50-second time window under first normal and then jammed network conditions.

transmit random bits into the network. Figure 12.7 (b) shows the location of the jammer, and the fact that the jammer destroyed the connections between several nodes and the sink. Specifically, in this example, nodes $\{9, 10, 33, 52, 53\}$ lost their connections because they were directly affected by the jammer, while nodes $\{1, 2, 3, 4\}$ lost their connections because their parents were jammed.

We next take a closer look at how the packet delivery quality between each node and the sink evolved with time, both with and without jamming. We present time series for the number of packets delivered to the sink in a window of 10 packet from six randomly chosen nodes in Figure 12.8. In these results, each node generated and sent packets at a rate of 1 packet every 5 seconds. Under a perfect network condition, we expect each node to deliver 10 packets in a 10-packet window, but even under normal network conditions prior to jamming, the traces exhibit short-term temporal fluctuations.

We note that, after introducing the jammer, nodes 9 and 53 were not able to deliver packets to the sink. As discussed earlier, that is because either these nodes were jammed, or all their possible parent nodes were jammed. As a result, the packet delivery ratio became 0. We note that although node 4 lost its former parent due to jamming, it later found a replacement parent on channel 1 and hence its packet delivery became normal after a short interruption.

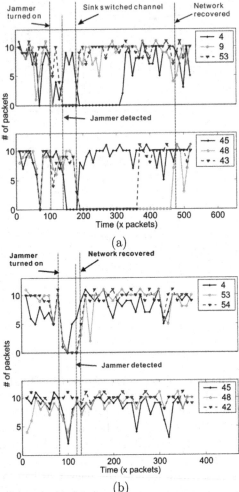

FIGURE 12.9. (a) Packet delivery time series for a 50-second time window for the autonomous channel surfing strategy. (b) Packet delivery time series for the broadcast-assist strategy.

12.6.2   Coordinated Channel Switching Results

The first set of channel surfing results that we provide are for the two coordinated channel switching protocols described in Section 12.4.1, in which the entire network changes its operating channel to a new channel.

Autonomous Channel Switching

We conducted an experiment to study how autonomous channel switching repairs a jammed network. In this experiment, we introduced the jammer after the network had run for 10 minutes. As soon as jamming was detected, the autonomous channel switching strategy worked to repair the network topology so that all of the nodes switched channels, forming a new routing tree on channel-2 (not illustrated due to space limitations).

The packet delivery time series for the experiment are presented in Figure 12.9. This plot presents the time series of number of packets delivered to the sink in a 10-packet window, from nodes $\{4, 9, 53, 45, 48, 43\}$. These six nodes represent some jammed (i.e. 9 and 53) and unaffected nodes. The autonomous algorithm has different impacts on different nodes, depending on their locations in the topology:

1. Before the sink switches to channel 2, those nodes that are not affected by the jammer will still continue delivering packets on channel 1, thus maintain good delivery ratio. Please refer to the traces for nodes 4, 48, and 43 between the 100th interval (when the jammer was turned on) and the 174th (when the sink switched channel).

2. Nodes that switch channel before the sink, either because they are jammed (i.e. node 9) or because they detect the loss of their neighbors before the sink (i.e. node 45), are able to resume packet delivery as soon as the sink switches. Please refer to the traces for nodes 9, 45, and 53 after the 174th interval, when the sink switched channel. Of course, these nodes usually suffer disrupted network services before the sink switches.

3. After the sink switches to channel 2, those nodes that were working on channel 1 suffer from low delivery ratios until they detect some of their neighbors have evaded, and they decide to switch. For example, nodes 4, 48, and 43 take a long time to reconnect to the sink. Among these three nodes, node 4 switches to channel 2 faster than the other two because it is closer to the nodes using channel 2.

Looking at traces from different nodes in the network, we can conclude that how fast a node adopts a new channel in a jammed scenario depends on several factors: (1) whether the node is jammed or not; (2) whether the node switches channel before the sink; and (3) if the node was unaffected by the jammer, its topological distance from the sink. We note that, if the percentage of nodes that are directly affected by the jammer is larger, then the network can switch to a new channel faster. Otherwise, it may take a long latency for some nodes to join the rest on the new channel. In this particular example, node 48 required 290 packet intervals to reconnect after jamming occurs.

Another important statistic is how many times a node switches channels throughout the experiment. A good channel surfing scheme should try to minimize this amount. Autonomous channel switching incurs only 1 channel switch for every node. This is due to its simplistic nature as well as the assurance that a node does not probe the next channel too soon. The latter is guaranteed by the following two facts: (1) in the case of a disappearing child, the underlying routing infrastructure can help a node differentiate a child that has evaded to the next channel from a child that has chosen another parent node; and (2) in the case of a disappearing parent, a node simply chooses another parent without probing the next channel, and only probes when there is no legitimate parent candidate.

Broadcast-assist Channel Switching

Next, we conducted experiments to study the effectiveness of the broadcast-assist channel switching method. Just like autonomous channel switching, we observed that the broadcast-assist channel switching method completely restores connectivity for every sensor to the network sink on the new channel. As expected, compared to autonomous channel switching, broadcast-assist channel switching can significantly reduce the transition latency for the entire network to escape to the new channel, as indicated by the packets-delivered time series shown in Figure 12.9. Here we selected a sampling of six nodes, and observed that regardless of a node's position, they can resume operations on the new channel quickly and almost at the same time. The switching latency is roughly the sum of jamming detection latency, probing time, and the broadcast latency. Specifically, the transition phase only took 46 packet intervals, and 39 out of the 46 intervals were used for the jammed nodes to detect they are jammed as well as for the boundary nodes to find out their children nodes are missing. We would like to emphasize that we purposefully let a node wait for a rather long time period (i.e. 39 packet intervals) before probing the next channel after it detects a poor link quality between itself and its children. *We take the viewpoint that sensor networks will experience a LOT more temporary topological changes than longer-term jamming/interference, and that we therefore must reduce the false positives of channel probing to achieve more stable network operations.* Also, we would like to point out that even with such a conservative approach, we could improve the recovery process by having jammed nodes buffer packets during the network transition periods.

We also measured the total number of channel switches for these six nodes versus time. As expected, we saw that, prior to introducing the jammer, no nodes switched channels because our algorithms were tuned to have very low false positives when determining whether to probe next channel. As soon as the jammer started, since nodes 53 and 54 were directly affected, they switched channels, and stayed there afterwards, thus switching only once. Node 42, however, reported 3 channel switches because it was a boundary node that first switched to channel-2 probing its child, then switched back to the original channel to broadcast the switch notice, and finally switched back to the new channel to resume network operations. Other boundary nodes exhibited similar behavior, while more distant nodes only switched once as a response to the switch notice. Overall, the broadcast-assist channel switch incurs very few channel switches for every node. Finally, we note that we conducted experiments with multiple simultaneous jammers in different positions, and observed that the broadcast-assist method was able to repair the network in these cases with latencies on the order of 50 packet intervals.

12.6.3 Spectral Multiplexing Results

We now examine results for spectral multiplexing. In these methods, we only switch channels for the jammed nodes while a subset of nodes multiplex between the original and new channels. We note that it is unfair to compare coordinated channel switching with spectral multiplexing under the same network configuration because these two strategies are complimentary to each other, each designed to deal with different situations. Specifically, the coordinated strategy is suitable for cases where a large region of the network is jammed, while the multiplexing strategy is suitable for cases with much smaller jammed regions.

Thus, for spectral multiplexing we block fewer nodes to have a smaller jammed region. One further difference between these two types of strategies is that spectral multiplexing typically requires more buffer space for storing packets during multiplexing. Given the limited buffer space on the motes, we chose to adopt a lower data rate (1 packet every 10 seconds) than the rate in earlier experiments.

Synchronous Spectral Multiplexing

In the synchronous spectral multiplexing experiment, we used the same general layout as shown in Figure 12.7(a). After the network had run for 40 packet intervals, the jamming/interference process began. The jammed region consisted of nodes 52, 53, and 10, and these three nodes promptly switched to channel 2. After the jammed nodes evaded to the new channel, their former parents detected their disappearance, and became the boundary nodes. Nodes 41 and 45 were such examples because they used to be the parents for nodes 10 and 52, respectively. The boundary nodes then announced themselves on the new channel, and waited to be selected as parents by the nodes on channel 2. In this experiment, all three jammed nodes chose node 41 as their parent. Thus, node 41 started working on two channels, while node 45 went back to work on channel 1. Overall, node 41 had four child nodes, three on channel 2 (nodes 52, 53, and 10), and one on channel 1 (node 44). In this experiment, the slot duration was 6.1 seconds, so that the number of packets buffered per channel was roughly 3 to 6 packets.

Figure 12.10(a) presents the time series of the number of packets delivered to the sink from six nodes in a 5-packet window. These six nodes include the three jammed nodes (nodes 52, 53, and 10), one boundary node (node 41), one potential boundary node (node 45), and one child of the boundary node that worked on channel 1 (node 44). The plot shows that the jammed nodes resumed their normal packet delivery performance. All other nodes were not affected much by the jammer. The plot shows that the jammed nodes

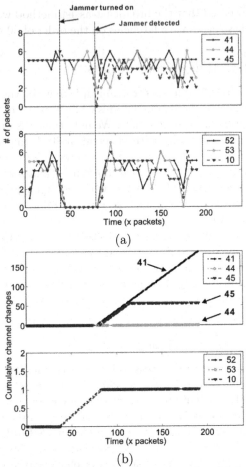

FIGURE 12.10. Statistics for the synchronous spectral multiplexing strategy: (a) number of packets delivered to the sink in a 50-second window vs. time, (b) total number of channel switches vs. time.

resumed their normal packet delivery performance. All other nodes were not affected much by the jammer. The interval during which the jammed nodes had disrupted services was roughly 48 packet intervals, which is similar to the recovery time in the coordinated channel surfing strategy. Again, during the total recovery latency of 50 packet intervals, the boundary nodes waited for around 39 packet intervals before switching to the new channel to search for their children. After the boundary nodes found their children on the new channel, it only took them 11 packet intervals to start working on dual channels.

We report the total number of channel switches for these six nodes in Figure 12.10(b). These numbers agree with the discussion above regarding how different nodes responded to the emulated jamming in the experiment

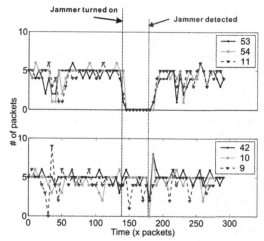

FIGURE 12.11. Packet delivery time series for asynchronous spectral multiplexing.

(e.g. node 41 continually switches channels). Interestingly, as noted earlier, node 45 started out as a dual-mode node and it initially switched channels frequently. However, after a short period of time, it was not selected as a dual-mode parent for any nodes on channel 2. It then returned to channel 1 and no longer switched channels.

Asynchronous Spectral Multiplexing

Our asynchronous experiment used a similar setup as the synchronous multiplexing experiment. The jammed region consisted of nodes 53, 54, and 11. After the jammed nodes evaded to channel 2, their former parents, i.e. nodes 42 and 10, also switched to channel 2, and announced their willingness to work on channel 2. Node 42 was chosen to be the parent by all three jammed nodes, and started flipping between both channels in a round-robin manner, while node 10 continued on channel 1. From then on, node 42 had three children (the three jammed nodes on channel 2).

Figure 12.11(a) presents the time series for the number of packets delivered to the sink from the above-mentioned nodes in a 5-packet window: the three jammed nodes (nodes 53, 54, and 11), one boundary node (node 42), and one potential boundary node (node 10). It is clear from this figure that the asynchronous multiplexing scheme can quickly recover the connections between the jammed nodes and the sink without affecting other nodes in the network. As in the case of synchronous multiplexing, this scheme also incurred roughly a 50-packet service disruption period for the jammed nodes. We also recorded the total number of channel switches for these six nodes and observed trends analogous to those reported for synchronous multiplexing.

12.6.4 Channel Surfing Discussion

As we noted, we designed the channel surfing strategies so that the protocols do not rely on extra beacons or messages. Therefore, there is no additional overhead needed if the network is in a normal state.

Further, though the results presented above have shown that the four channel surfing strategies fare comparably, we emphasize that it is better to evaluate these strategies according to the network and interference scenarios for which they are most appropriate. The autonomous channel switching strategy relies on each individual node to detect whether its lost neighbors have been jammed and have escaped to the next channel. It does not introduce any additional protocol overhead, which can help the protocol design of a sensor network simple and clean. Its less proactive nature, however, leads to a longer delay in switching the entire network. Broadcast-assist Channel Switching, which requires all network nodes to switch their channel, is most suitable for cases where a large region is jammed, and the jamming occurs on a longer time scale (e.g. from a long-duration unintentional interference source). Spectrum multiplexing, however, is more effective for transient jamming where a few nodes are affected for a short duration. Here, we should determine whether to adopt synchronous or asynchronous spectral multiplexing based on the underlying traffic model. For instance, synchronous spectral multiplexing is more suitable for regular traffic patterns, while asynchronous spectral multiplexing can better cope with irregular (e.g. bursty) traffic.

12.6.5 Channel Following Jammers

We were also interested in more challenging interference scenarios, such as the scenario in which the jammer follows the network as it channel surfs. We conducted experiments with this new jammer model, and found that all the four proposed schemes could restore network connectivity in such cases. Due to space limits, we do not provide results for all four schemes here, but rather we show the experimental results for the broadcast-assist channel switch scheme.

The network setup was the same as in Figure 12.7(a). We started the network on channel 1, and then introduced the jammer approximately at the 40th packet into the experiment, as depicted in the time series in Figure 12.12 (a). Shortly thereafter, the network adapted, with all nodes switching to the second channel after an overall latency of approximately 47 packet intervals. We allowed the network to run in the new channel for 50 packet intervals to find the new channel the network moved to, and thus the jammer switched to channel 2 at the 140th packet interval. The network again adapted to the interference, switching to the third channel after a total latency of roughly 51 packet intervals. Examining the packet delivery time series for node 51 illustrates an interesting phenomena regarding the effec-

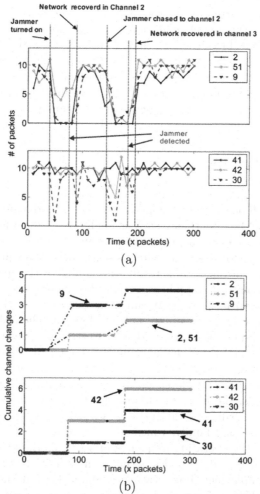

FIGURE 12.12. (a) Packet delivery time series for the broadcast-assist strategy when the jammer follows the network's channel surfing. (b) Time series illustrating the amount of times a node has changed its channel during the network's operation.

tiveness of a jammer: when the jammer was on channel 1 it was not entirely effective in disrupting the operation of node 51 while when the jammer was on channel 2, it was more effective at disrupting the communications from node 51. We believe this is due to the irregularity of the Mica2 radio. In addition to packet delivery traces, we recorded the amount of times different nodes switched channels during the experiment, as illustrated in Figure 12.12 (b). This curve points to the fact that the protocol does not require an excessive amount of channel switching, typically requiring either one or three channel change attempts prior to settling in on the new channel. We draw the reader's attention to the channel change trace for node 41 and node 42. During the first channel change, these nodes switched back to the original channel to broadcast the change channel command, thus requiring a total of 3 channel changes. However, after the jammer follows them to the new channel, both the radio dynamics and the underlying network topology have changed, and in this case only node 42 was involved in switching back to channel 2 to announce the change channel command.

12.7   Conclusion

It is foreseeable that, as wireless sensor networks become increasingly deployed, they will be subjected to increased levels of radio interference. Such interference may be intentional, such as might arise from a jammer, or may be incidental, as may occur due to the presence of other wireless networks. In either case, most commercial wireless sensor networks will be susceptible to radio interference and, as a result, the ability of the sensor system to feed data to monitoring applications may be undermined. It is therefore critical to develop methods that can make sensor networks coexist with each other and even survive external interference. These defense mechanisms, however, must be distributed, easy to scale, and have low false positives. We have tackled this challenge in this chapter by presenting a family of channel surfing strategies that may be used to restore connectivity in the presence of radio interference. We presented two families of channel surfing strategies: the first, which we refer to as channel switching, consists of techniques whereby the entire sensor network changes its operating frequency; the second family, which we refer to as spectral multiplexing, aims to change the operating frequency in a neighborhood local to the interference, with boundary nodes acting as a radio-bridge across different channels. We implemented our strategies on a testbed of Mica2 motes, and have reported their performance for several interference scenarios. We found that our broadcast-assist strategy, as well as our multiplexing schemes, can effectively repair the network with short latency.

Part IV

Preserving Privacy in Wireless Networks

13

Enhancing Source-Location Privacy in Sensor Network Routing

13.1 Introduction

One of the most notable challenges looming on the horizon that threatens successful deployment of sensor networks is privacy. Providing privacy in sensor networks is complicated by the fact that sensor networks consist of low-cost radio devices that employ readily-available, standardized wireless communication technologies. As an example, Berkeley Motes employ a tunable radio technology that is easily observable by spectrum analyzers, while other examples exist of sensor devices employing low-power versions of 802.11 wireless technologies. As a result of the open-architecture of the underlying sensor technology, adversaries will be able to easily gain access to communications between sensor nodes either by purchasing their own low-cost sensor device and running it in a monitor mode, or by employing slightly more sophisticated software radios capable of monitoring a broad array of radio technologies.

Privacy may be defined as the guarantee that information, in its general sense, is observable or decipherable by only those who are intentionally meant to observe or decipher it. The phrase "in its general sense" is meant to imply that there may be types of information besides the message content that are associated with a message transmission. Consequently, the privacy threats that exist for sensor networks may be categorized into two broad classes: content-oriented security/privacy threats, and contextual privacy threats. Content-oriented security and privacy threats are issues that arise due to the ability of the adversary to observe and manipulate the exact content of packets being sent over the sensor network, whether these packets correspond to actual sensed-data or sensitive lower-layer control information. Although issues related to sensor security are important, we believe many of the core problems associated with sensor security are on the road to eventual resolution due to an abundance of recent research by the technical community, c.f. [133, 144, 151].

Contextual privacy issues associated with sensor communication, however, have not been as thoroughly addressed. In contrast to content-oriented security, the issue of contextual privacy is concerned with protecting the *context* associated with the measurement and transmission of sensed data. For many scenarios, general contextual information surrounding the sensor application, especially the location of the message originator, are sensitive and must be protected. This is particularly true when the sensor network monitors valuable assets since protecting the asset's location becomes critical.

Many of the privacy techniques employed in general network scenarios are not appropriate for protecting the source location in a sensor network [64, 152–154]. This is partially due to the fact that the problems are different, and partially due to the fact that many of the methods introduce overhead which is too burdensome for sensor networks. One notable challenge that arises in sensor networks is that the shared wireless medium makes it feasible for an adversary to locate the origin of a radio transmission, thereby facilitating hop-by-hop traceback to the origin of a multi-hop communication.

To address source-location privacy for sensor networks, this chapter provides a formal model for the source-location privacy problem and examines the privacy characteristics of different sensor routing protocols. We introduce two metrics for quantifying source-location privacy in sensor networks, the safety period and capture likelihood. In our examination of popular routing techniques used in today's sensor networks, we also considered important systems issues, like energy consumption, and found that most protocols cannot provide efficient source-location privacy. We propose new techniques to enhance source-location privacy that augment these routing protocols. It is important that this privacy enhancement does not come at a cost of a significant increase in resource consumption. We have devised a strategy, called phantom routing, that has proven flexible and capable of preventing the adversary from tracking the source location with minimal increase in energy overhead.

The rest of this chapter is organized as follows. We begin in Section 13.2 by presenting an overview of our basic asset monitoring sensor networks as well as introduce a formal model and simulation models that will be used to study the source-location privacy problem. Following the setup of the problem, in Section 13.3, we examine several existing routing schemes to protect the source's location. Particularly, the source is not mobile. Then, we study routing and the location privacy of a mobile source in Section 13.4. We present conclusions in Section 13.5.

13.2   Asset Monitoring Sensor Networks

One important class of future sensor-driven applications will be applications that monitor a valuable asset. For example, sensors will be deployed in natural habitats to monitor endangered animals, or may be used in tactical military deployments to provide information to networked operations. In these asset monitoring applications, it is important to provide confidentiality to the source sensor's location.

In order to facilitate the discussion and analysis of source-location privacy in sensor networks, we need to select an exemplary scenario that captures most of the relevant features of both sensor networks and potential adversaries in asset monitoring applications. Throughout this chapter, we use a generic asset monitoring application, which we have called the *Panda-Hunter Game*, as well as refer to a formal model for asset monitoring applications that can benefit from source-location privacy protection. In this section we begin by introducing the Panda-Hunter Game and the formal model, and then discuss how to model the Panda-Hunter Game using a discrete, event-driven simulation framework.

13.2.1   The Panda-Hunter Game

In the Panda-Hunter Game, a large array of panda-detection sensor nodes have been deployed by the Save-The-Panda Organization to monitor a vast habitat for pandas [155]. As soon as a panda is observed, the corresponding *source* node will make observations, and report data periodically to the *sink* via multi-hop routing techniques. The game also features a hunter in the role of the adversary, who tries to capture the panda by back-tracing the routing path until it reaches the source. As a result, a privacy-cautious routing technique should prevent the hunter from locating the source, while delivering the data to the sink.

In the Panda-Hunter Game, we assume there is only a single panda, thus *a single source*, and this source can be either stationary or mobile. During the lifetime of the network, the sensor nodes will continually send data, and the hunter may use this to his advantage to track and hunt the panda. We assume that the source includes its ID in the encrypted messages, but only the sink can tell a node's location from its ID. As a result, even if the hunter is able to break the encryption in a reasonably short time frame, it cannot tell the source's location. In addition, the hunter has the following characteristics:

- **Non-malicious:** The adversary does not interfere with the proper functioning of the network, otherwise intrusion detection measures might flag the hunter's presence. For example, the hunter does not modify packets in transit, alter the routing path, or destroy sensor devices.

- **Device-rich:** The hunter is equipped with devices, such as antenna and spectrum analyzers, so that it can measure the angle of arrival of a message and the received signal strength. From these two measurements, after it hears a message, it is able to identify the immediate sender and move to that node. We emphasize, though, that the hunter cannot learn the origin of a message packet by merely observing a relayed version of a packet. In addition, the hunter can detect the panda when it is near.

- **Resource-rich:** The hunter can move at any rate and has an unlimited amount of power. In addition, it also has a large amount of memory to keep track of information such as messages that have been heard and nodes that have been visited.

- **Informed:** To appropriately study privacy, we must apply Kerckhoff's Principle from security to the privacy setting [96]. In particular, Kerckhoff's Principle states that, in assessing the privacy of a system, one should always assume that the enemy knows the methods being used by the system. Therefore, we assume that the hunter knows the location of the sink node and knows various methods being used by the sensor network to protect the panda.

13.2.2   A Formal Model

In order to understand the issue of location privacy in sensor communication, we now provide a formal model for the privacy problem. Our formal model involves the definition of a general asset monitoring network game, which contains the features of the Panda-Hunter game analyzed in this chapter.

Definition 3. *An asset monitoring network game is defined as a six-tuple* $(\mathcal{N}, S, A, \mathcal{R}, \mathcal{H}, \mathcal{M})$, *where*

1. $\mathcal{N} = \{n_i\}_{i \in I}$ *is the network of sensor nodes* n_i, *which are indexed using an index set* I.

2. S *is the network sink, to which all communication in the sensor network must ultimately be routed to.*

3. A *is an asset that the sensor network monitors. Assets are characterized by the mobility pattern that they follow.*

4. $\mathcal{R}$ *is the routing policy employed by the sensors to protect the asset from being acquired or tracked by the hunter* $\mathcal{H}$.

5. $\mathcal{H}$ *is the hunter, or adversary, who seeks to acquire or capture the asset* A *through a set of movement rules* $\mathcal{M}$.

The game progresses in time with the sensor node that is monitoring the asset periodically sending out messages.

The purpose of the network is to monitor the asset, while the purpose of the routing strategy is two-fold, to deliver messages to the sink and to enhance the location-privacy of the asset in the presence of an adversarial hunter following a movement strategy. We are therefore interested in privacy measures and network efficiency metrics.

Definition 4. *The privacy associated with a sensor network's routing strategy* $\mathcal{R}$ *can be quantified through two differing performance metrics:*

1. *The safety period* Φ *of a routing protocol* $\mathcal{R}$ *for a given adversarial movement strategy* $\mathcal{M}$ *is the number of new messages initiated by the source node that is monitoring an asset, before the adversary locates the asset.*

2. *The capture likelihood* L *of a routing protocol* $\mathcal{R}$ *for a given adversarial movement strategy* $\mathcal{M}$ *is the probability that the hunter can capture the asset within a specified time period.*

On the other hand, the network's performance may be quantified in terms of its energy consumption, and the delivery quality. A sensor node consumes energy when it is sending messages, receiving messages, idling, computing, or sensing the physical world. Among all the operations, sending and receiving messages consume the most energy [156, 157]. We measure the energy consumed in a sensor network by the total number of messages that are sent by all the nodes within the entire network until the asset is captured. We assume that messages are all the same length, each sensor transmits with the same transmission power, and hence each transmission by each sensor requires an equal amount of energy. Consequently, the greater the amount of messages required by a strategy, the more energy that strategy consumes. We use two metrics to measure the delivery quality. One is the average message latency, and the other is the event delivery ratio.

In order to illustrate the formal model of the asset monitoring game, we examine a special case of the Panda-Hunter Game. Suppose that we have a sensor network $\mathcal{N} = \{n_i\}$, where nodes n_i are located on a two-dimensional integer grid and that one of these nodes is designated as the network sink. Network devices might monitor a stationary panda, i.e. the asset A, located at a particular sensing node n_A. This node will periodically transmit sensor messages to the sink S following a routing policy $\mathcal{R}$. One possible routing policy $\mathcal{R}$ might be to employ shortest-path routing in which a single route is formed between the source and sink S according to a gradient-based approach. A hunter $\mathcal{H}$, might start at the network sink S, and might follow a movement strategy $\mathcal{M}$. One possible movement strategy could involve $\mathcal{H}$ repeatedly determining the position of the node that relayed the sensor message and moving to that relay node. Another movement strategy might involve $\mathcal{H}$ initially moving two hops, in order to get a head start, and then continue by moving one hop at a time. The safety period Φ corresponds to the amount of messages transmitted by the source which, in the case of the first movement strategy, corresponds directly to the amount of time it takes the hunter to reach the panda. On the other hand, there is a possibility, in the second movement strategy, that the hunter might skip past the panda (when the panda is one hop from the sink), in which case the hunter will miss the panda entirely and thus $L \neq 1$. Clearly, both the safety period Φ and the capture likelihood L depend on the location of the panda, the mobility of the panda, the routing strategy $\mathcal{R}$ and the movement rules $\mathcal{M}$ for the hunter.

13.2.3 Simulation Model

We have built a discrete event-based simulator to study the privacy protection of several routing techniques. We are particularly interested in large-scale sensor networks where there is a reasonably large separation between the source and the sink. In order to support a large number of nodes in our simulations, we have made a few approximations. Unless otherwise noted, for the simulation results provided in this chapter, we have a network $\mathcal{N}$ of 10,000 randomly located nodes, and the hunter had a hearing radius equal to the sensor transmission radius.

In reality, wireless communication within one hop involves channel sensing (including backoffs) and MAC-layer retransmissions due to collisions. Our simulator ignores the collisions. We emphasize that this should not have a noticeable effect on our accuracy for the following reasons. First, when more reliable MAC protocols are employed, the probability of collision decreases considerably, and channel sensing time may go up correspondingly. Second, sensor networks usually involve light traffic loads with small packets, which result in a lower likelihood of collisions. As a result, our simulator focuses on the channel sensing part. We employ a simple channel sensing model: if a node has m neighbors that may send packets concurrently, the gap before its transmission is a uniformly distributed random number between 1 and m clock ticks. Further, we argue that, although the absolute numbers we report in this chapter may not directly calibrate to a real network, the observed performance trends should hold.

Next, let us look at how we implement the Panda-Hunter game in our simulator. In the game, the panda pops up at a random location. Section 13.3 considers the scenario where the panda stays at the source until it is caught, while Section 13.4 investigates how the routing techniques perform for a moving panda. Once the hunter gets close to the panda (i.e., within Δ hops from the panda), the panda is considered captured and the game is over. As soon as the panda appears at a location, the closest sensor node, which becomes the source, will start sending packets to the sink reporting its observations. The simulator uses a global clock and a global event queue to schedule all the activities within the network, including messages sends, receives and data collections. The source generates a new packet every T clock ticks until the simulation ends, which occurs either when the hunter catches the panda or when the hunter cannot catch the panda within a threshold amount of time (e.g. the panda has returned to its cave).

13.3   Privacy Protection for a Stationary Source

Rather than build a completely new layer for privacy, we take the viewpoint that existing technologies can be suitably modified to achieve desirable levels of privacy. We will therefore examine several existing routing schemes $\mathcal{R}$ to protect the source's location, while simultaneously exploring how much energy they consume. Specifically, we explore two popular classes of routing mechanisms for sensor networks: flooding and single-path routing. For each of these techniques, we propose modifications that allow for enhanced preservation of the source's location or allow us to achieve improved energy conservation. After exploring each of these two classes, we combine our observations to propose a new technique, which we call *phantom routing*, which has both a flooding and single-path variation. Phantom routing is a powerful and effective privacy enhancing strategy that carefully balances the tradeoffs between privacy and energy consumption.

13.3.1   Baseline Routing Techniques

In sensor networks, flooding-based routing and single-path routing are the two most popular classes of routing techniques. In this study, we first examine baseline routing strategies $\mathcal{R}$ from these two classes, and examine their capabilities in protecting the source-location privacy as well as in conserving energy in great depth.

Flooding-based Routing

Many sensor networks employ flooding to disseminate data and control messages [132, 158–160]. In flooding, a message originator transmits its message to each of its neighbors, who in turn retransmit the message to each of their neighbors. Although flooding is known to have performance drawbacks, it nonetheless remains a popular technique for relaying information due to its ease of implementation, and the fact that minor modifications allow it to perform relatively well [161, 162].

In our baseline implementation of flooding, we have ensured that every node in the network only forwards a message once, and no node retransmits a message that it has previously transmitted. When a message reaches an intermediate node, the node first checks whether it has received that message before. If this is its first time, the node will broadcast the message to all its neighbors. Otherwise, it just discards the message. Realistically, this would require a cache at each sensor node. However, the cache size can be easily kept very small because we only need to store the sequence number of each message. We assume that each intermediate sensor node can successfully decrypt just the portion of the message corresponding to the sequence number to obtain the sequence number. Such an operation can easily be done using the CTR-mode of encryption. It is thus reasonable to expect that each sensor device will have enough cache to keep track of enough messages to determine whether it has seen a message before.

Probabilistic flooding [161, 162] was first proposed as an optimization of the baseline flooding technique to cut down energy consumption. In probabilistic flooding, only a subset of nodes within the entire network participate in data forwarding, while the others simply discard the messages they receive. The probability that a node forwards a message is referred to as the *forwarding probability* ($P_{forward}$), and plain flooding can be viewed as probabilistic flooding with $P_{forward} = 1$.

In our simulation, we implement probabilistic flooding as follows. Every time a node receives a new message (it discards the message that it has received before no matter whether it has forwarded it or not), it generates a random number q that is uniformly distributed between 0 and 1. If $q < P_{forward}$, the node will forward/broadcast this message to its neighbors. Otherwise, it will just discard that message. The parameter, $P_{forward}$, is important to the overall performance of this approach. A small value can help reduce the energy consumption though at the expense of lower network coverage and connectivity, while a large value can ensure a higher network coverage and connectivity but will have a correspondingly higher energy consumption.

Single-Path Routing

Unlike flooding, a large number of energy-efficient routing techniques allow a node to forward packets only to one of (or a small subset of) its neighbors. This family of routing techniques is referred to as *single-path routing* in this chapter (e.g., GPSR [118], trajectory-based routing [163], directed diffusion [132], etc). Single-path routing techniques usually require either extra hardware support or a pre-configuration phase. For example, in [118], Karp and Kung propose to use the location information of a node, its neighbors and the destination to calculate a greedy single routing path. In [163], Niculescu and Nath propose trajectory-based routing, which uses the location information associated with a node and its neighbors to create a routing path along a specified trajectory. Such location information can be obtained by either using GPS or other means. In Directed Diffusion [132], an initial phase sets up the "gradients" from each sensor node towards the sink. Later in the routing phase, each intermediate forwarding node can use its neighbors' gradients to implement single-path routing. Whenever

---

Algorithm:*Adversary Strategy I: Patient Adversary* $\mathcal{H}$

next_location = sink;
while *(next_location != source)* **do**
    Listen(next_location);
    msg = ReceiveMessage();
    if *(IsNewMessage(msg))* **then**
        next_location = CalculateImmediateSender(msg);
        MoveTo(next_location);
    end
end

Algorithm 8: *The adversary waits at a location until it receives a new message.*

the source or the sink changes, a re-configuration stage is required in order to reset the routes.

In this study, we try not to assume extra hardware for a normal sensor node. Instead, we use an initial configuration phase to set up the gradients, i.e. hop count between each node and the sink. In the configuration phase, the sink initiates a flood, setting the initial hop count to 0. Any intermediate node will receive the packet many times. It makes sure that it only processes the packet from all of its neighbors once, discarding duplicates. Every time it receives the message, it increments the hop in the message, records it in its local memory, and then broadcasts to its neighbors. After the initial phase, among all the hop counts it has recorded, a sensor node chooses the minimum value as the number of hops from the sink, and updates its neighbors with that number. Then, every sensor node maintains a neighbor list, which is rank-sorted in ascending order according to each neighbor's hop count to the sink. The head of the list, which has the shortest distance to the sink, is said to have the maximum gradient towards the sink. In the baseline single-path routing protocol, as soon as the source generates a new packet, it forwards the packet to the neighbor with the maximum gradient. Every node along the routing path will repeat this process until the packet reaches the sink. Our version of single-path routing thus corresponds to shortest-path routing, and we use these two terms interchangeably.

Adversary Model and Performance Comparison

Before we delve into the location-privacy protection capability of routing techniques, we define one class of hunter $\mathcal{H}$. In Algorithm 8, the hunter follows a simple but natural adversary model, where the adversary starts from the sink, waits at a location until it hears a new message, and then moves to the immediate sender of that message. It repeats this sequence until it reaches the source location. In this model, the adversary assumes that as long as he is patient enough, he will obtain some information that can direct him to the source. We thus refer to this $\mathcal{H}$ model as a *patient adversary*.

Figures 13.1(a)-(d) provide the performance of these baseline routing techniques for a patient adversary for different source-sink distances. In this set of results, we have 10,000 nodes uniformly randomly distributed over a 6000 × 6000 (m^2) network field. The average number of neighbors is 8.5. Among 10,000 nodes, less than 1% are weakly connected with less than 3 neighbors.

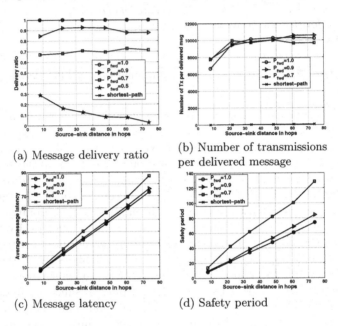

(a) Message delivery ratio

(b) Number of transmissions per delivered message

(c) Message latency

(d) Safety period

FIGURE 13.1. Performance of baseline routing techniques.

Delivery Quality

As expected, baseline flooding and shortest-path routing both give good delivery quality, namely, 100% delivery ratio (Figure 13.1(a)) and lowest message latency (Figure 13.1(c)). On the other hand, probabilistic flooding may have a poorer delivery quality. In particular, we find that probabilistic flooding techniques with $P_{forward} < 0.7$ result in a low message delivery ratio, especially when the source and the sink are far apart. Figure 13.1(a) shows that for $P_{forward} = 0.5$, the message delivery ratio can drop below 5%. As a result, we focus our attention on probabilistic flooding techniques with $P_{forward} \geq 0.7$ in the discussion below.

Energy Consumption

We use the number of transmissions to measure energy consumption, and instead of using the total energy consumed, we report energy consumption per successfully delivered message since some of the messages may not reach the sink (for probabilistic flooding) and this metric captures the wasted energy. For baseline flooding, every message can successfully reach the sink, and each message incurs n transmissions, where n is the number of sensor nodes in the network. Similarly, single-path routing can deliver all the messages, while each message incurs h transmissions where h is the number of hops in the shortest source-sink path. The number of transmissions per successfully delivered message is more complicated for probabilistic flooding schemes. Each successfully delivered message incurs $nP_{forward}$ transmissions, yet there is no guarantee that each message reaches the sink. This behavior has been studied thoroughly by the community [161, 162].

The effective energy usage is reported in Figure 13.1(b). Shortest-path routing incurs a much lower energy consumption (h as we discussed above). Three flooding-based techniques have similar energy consumption figures for each successfully delivered message (n as we discussed above). We would like to point out that those data points below $n = 10,000$ for nearby source-sink configurations are because we stopped the simulation as soon as the panda was caught and the flooding of messages had not yet finished.

Privacy Protection

Although single-path protocols have desirable energy consumption since they reduce the number of messages sent/received, they are rather poor at protecting the source location privacy (Figure 13.1(d)). Since only the nodes that are on the routing path forward messages, the adversary can track the path easily, and can locate the source within h moves. The safety period Φ of baseline single-path routing protocols is the same as the length of the shortest routing path because the adversary can observe every single message the source transmits.

At first glance, one may think that flooding can provide strong privacy protection since almost every node in the network will participate in data forwarding, and that the adversary may be led to the wrong source. Further inspection, however, reveals the contrary. We would like to emphasize that *flooding provides the least possible privacy protection as it allows the adversary to track and reach the source location within the minimum safety period.* Figure 13.1(d) shows that flooding and shortest-path routing lead to the same minimal privacy level. Specifically, the safety period is the same as the hop count on the shortest path.

The poor privacy performance of flooding can be explained by considering the set of all paths produced by the flooding of a single message. This set consists of a mixture of different paths. In particular, this set contains the shortest source-sink path. The shortest path is more likely to reach the hunter first, and thus the hunter will always select the shortest path out of all paths produced by flooding.

In addition to its energy efficiency, probabilistic flooding can improve the privacy protection as well. Imagine there exists a path $\{1, 2, 3, 4, sink\}$, and the adversary is waiting for a new message at node 4. In flooding, the subsequent message will certainly arrive at node 4. However, in probabilistic flooding, the subsequent message may not arrive at node 4 because neighboring nodes may not forward, or take longer to arrive. As a result, the source will likely have to transmit more messages in order for the adversary to work his way back to the source. The more messages the adversary misses, the larger the safety period for the panda, and hence source location protection is provided.

The primary observation is that it is hard for probabilistic flooding techniques to strike a good balance between privacy protection and delivery ratio. For instance, in our study, probabilistic flooding with $P_{forward} = 0.7$ can improve the safety period of baseline flooding roughly by a factor of 2. At the same time, however, it has a message delivery ratio of 70%, which may not be enough for some applications. On the other hand, $P_{forward} = 0.9$ can give a good delivery ratio, but its privacy level is only marginally improved compared to baseline flooding.

13.3.2   Routing with Fake Sources

Baseline flooding and single-path routing cannot provide privacy protection because the adversary can easily identify the shortest path between the source and the sink. This behavior may be considered a result of the fact that there is a single source in the network, and that messaging naturally pulls the hunter to the source. This suggests that one approach we can take to alleviate the risk of a source-location privacy breach is to devise new routing protocols $\mathcal{R}$ that introduce more sources that inject fake messages into the network.

In order to demonstrate the effectiveness of fake messaging, we assume that these messages are of the same length as the real messages, and that they are encrypted as well. Therefore, the adversary cannot tell the difference between a fake message and a real one. As a result, when a fake message reaches the hunter, he will think that it is a legitimate new message, and will be guided towards the fake source.

One challenge with this approach is how to inject fake messages. We need to first decide how to create the fake sources, and when and how often these fake sources should inject false messages. Specifically, we want these fake sources to start only after the event is observed, otherwise the use of fake sources would consume precious sensor energy although there is no panda present to protect.

First, let us look at one naive injection strategy that does not require any additional overhead, which we refer to as the *Short-lived Fake Source* routing strategy. This strategy uses the constant P_{fake} to govern the fake message rate, and choose $P_{fake} \propto \frac{1}{n}$. For any node within the network, after it receives a real message, it generates a random number q that is uniformly distributed between 0 and 1. If $q < P_{fake}$, then this node will produce a fake packet and flood it to the network. In this strategy, the fake source changes from one fake message to another. Although this strategy is easy

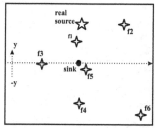 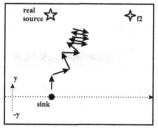

(a) Different locations of fake sources

(b) Pulls from both real source and fake source

FIGURE 13.2. Routing with fake sources.

to implement, it does not improve the privacy level of baseline flooding because the fake sources are short-lived. Even if the hunter is guided by one fake message towards a wrong location, there are no subsequent fake messages around that location to draw him even further away, so he can catch the next real message. As a result, we need a persistent fake source to mislead the hunter.

Thus, we introduce a *Persistent Fake Source* routing strategy. The basic idea of this method is that once a node decides to become a fake source, it will keep generating fake messages regularly so that the hunter can be misled. It is intuitive that a fake source close to the real source, or on the way from the sink to the source, can only help lead the adversary towards the real source, thus providing a poor privacy protection (such as $f1$ in Figure 13.2(a)). As a result, locations $f2, f3, f4, f5, f6$ are better alternatives in terms of protecting privacy. Among these locations, we would like to point out that the distances of the fake sources to the sink should be considered as well when choosing a fake source. For example, if a fake source is too far away from the sink compared to the real source, such as $f6$ in our example, then it would not be as effective in pulling the adversary. On the other hand, if a fake source is too close to the sink, it can draw the hunter quickly towards its location, and as we mention below, a hunter can easily detect the fake source in such cases. As a result, we conclude that the fake sources should be comparable to the real source with respect to their distances to the sink. Hence, $f2, f3$, and $f4$ are good candidates.

The above discussion assumes that we have the global picture of the network deployment. There are many ways of implementing this in a distributed manner, and in this study, we discuss a simple way where we assume that each node knows the hop count between itself and the sink, and that the sink has a sectional antenna. The first assumption can be achieved by a simple flood from the sink, as described in Section 13.3.1. The second assumption is valid because sinks usually are much more powerful than normal sensor nodes. Suppose the source is h hops away from the sink and seeks to create a fake source on the opposite side of the sink with a similar distance to the sink. Then the source can embed that information into the data packets. As soon as the sink receives the hop count from the

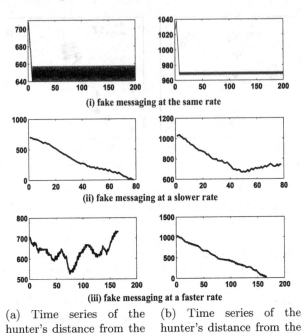

(i) fake messaging at the same rate

(ii) fake messaging at a slower rate

(iii) fake messaging at a faster rate

(a) Time series of the   (b) Time series of the
hunter's distance from the   hunter's distance from the
real source   fake source

FIGURE 13.3. Fake messaging rates.

source, it will send a message to one of its neighbors that are in the direction of $-y$ (using the sectional antenna). This node will further pass the
message to one of its neighbors whose hop count is larger than its own. If
the current node that has the message does not have any neighbors with
a larger hop count then we backtrace one step. We repeat this procedure
until the message reaches a node whose hop count is comparable to h, and
it becomes a fake source. This simple method also allows us to control the
number of fake sources.

After a fake source is chosen, the rate of fake messaging can have a significant impact. Figure 13.3 presents the time series of the hunter's distance
from the real source and the fake source for different fake messaging rates
corresponding to $f2$ in the scenario in Figure 13.2(b). If the fake messages
are injected into the network at the same rate as the real messages (as
shown in Figure 13.3(i)), then the hunter oscillates between the real source
and the fake source, and cannot make progress towards either of them. If
the fake messages are injected at a slower rate, as shown in (ii), then the
hunter will be drawn towards the real source easily. On the other hand, if
the fake messaging rate is higher than the real messaging rate, then the
hunter will be kept at the fake source (Figure 13.3(iii)).

The Perceptive Hunter Adversary Model: From the discussion
above, one can quickly conclude that, if we have a large energy budget,
we can always let fake sources inject messages at a comparable or faster

speed than the real messages to protect privacy. However, this scheme cannot work for a more sophisticated hunter. By using the fact that the hunter knows that fake sources are used (Kerckhoff's Principle), the hunter may detect that he has arrived at a fake source because he cannot detect the panda. As a result, if the fake source is too close to the sink, or injects fake messages too fast, then it will be identified as a fake source quickly. Hence, it may appear appealing for the fake source to inject messages at the same rate as the real source. For the scenario in Figure Figure 13.2(b), we present the results in Figure 13.3(i), where it is seen that the hunter cannot reach either source, but just oscillate between the two. In the figure, the arrows depict the heard messages that can pull the adversary towards both the real source and the fake source. The hope is that the hunter is trapped by the two conflicting pulls into a "zigzag" movement and will not reach the real source. However, the adversary can detect the zigzag movement rather easily, with the help of its cache that stores the history of locations it has recently visited. At this point, the hunter can conclude that he might be receiving fake messages. As a response, the hunter can choose a random direction and only follow messages from that direction. In our example, let us assume that the adversary chooses to follow the messages from its right, and it can reach the fake source. As soon as it reaches the fake source, it stops because the subsequent messages it receives are from the location it is at, and it can conclude it is sitting at a message source. On the other hand, the hunter is assumed to be able to detect the panda if it is at the real source. As a result, it can conclude that it has reached a fake source. Thus, it *learns* that it should only follow messages coming from its left, and can attempt to trace back to the real source. The lessons learned from the study of fake sources is that, though at an enormous energy cost, fake messaging is nonetheless not effective in protecting the privacy of source locations.

13.3.3 *Phantom Routing Techniques*

In the previous sections, we examined the privacy protection capabilities of baseline routing techniques and fake messaging techniques. Both approaches are not very effective in protecting privacy. In both approaches, the sources (either the real one or the fake ones) provide a fixed route for every message so that the adversary can easily back trace the route. Based on this observation, we introduce a new family of flooding and single-path routing protocols for sensor networks, called *phantom routing techniques*. The goal behind phantom techniques is to entice the hunter away from the source towards a phantom source.

In phantom routing, the delivery of every message experiences two phases: (1) the random walk phase, which may be a pure random walk or a directed walk, meant to direct the message to a phantom source, and (2) a subsequent flooding/single-path routing stage meant to deliver the message to

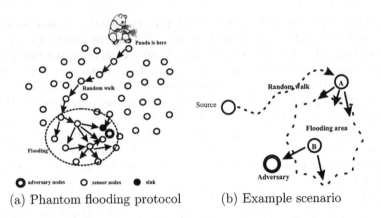

(a) Phantom flooding protocol   (b) Example scenario

FIGURE 13.4. Illustration of Phantom Flooding.

the sink. When the source sends out a message, the message is unicasted in a random fashion for a total of h_{walk} hops. After the h_{walk} hops, in phantom flooding the message is flooded using baseline (probabilistic) flooding. In phantom single-path routing, after the h_{walk} hops the message transmission switches to single-path routing. A depiction of the phantom flooding protocol is illustrated in Figure 13.4(a).

We now discuss the random walk phase in more detail. The ability of a phantom technique to enhance privacy is based upon the ability of the random walk to place the phantom source (after h_{walk} hops) at a location far from the real source. The purpose of the random walk is to send a message to a random location away from the real source. However, if the network is more or less uniformly deployed, and we let those nodes randomly choose one of their neighbors with equal probability, then there is a large chance that the message path will loop around the source spot, and branch to a random location not far from the source.

To further quantify this notion, suppose the network of sensors $\mathcal{N}$ is arrayed on a two-dimensional integer grid with the source and asset A located at $(0,0)$. Suppose the random walk chooses randomly from moving north, south, east, or west, i.e. from $\{(1,0),(-1,0),(0,1),(0,-1)\}$, with equal probability and that the random walk may visit a node more than once. We now estimate the probability that, after h_{walk} hops, the phantom source is within a distance $d < h_{walk}$ of the true source. The movement consists of h_{walk} steps, where each step is an independent random variable X_j with vector values $\{(1,0),(-1,0),(0,1),(0,-1)\}$. The location of the random walk, after h_{walk} steps, is given by

$$D_{h_{walk}} = X_1 + X_2 + \cdots + X_{h_{walk}}.$$

Then, by the central limit theorem, $D_{h_{walk}}/\sqrt{h_{walk}}$ converges in distribution to a bivariate Gaussian with mean $\mathbf{0} = (0,0)$, and covariance matrix $(1/2)\mathbf{I}$. Consequently, $D_{h_{walk}} \sim \mathcal{N}(\mathbf{0}, \frac{h_{walk}}{2}\mathbf{I})$. Let $B = B(\mathbf{0}, d)$ be a ball of radius d centered at $(0,0)$. The asymptotic probability of the phantom source's location $D_{h_{walk}}$ being within a distance d of the real source, after h random walk steps, is given by

$$
\begin{aligned}
P\left(D \in B\right) &= \frac{1}{h\pi} \int_B e^{-\frac{(x^2+y^2)}{h_{walk}}} dx\, dy \\
&= \frac{1}{h\pi} \int_0^d \int_0^{2\pi} e^{-r^2/h_{walk}} r\, d\theta\, dr \\
&= 1 - e^{-d^2/h_{walk}}. \tag{13.1}
\end{aligned}
$$

From this formula, we may examine the likelihood of the phantom's source being within 20% of h_{walk} from the true source after h_{walk} steps, i.e. $d = h_{walk}/5$. The probability is $p = 1 - e^{-h_{walk}/25}$. As we increase h_{walk}, the probability tends to 1, indicating that relative to the amount of energy spent moving a message around, we remain clustered around the true source's location. That is, purely random walk is inefficient at making the phantom source far from the real source, and therefore for reasonable h_{walk} values the location-privacy is not significantly enhanced. These results have been corroborated by simulations involving more general network arrangements, but are not presented due to space considerations.

In order to avoid random walks cancelling each other, we need to introduce bias into the walking process, and therefore we propose the use of a *directed walk* to provide location-privacy. There are two simple approaches to achieving directed walk (without equipping sensor nodes with any extra hardware) that we propose:

- *A sector-based directed random walk.* This approach requires each sensor node to be able to partition the the 2-dimensional plane into two half planes. This can be achieved without using a sectional antenna. Instead, we assume that the network field has some landmark nodes. For example, after the network is deployed, we can mark the west-most node. Then we let that node initiate a flood throughout the network. For a random node i in the network, if it forwards a packet to its neighbor j before it receives the same packet from j, then it can conclude that j is to the east; otherwise, j is to the west. Using this simple method, every node can partition its neighbors into two sets, S_0 and S_1. Before the source starts the directed random walk, it flips a coin and determines whether it is going to use S_0 or S_1. After that, within the first h_{walk} hops, every node that receives the packet randomly chooses a neighbor node from the chosen set for that packet.

- *A hop-based directed random walk.* This approach requires each node to know the hop count between itself and the sink. This can be achieved by the sink initiating a flood throughout the network. After a node first receives the packet, it increments the hop count, and passes the packet on to its neighbors. After the flood phase, neighbors update each other with their own hop counts. As a result, node i can partition its neighbors into two sets, S_0 and S_1, where S_0 includes all the neighbors whose hop counts are smaller than or equal to i's hop count and S_1 includes all the neighbors with a larger hop count. Just as in the sector-based directed random walk, once the two sets are formed, each new message can choose a random set, and every node in the walk can choose a random neighbor from its corresponding set.

We now discuss the ability of phantom techniques to increase the safety period, and hence the location-privacy of sensor communications. Phantom flooding can significantly improve the safety period because every message may take a different (shortest) path to reach any node within the network. As a result, after the adversary hears message i, it may take a long time before it receives $i + 1$. When it finally receives message $i + 1$, the immediate sender of that message may lead the adversary farther away from the source. In the illustration shown in Figure 13.4(b), the adversary is already pretty close to the source before it receives the next new message. This new message goes through the random walk phase and reaches node A, and then goes through the flooding phase. The adversary receives this message from node B, and according to its strategy, it will be duped to move to node B, which is actually farther away from the source compared to the current location of the source.

Both phantom flooding and phantom single-path routing exhibit increased privacy protection because of the path diversity between different messages. We conducted a simulation to examine the privacy enhancement for both types of phantom routing. In this simulation, the source-sink separation was fixed at 60 hops, and we used a sector-based directed walk with different walk lengths h_{walk}. The results are presented in Figure 13.5. A value of $h_{walk} = 0$ corresponds to baseline cases. Phantom techniques clearly demonstrate a much better safety period compared to their baseline counterparts. More importantly, the improvement of phantom schemes keeps increasing with a larger h_{walk}. This is due to the fact that a larger h_{walk} creates a more divergent family of locations for the phantom source, and the probability of sending messages over precisely the same path decreases dramatically.

It is interesting to note that the safety period for phantom shortest-path is larger than for phantom flooding ($p = 1.0$). This behavior is due to the fact that, when we perform routing after the random walk, there is a high likelihood that the resulting single-paths from subsequent phantom sources will not significantly intersect and hence the hunter may miss messages. On the other hand, the resulting floods from subsequent phantom sources will still result in packets arriving at the hunter, allowing him to make progress.

The energy consumed by the phantom techniques is governed by two factors: (1) the walk distance h_{walk}, and (2) the type of flooding/single-path

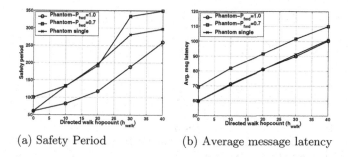

(a) Safety Period   (b) Average message latency

FIGURE 13.5. Performance of different phantom routing techniques (source-sink separation is 60 hops).

routing stage used. The random walk stage automatically introduces h_{walk} transmissions that were not present in the baseline cases. Typically, however, the predominant energy usage for flooding-based techniques comes from the flooding phase, and usually $h_{walk} \ll n$. As a result, the increased energy consumption is negligible (in fact, it does not even change the energy consumption of baseline flooding). Further, for single-path routing techniques, it introduces at most $2h_{walk}$ extra transmissions to the shortest path between the source and the sink, and the total energy consumption of this approach is still minimal.

Phantom techniques also introduce additional latency because every message is directed to a random location first. We conducted simulations to examine the increase in latency for phantom flooding and phantom single-path routing, as presented in Figure 13.5(b). Examining this plot we see that the additional latency increases roughly linearly with h_{walk} for each phantom technique. Combining the latency results and the safety period results, it is interesting to note that for a minor increase in latency, the safety period increases dramatically. For example, for $h_{walk} = 20$, the latency increased roughly 30% while the privacy almost quadrupled!

The Cautious Hunter Adversary Model: We now introduce a new model for the hunter $\mathcal{H}$, which we call the *cautious adversary* model. Since phantom techniques might leave the hunter stranded far from the true source location, the cautious adversary seeks to cope by limiting his listening time at a location. If he has not received any new message within a specified interval, he concludes that he might have been misled to the current location, and he goes back one step and resumes listening from there. We illustrate the cautious adversary model in Algorithm 9. We conducted an experiment with different source-sink separations using phantom single-path routing with $h_{walk} = 10$ hops. In our study, the cautious adversary waited at a location for a period of time corresponding to 4 source messages before deciding to retreat one step. The results are presented in Table 13.1. The cautious adversary model does not provide any benefit over the patient adversary model, as the safety period is higher and the capture likelihood is less. This is because the hunter does not make significant forward progress.

Algorithm:*Adversary Strategy II: Cautious Adversary* $\mathcal{H}$

prev_location = sink;
next_location = sink;
while *(next_location != source)* **do**
    reason = TimedListen(next_location, interval);
    if *(reason == MSG_ARRIVAL)* **then**
        msg = ReceiveMessage();
        if *(IsNewMessage(msg))* **then**
            next_location = CalculateImmediateSender(msg);
            MoveTo(next_location);
        end
    else
        next_location = prev_location;
        prev_location = LookUpPrevLocation(prev_location);
        MoveTo(next_location);
    end
end

Algorithm 9: *The adversary waits at a location for a period of time and returns to its previous location if no message arrives within that period of time.*

TABLE 13.1. Comparison of phantom single-path routing for two adversarial models.

Source-Sink Separation 8 hops		
	Capture Likelihood (L)	Safety Period (Φ)
Patient hunter	1	32
Cautious hunter	0.90	54
Source-Sink Separation 34 hops		
	Capture Likelihood (L)	Safety Period (Φ)
Patient hunter	1	90
Cautious hunter	0.60	301

Waiting time is 60 clock ticks and $h_{walk} = 10$ hops.

Consequently, it is better for the hunter to stay where he is and be patient for message to arrive.

13.4   Privacy Protection for a Mobile Source

In this section we study routing and the location privacy of a mobile asset A. Particularly, in the context of the Panda-Hunter Game, the panda is now mobile. The observations regarding privacy for stationary assets do not directly apply to a mobile asset scenario. Instead, a set of new questions arise. For example, since a mobile panda corresponds to a mobile source, there is a dynamically changing shortest routing path, and therefore it is natural to ask whether the moving panda alone is sufficient to protect its location privacy? Is a faster panda more safe or vice versa? How do flooding-

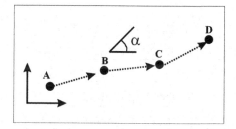

FIGURE 13.6. A simple movement pattern.

TABLE 13.2. The impact of moving velocity on different routing techniques.

Routing techniques	$\delta/T = 2$		$\delta/T = 6$		$\delta/T = 18$	
	L	Φ	L	Φ	L	Φ
flooding	1.0	54	1.0	50	1.0	47
phantom-flood	1.0	92	1.0	75	1.0	78
single-path	0.43	51	0.80	50	1.0	51
phantom-single	0.40	134	0.67	169	1.0	107

In this experiment the hop count between the source and the sink is 48. The source emits a new message every 15 clock ticks.

based techniques fare for a mobile panda compared to a static one? How about single-path routing techniques?

The panda's mobility is defined by its movement pattern and its velocity. The purpose of this chapter is not to define a sophisticated movement pattern, nor to study a comprehensive set of movement patterns. Rather, we employ a rather simple movement model, illustrated in Figure 13.6, to study privacy. In this model, the panda knows the coordinates and knows which direction it is moving along. The parameter α governs the direction of movement. Specifically, if u is its current location, and v is its next location, then the angle of $\overrightarrow{uv}$ should be be within the range $[0, \alpha]$. For instance, in Figure 13.6, the Panda traverses $A, B, C,$ and D, and the direction of any link is within $[0, \alpha]$. Since our simulator has a finite network field, after the panda reaches the boundary of the network, it cannot find any sensor node in the specified direction, retreats a few steps, and resumes its normal pattern. In addition to its direction, it has the other parameters which describe its velocity: δ is the stay time at each location, and d denotes the distance for each of its movements. In the simulation, the sensor node that is closest to the Panda will become the new source, and will send $\lfloor \frac{\delta}{T} \rfloor$ (where T is the reporting interval) new messages before the Panda moves on.

The Impact of Velocity: We first conducted simulations to evaluate the effect of the panda's velocity on source-location privacy. In this experiment, the source-sink hop count is 48, the source sends out a message every 15 clock ticks and $P_{forward}$ is 1.0 in the flooding techniques studied. The results are presented in Table 13.2. Here, the first observation is that, for all routing techniques, a fast moving panda (lower δ values) is safer than a slow panda. The second observation is that, among different techniques, the velocity of

TABLE 13.3. The impact of the hunter's hearing range on capture likelihood.

Phantom Single-Path Routing			
δ/T	$r_H/r = 1$	$r_H/r = 2$	$r_H/r = 3$
1	0.23	0.43	0.60
2	0.40	0.77	0.93
6	0.67	0.90	0.97
8	0.80	0.97	0.97
Single-path Routing			
δ/T	$r_H/r = 1$	$r_H/r = 2$	$r_H/r = 3$
1	0.23	0.50	0.73
2	0.43	0.77	0.90
6	0.80	0.97	0.97
8	0.87	0.97	1.0

the panda has a more noticeable impact on single-path routing techniques than it does on flooding-based routing techniques. For single-path routing, the capture likelihood L is closely related to the velocity of the panda. In particular, a faster moving panda makes it unlikely that the adversary can track the panda. On the other hand, flooding for the same mobility allows the panda to be caught, though with an increased safety period Φ. This observation can be explained as follows. In single-path routing, subsequent shortest paths might not have significant overlap due to the panda's movement, and hence the hunter may not even see a subsequent message. On the other hand, flooding guarantees that the hunter will see the message, though not from the shortest source-sink path, and he may still follow the panda's movement. That is, a reasonably fast moving panda alone is sufficient to protect its location when using single-path routing. The third observation is that panda mobility can improve the privacy protection of phantom techniques more than it does to other schemes. These observations are due to the fact that the source mobility serves to further decorrelate the source's location from the phantom source's location, resulting in enhanced location privacy.

The Impact of the Hunter's Hearing Range: So far, we have assumed that the hunter's hearing range (r_H) is the same as any normal sensor node (r). Next, let us look at the impact of different hearing ranges on the privacy level of a network. For this purpose, we conducted a set of simulation studies for phantom single-path routing with a source-sink separation of 48 hops. The resulting capture likelihoods for different δ/T and r_H/r combinations are presented in Table 13.3. In general, we find that a larger hearing range helps the hunter since this translates into the hunter hearing messages sooner and allows him to make larger moves, effectively allowing him to move faster. We also see that ability for the hunter to capture pandas improves with larger hearing ranges, and that the relative improvement is more pronounced for faster pandas. It should be realized, however, that this corresponds to introducing a powerful adversary. We also measured the impact of hearing range for single-path routing, and observed that phantom single-path routing has improved privacy for larger hearing radii compared to baseline single-path routing.

13.5   Conclusion

Sensor networks will be deployed to monitor valuable assets. In many scenarios, an adversary may be able to backtrace message routing paths to the event source, which can be a serious privacy breach for many monitoring and remote-sensing application scenarios. In this chapter, we have studied the ability of different routing protocols to obfuscate the location of a source sensor. We examined several variations of flooding-based and single-path routing techniques, and found that none of these protocols are capable of providing source location privacy. To achieve improved location privacy, we proposed a new family of routing techniques, called phantom routing, for both the flooding and single-path classes that enhance privacy protection. Phantom routing techniques are desirable since they only marginally increase communication overhead, while achieving significant privacy amplification. Going forward we plan to investigate stronger adversarial models, as well as multiple asset tracking scenarios and their impact on location privacy in sensor networks.

14

Temporal Privacy in Wireless Networks

14.1 Introduction

Since wireless sensor networks employ a broadcast medium, an adversary may monitor sensor communications to piece together knowledge of the context surrounding sensor messages. In particular, by applying wireless localization algorithms and some level of diligence, an adversary will be able to infer the sensor network deployment, i.e. an association of sensor IDs with their physical locations. This information, combined with knowledge of the routing algorithms employed and the location of the base-station (data sink), can allow the adversary to track the spatio-temporal evolution of a sensed-event from the remote location of the network sink by merely monitoring the arrival of incoming packets [164]. This spatio-temporal information is available regardless of whether the adversary can decipher encrypted packet payloads, and represents a breach of the spatio-temporal privacy associated with the sensor network's operation. This breach of privacy can be put to very malicious use. For example, in an asset tracking sensor network, an adversary can use the spatio-temporal characteristics of the network traffic to determine the speed and direction of motion of an asset and track it, merely by observing packets arriving at the network sink.

In order to protect against such a privacy breach, there are two types of information that can be protected: the spatial information surrounding the flow of sensor messages, and the temporal context surrounding the creation of sensor readings. Protecting the spatial context of sensor routing involves obscuring the location of the source sensor [165, 166], as well as the location of the network sink [167, 168]. However, should an adversary compromise the defense mechanisms meant to protect a sensor network's spatial context and learn the location of the originating sensor and the network sink, then the spatio-temporal context of a sensor's message flow may still be protected by employing mechanisms that protect the *temporal context* of the sensor's message. In this chapter we focus on the problem of protecting the temporal context associated with a sensor's measurement of underlying physical phenomena. Specifically, for the typical delay-tolerant application, we propose the use of additional store-and-forward buffering at intermediate nodes along the routing path between a source sensor and the sink in order to obfuscate the time of creation associated with the flow of sensor messages.

We begin the chapter in Section 14.2 by describing our sensor network model, overview the problem of temporal privacy and how additional buffering can enhance privacy. We then examine the two conflicting aspects of buffering: in Section 14.3, we formulate temporal privacy from an information-theoretic perspective, and in Section 14.4, we examine the stress that additional delay places on intermediate buffers. Then, in Section 14.5, we present an adaptive buffering strategy that effectively manages these trade-offs through the preemptive release of packets as buffers attain their capacity. We evaluate our temporal privacy solutions in Section 14.6 through simulations involving a large-scale network, where the adversary's mean square error (a metric directly related to our information-theoretic formulation of privacy) is used to quantify the temporal privacy. Finally, we conclude the chapter in Section 14.7.

14.2 Overview of Temporal Privacy in Sensor Networks

We start our overview by describing a couple scenarios that illustrate the issues associated with temporal privacy. To begin, consider a sensor network that has been deployed to monitor an animal habitat [165, 169]. In this scenario, animals ("assets") move through the environment, their presence is sensed by the sensor network and reported to the network sink. The fact that the network produces data and sends it to the sink provides an indication that the animal was present at the source at a specific time. If the adversary is able to associate the origin time of the packet with a sensor's location, then the adversary will be able to track the animal's behavior– a dangerous prospect if the animal is endangered and the adversary is a hunter! This same scenario can be easily translated to a tactical environment, where the sensor network monitors events in support of military networked operations. In asset tracking, if we add temporal ambiguity to the time that the packets are created then, as the asset moves, this would introduce spatial ambiguity and make it harder for the adversary to track the asset.

The situations where temporal privacy is important are not always associated with protecting spatio-temporal context, but instead there are scenarios where we are solely interested in masking the time at which an event occurred. For example, sensor networks may be deployed to monitor inventory in a warehouse. In this scenario, a sensor would create audit logs associated with the removal/relocation of items (bearing RFID tags) within the warehouse and route these audit messages to the network sink. Here, an adversary located near the sink (perhaps outside the warehouse) could observe packets arriving and use this information to infer the stock levels or the volume of transactions going through a warehouse at a specific time. Such information could be of great benefit to a rival corporation that is interested in knowing its competitor's sales and inventory profile. Here, if we add temporal ambiguity to the delivery of the audit messages, then the warehouse would still be able to verify its inventory against purchase orders, but the competitor would have outdated information about the inventory activity.

For both scenarios, temporal privacy amounts to preventing an adversary from inferring the time of creation associated with one or more sensor packets arriving at the network sink. In order to protect the temporal context of the packet's creation, it is possible to introduce additional, random delay to the delivery of packets in order to mask a sensor reading's time of creation. However, although delaying packets might increase temporal privacy, this strategy also necessitates the use of buffering either at the source or within the network and places new stress on the internal store-and-forward network buffers.

We now provide a generic model for both the sensor network and the adversary that captures the most relevant features of the temporal privacy problem studied in this chapter. The abstract sensor network model that we will use involves:

- **Delay-Tolerant Application:** A sensor application that is delay-tolerant in the sense that observations can be delayed by reasonable amounts of time before arriving at the monitoring application, thereby allowing us to introduce additional delay in packet delivery.

- **Payload Encrypted:** The payload contains application-level information, such as the sensor reading, application sequence number, and the time-stamp associated with the sensor reading. In order to guarantee the confidentiality of the sensor application's data, conventional encryption is employed.

- **Headers are Cleartext:** The headers associated with essential network functionality are not encrypted. For example, the routing header associated with [142], and used in the TinyOS 1.1.7 release (described in `MultiHop.h`) includes the ID of the previous hop, the ID of the origin (used in the routing layer to differentiate between whether the packet is being generated or forwarded), the routing-layer sequence number (used to avoid loops, not flow-specific and hence cannot help the adversary in estimating time of creation), and the hop count.

On the otherhand, the assumptions that we have for the adversary are

- **Protocol-Aware:** By Kerckhoff's Principle [170], we assume the adversary has knowledge of the networking and privacy protocols being employed by the sensor network. In particular, the adversary knows the delay distributions being used by each node in the network.

- **Able to Eavesdrop:** We assume that the adversary is able to eavesdrop on communications in order to read packet headers, or control traffic. We emphasize that the adversary is not able to decipher packet contents by decrypting the payloads, and hence the adversary must infer packet creation times solely from network knowledge and the time it witnesses a packet.

- **Deployment-Aware:** We assume that the adversary has a single point-of-presence, and is thus at the sink. Further, to empower the adversary, we assume he is aware of the identity of all sensor nodes. Since the adversary can monitor communications, we assume that the adversary knows the source identity associated with each transmission. Further, since the adversary is aware of the routing protocols employed and can eavesdrop, the adversary is able to build its own source-sink routing tables.

- **Non-intrusive:** The adversary does not interfere with the proper functioning of the network, otherwise intrusion detection measures might flag the adversary's presence. In particular, the adversary does not inject or modify packets, alter the routing path, or destroy sensor devices.

These security assumptions are intended to give the adversary significant power and thus, if our temporal privacy techniques are robust under these assumptions, they can be considered to be powerful under more general threat conditions. Additionally, we note that we have separated out issues associated with obscuring the location of the source's origin, and solely focus on temporal privacy. In practice, however, the combination of temporal privacy methods with location-privacy methods will yield a more complete solution to protecting contextual privacy in sensor networks.

14.2.1   The Baseline Adversary Model

Our sensor network model assumes multiple source nodes that create packets and send these packets to a common sink via multi-hop networking. As an important player of the game, the adversary stays at the sink, observes packet arrivals, and estimates the creation times of these packets. We note that, while it may seem like the adversary would be better off being mobile or that the adversary is located at several random places within the network, it is not so. *Since all activities in a sensor network are reported to the sink, being closer to the sink enables the adversary to maximize his chances of observing as many traffic flows as possible.*

In order to better focus on temporal privacy of a sensor network, we assume a rather powerful adversary that can acquire the following information about the underlying network:

1. The hop count h_i, of flow i. This can be inferred by the adversary by looking at hop-count information in the packet headers or by correlating the source-id with his knowledge of the network topology.

2. The transmission delay that a node takes to transmit a packet τ.

For an observed packet arrival time z, our adversary estimates the creation time of this packet as $x' = z - h\tau$. In the literature, the *square error* is often used to quantify the estimation error, i.e. $(x' - x)^2$ where x is the true creation time. Similarly, for a series of packet arrivals from the same flow $z_1, z_2, \ldots, z_m$, our adversary estimates their creation times as $x'_1, x'_2, \ldots, x'_m$, and $x'_i = z_i - y$. The total estimation error for m packets is then calculated as the *mean square error* $MSE = \sum (x'_i - x_i)^2/m$. The network that has a higher estimation error consequently better preserves the temporal privacy of the source.

14.3   Temporal Privacy Formulation

We start by first examining the theoretical underpinnings of temporal privacy. Our discussion will start by first setting up the formulation using a simple network of two nodes transmitting a single packet, and then we extend the formulation to more general network scenarios.

14.3.1   Temporal Privacy: Two-Party Single-Packet Network

We begin by considering a simple network consisting of a source S, a receiver node R, and an adversarial node E that monitors traffic arriving at R. The goal of preserving temporal privacy is to make it difficult for the adversary to infer the time when a specific packet was created. Suppose that the source sensor S observes a phenomena and creates a packet at some time X. In order to obfuscate the time at which this packet was created, S can choose to locally buffer the packet for a random amount of time Y before transmitting the packet. Disregarding the negligible time it takes for the packet to traverse the wireless medium, both R and E will witness that the packet arrives at a time $Z = X + Y$. The legitimate receiver can decrypt the payload, which contains a timestamp field describing the correct time of creation. The adversary's objective is to infer the time of creation X, and since it cannot decipher the payload, it must make an inference based solely upon the observation of Z and (by Kerckhoff's Principle) knowledge of the buffering strategy employed at S.

The ability of E to infer X from Z is controlled by two underlying distributions: first, is the a priori distribution $f_X(x)$, which describes the knowledge the adversary had for the likelihood of the message creation prior to observing Z; and second, the delay distribution $f_Y(y)$, which the source employs to mask X. In classical security and privacy, the amount of information that E can infer about X from observing Z is measured by the mutual information [170, 171]:

$$I(X; Z) = h(X) - h(X|Z) = h(Z) - h(Z|X) \tag{14.1}$$
$$= h(Z) - h(Y), \tag{14.2}$$

where $h(X)$ is the differential entropy of X. We note that, due to the relationship between mutual information and mean square error [172], large $I(X; Z)$ implies that a well-designed estimator of X from Z will have small MSE. For certain choices of f_X and f_Y, we may directly calculate $I(X; Z)$. For example, if $X \sim Exp(\lambda)$ (i.e. exponential with mean $1/\lambda$), and $Y \sim Exp(\lambda)$, then $Z \sim Erlang(2, \lambda)$, and $h(Z) = -\psi(2) + \ln \Gamma(2) - \ln(\lambda) + 2$, where $\psi(w)$ is the digamma function and $\Gamma(w)$ is the gamma function. For this case, $h(Y) = 1 - \ln \lambda$, and hence $I(X; Z) = 1 - \psi(2) \approx 1.077$. In other words, roughly 1 nat of information about X is learned by observing Z. For more general distributions, the entropy-power inequality [171] gives a lower bound

$$I(X; Z) \geq \frac{1}{2 \ln 2} \left(2^{2h(X)} + 2^{2h(Y)} \right) - h(Y). \tag{14.3}$$

In general, however, the distribution for X is fixed and determined by an underlying physical phenomena being monitored by the sensor. Since the objective of the temporal privacy-enhancing buffering is to hide X, we may formulate the temporal privacy problem as

$$\min_{f_Y(y)} I(X; Z) = h(X + Y) - h(Y),$$

or in other words, choose a delay distribution f_Y so that the adversary learns as little as possible about X from Z.[1]

---

[1] The astute reader will note the similarity with the information-theoretic formulation of communication, where the objective is to maximize mutual information.

14.3.2 Temporal Privacy: Two-Party Multiple-Packet Network

We now extend the formulation of temporal privacy to the more general case of a source S sending a stream of packets to a receiver R in the presence of an adversary E. In this case, the sender S will create a stream of packets at times $X_1, X_2, \ldots, X_n, \ldots$, and will delay their transmissions by $Y_1, Y_2, \ldots, Y_n, \ldots$. The packets will be observed by E at times $Z_1, Z_2, \ldots, Z_n, \ldots$. In going to the more general case of a packet stream, several new issues arise. First, as noted earlier in Section 14.2, when we delay multiple packets it will be necessary to buffer these packets. For now we will hold off on discussing queuing issues until Section 14.4. The next issue involves how the packets should be delayed. There are many possibilities here. For example, one possibility would have packets released in the same order as their creation, i.e. $Z_1 < Z_2 < \ldots < Z_n$, which would correspond to choosing Y_j to be at least the wait time needed to flush out all previous packets. Such a strategy does not reflect the fact that most sensor monitoring applications do not require that packet ordering is maintained. Therefore, a more natural delay strategy would involve choosing Y_j independent of each other and independent of the creation process $\{X_j\}$. Consequently, there will not be an ordering of $(Z_1, Z_2, \ldots, Z_n, \ldots)$.

In our sensor network model, however, we assumed that the sensing application's sequence number field was contained in the encrypted payload, and consequently the adversary does not directly observe $(Z_1, Z_2, \ldots, Z_n, \ldots)$, but instead observes the sorted process $\{\tilde{Z}_j\} = \Upsilon(\{Z_j\})$, where $\Upsilon(\{Z_j\})$ denotes the permutations needed to achieve a temporal ordering of the elements of the process $\{Z_j\}$, i.e. $\{\tilde{Z}_j\} = (\tilde{Z}_1, \tilde{Z}_2, \ldots, \tilde{Z}_n, \ldots)$ where $\tilde{Z}_1 < \tilde{Z}_2 < \cdots$. The adversary's task thus becomes inferring the process $\{X_j\}$ from the sorted process $\{\tilde{Z}_j\}$. The amount of information gleaned by the adversary after observing $\tilde{Z}^n = (\tilde{Z}_1, \cdots, \tilde{Z}_n)$ is thus $I(X^n; \tilde{Z}^n)$, and the temporal-privacy objective of the system designer is to make $I(X^n; \tilde{Z}^n)$ small.

Although it is analytically cumbersome to access $I(X^n; \tilde{Z}^n)$, we may use the data processing inequality [171] on $X^n \to Z^n \to \tilde{Z}^n$ to obtain the relationship $0 \le I(X^n, \tilde{Z}^n) \le I(X^n, Z^n)$, which allows us to use $I(X^n, Z^n)$ in a pinching argument to control $I(X^n, \tilde{Z}^n)$. Expanding $I(X^n, Z^n)$ as

$$I(X^n, Z^n) = h(Z^n) - h(Y^n) \tag{14.4}$$

$$\le \sum_{j=1}^{n} (h(Z_j) - h(Y_j)) \tag{14.5}$$

$$= \sum_{j=1}^{n} I(X_j, Z_j), \tag{14.6}$$

we may thus bound $I(X^n, Z^n)$ using the sum of individual mutual information terms.

As before, the objective of temporal privacy enhancement is to minimize the information that the adversary gains, and hence to mask $\{X_j\}$, we should minimize $I(X^n, Z^n)$. Although there are many choices for the delay process $\{Y_j\}$, the general task of finding a non-trivial stochastic process $\{Y_j\}$ that minimizes the mutual information for a specific temporal process $\{X_j\}$ is challenging and further depends on the sensor network design constraints (e.g. buffer storage). In spite of this, however, we may seek to optimize within a specific type of process $\{Y_j\}$, and from this make some general observations.

As an example of this, let us look at an important and natural example. Suppose that the source sensor creates packets at times $\{X_j\}$ as a Poisson process of rate λ, i.e. the interarrival times A_j are exponential with mean $1/\lambda$, and that the delay process $\{Y_j\}$ corresponds to each Y_j being an exponential delay with mean $1/\mu$. We note that our choice of a Poisson source is intended for explanation purposes, and that more general packet creation processes can be handled using the same machinery. One motivation for choosing an exponential distribution for the delay is the well-known fact that the exponential distribution yields maximal entropy for non-negative distributions. We note that $X_j = \sum_{k=1}^{j} A_k$ (and hence the X_j are j-stage Erlangian random variables with mean j/λ). Using the result of Theorem 3(d) from [173], we have that

$$
\begin{aligned}
I(X_j; Z_j) &= I(X_j; X_j + Y_j) \\
&= \ln\left(1 + \frac{j\mu}{\lambda}\right) - D\left(f_{X_j + Y_j} \| f_{\overline{X}_j + Y_j}\right) \\
&\leq \ln\left(1 + \frac{j\mu}{\lambda}\right).
\end{aligned}
$$

Here, the $D(f\|g)$ corresponds to the divergence between two distributions f and g, while $\overline{X}$ is the mixture of a point mass and exponential distribution with the same mean as X, as introduced in [173]. Since divergence is non-negative and we are only interested in pinching $I(X^n; \tilde{Z}^n)$, we may discard this auxiliary term. Using the above result, we have that

$$
I(X^n, Z^n) \leq \sum_{j=1}^{n} \ln\left(1 + \frac{j\mu}{\lambda}\right). \tag{14.7}
$$

Our objective is to make

$$
0 \leq I(X^n; \tilde{Z}^n) \leq I(X^n, Z^n) \leq \sum_{j=1}^{n} \ln\left(1 + \frac{j\mu}{\lambda}\right)
$$

small, and from this we can see that by tuning μ to be small relative to λ (or equivalently, the average delay time $1/\mu$ to be large relative to the average interarrival time $1/\lambda$), we can control the amount of information the adversary learns about the original packet creation times. It is clear that choosing μ too small will place a heavy load on the source's buffer. We will revisit buffer issues in Section 14.4 and Section 14.5.

14.3.3 Temporal Privacy: Multihop Networks

In the previous subsection, we considered a simple network case consisting of two nodes, where the source performs all of the buffering. More general sensor networks consist of multiple nodes that communicate via multi-hop routing to a sensor network sink. For such networks, the burden of obfuscating the times at which a source node creates packets can be shared amongst other nodes on the path between the source and the sensor network sink.

To explain, we may consider a generic sensor network consisting of an abundant supply of sensor nodes, and focus on an N-hop routing path between the source and the network sink. By doing so, we are restricting our attention to a line-topology network $S \to F_1 \to F_2 \to \cdots \to F_{N-1} \to R$, where R denotes the receiving network sink, and F_j denotes the j-th intermediate node on the forwarding path.

By introducing multiple nodes, the delay process $\{Y_j\}$ can be decomposed across multiple nodes as

$$Y_j = Y_{0j} + Y_{1j} + \cdots + Y_{N-1,j},$$

where Y_{kj} denotes the delay introduced at node k for the j-th packet (we use Y_{0j} to denote the delay used by the source node S). Thus, each node k will buffer each packet j that it receives for a random amount of time Y_{kj}.

This decomposition of the delay process $\{Y_j\}$ into sub-delay processes $\{Y_{kj}\}$ allows for great flexibility in achieving both temporal privacy goals and ensuring suitable buffer utilization in the sensor network. For example, it is well-known that traffic loads in sensor networks accumulate near network sinks, and it may be possible to decompose $\{Y_j\}$ so that more delay is introduced when a forwarding node is further from the sink.

14.4 Queuing Analysis of Privacy-Enhancing Buffering

Although delaying packets might increase temporal privacy, such a strategy places a burden on intermediate buffers. In this section we will examine the underlying issues of buffer utilization when employing delay to enhance temporal privacy.

When using buffering to enhance temporal privacy, each node on the routing path will receive packets and delay their forwarding by a random

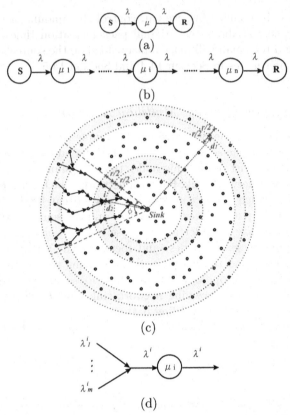

FIGURE 14.1. (a) Queue model for buffering at the source node S, (b) chain of queues along a routing path from S to receiver sink R, (c) the effect of flow convergence in a large sensor network, and (d) queuing model for the merging of traffic flows at an intermediate sensor node.

amount of time. As a result, sensor nodes must buffer packets prior to releasing them, and we may formulate the buffer occupancy using a queuing model. In order to start our discussion, let us again examine the simple two-node case where a source node S generates packets according to an underlying process and the packets are delayed according to an exponential distribution with average delay $1/\mu$, prior to being forwarded to the receiver R, as depicted in Figure 14.1 (a). If we assume that the creation process is Poisson with rate λ (if the process is not Poisson, the source may introduce additional delay to shape the traffic or, at the expense of lengthy derivations, similar results can be arrived at using embedded Markov chains), then the buffering process can be viewed as an $M/M/\infty$ queue where, as new packets arrive at the buffer, they are assigned to a new "variable-delay server" that processes each packet according to an exponential distribution with mean $1/\mu$. Following the standard results for $M/M/\infty$ queues, we have that the amount of packets being stored at an arbitrary time, $N(t)$, is Poisson distributed, with $p_k = P\{N(t) = k\} = \frac{\rho^k}{k!}e^{-\rho}$, where $\rho = \lambda/\mu$ is the system utilization factor. $\overline{N}$, the expected number of messages buffered at S, is ρ.

The slightly more complicated scenario involving more than one intermediate node allows for the buffering responsibility to be divided across the routing path, and is depicted by a chained path in Figure 14.1 (b). A tandem queuing network is formed, where a message departing from node i immediately enters an $M/M/\infty$ queue at node $i+1$. Thus, the interdeparture times from the former generate the interarrival times to the latter. According to Burke's Theorem [174], the steady-state output of a stable $M/M/m$ queue with input parameter λ and service-time parameter μ for each of the m servers is in fact a Poisson process at the same rate λ when $\lambda < \mu$. Hence, we may generally model each node i on the path as an $M/M/\infty$ queue with average input message rate λ, but with average service-time $1/\mu_i$ (to allow each node to follow its own delay distribution).

So far we have only considered a single routing path in a sensor network, but in practice the network will monitor multiple phenomena simultaneously, and consequently there will be multiple source-sink flows traversing the network. As a result, for the most general scenario, the topological structure of the network will have an impact on buffer occupancy. For example, nodes that are closer to network sink typically have higher traffic loads, and thus will be expected to suffer from a higher buffer occupancy than nodes further from the sink. We now explore this behavior, and the relationship between buffering for privacy-enhancement and the traffic load placed on intermediate nodes due to flow convergence in the sensor network.

Consider a sensor network deployment as depicted in Figure 14.1(c), where we have assumed (without loss of generality) that there is only one sink. Here, multiple sensors generate messages intended for the sink, and each message is routed in a hop-by-hop manner based on a routing tree (as suggested in the figure). Message streams merge progressively as they approach the sink. As before, let us assume for the sake of discussion that the senders in the network generate Poisson flows, then by the superposition property of Poisson processes the combined stream arriving at node i of m independent Poisson processes with rate λ_j^i is a Poisson process with rate $\lambda^i = \lambda_1^i + \lambda_2^i + \cdots + \lambda_m^i$. We depict this phenomena for node i in Figure 14.1(d), where m is the number of "routing" children for node i. Additionally, we let $1/\mu_i$ be the average buffer delay injected by node i. Then node i is an $M/M/\infty$ queue, with arrival parameter λ^i and departure parameter μ_i, yielding:

- $N_i(t)$, the number of packets in the buffer at node i, is Poisson distributed.

- $p_{ik} = P\{N_i(t) = k\} = \frac{\rho_i^k}{k!}e^{-\rho_i}$, where $\rho_i = \lambda^i/\mu_i$.

- The expected number of messages at node i is $\overline{N_i} = \rho_i$.

As expected, if we choose our delay strategy at node i such that μ_i is much smaller than λ^i (as is desirable for enhanced temporal privacy), then the expected buffer occupancy $\overline{N_i}$ will be large. Thus, temporal privacy and buffer utilization are conflicting system objectives.

We now evaluate the impact of the depth of node i in the routing tree (the number of hops from the node i to the sink). For the sake of calculations, we shall assume that the density η of the sensor deployment is sufficient that a communicating sensor node will always find a path to the network sink. Additionally, let us denote the average geographical distance between parents and children in the routing tree by r. Then, to quantify the effect of flow convergence on the local traffic rate in the sensor network, let us assume that an outer annulus O_1 of distance d_1, angular spread φ and width r creates a total traffic of rate λ_{O_1} packets/second, as depicted in Figure 14.1 (c). Hence, in a spatial ensemble sense, each node carries an average traffic rate of $\overline{\lambda_{O_1}} = \lambda_{O_1}/(\varphi r d_1 \eta)$. This traffic flows toward the sink, and if we examine an annulus at distance $d_2 < d_1$ with width r and spread φ, the area of this annulus is $\varphi r d_2$, and there will be an average of $\varphi r d_2 \eta$ sensors in O_2 carrying a total rate of λ_{O_1}. Hence, on average, each sensor in this inner annulus will carry traffic of rate $\overline{\lambda_{O_2}} = \lambda_{O_1}/(\varphi r d_2 \eta)$. Comparing the average traffic load $\overline{\lambda_{O_2}}$ that a single sensor in an inner annulus O_2 carries with the average traffic load $\overline{\lambda_{O_1}}$ of a single sensor in annulus O_1, yields

$$\frac{\overline{\lambda_{O_2}}}{\overline{\lambda_{O_1}}} = \frac{d_1}{d_2}, \tag{14.8}$$

and hence traffic load increases in inverse relationship to the distance a node is from the sink.

The last issue that we need to consider is the amount of storage available for buffering at each sensor. As sensors are resource-constrained devices, it is more accurate to replace the $M/M/\infty$ queues with $M/M/k/k$ queues, where memory limitations imply that there are at most k servers/buffer slots, and each buffer slot is able to handle 1 message. If an arriving packet finds all k buffer slots full, then either the packet is dropped or, as we shall describe later in Section 14.5, a preemption strategy can be employed. For now, we just consider packet dropping. We note that packet dropping at a single node causes the outgoing process to lose its Poisson characteristics. However, we further note that by Kleinrock's Independence approximation (the merging of several packet streams has an affect akin to restoring the independence of interarrival times) [174], we may continue to approximate the incoming process at node i as a Poisson process with aggregate rate λ^i. Hence, in the same way as we used a tree of $M/M/\infty$ queues to model the network earlier, we can instead model the network as a tree of $M/M/k/k$ queues.

The $M/M/k/k$ formulation provides us with a means to adaptively design the buffering strategy at each node. If we suppose that the aggregate traffic levels arriving at a sensor node is λ, then the packet drop rate (the probability that a new packet finds all k buffer slots full) is given by the well-known Erlang Loss formula for $M/M/k/k$ queues:

$$E(\rho, k) = \frac{\rho^k}{k!} p_0 = \frac{\frac{\rho^k}{k!}}{\sum_{i=0}^{k} \frac{\rho^i}{i!}}, \tag{14.9}$$

where $\rho = \lambda/\mu$. For an incoming traffic rate λ, we may use the Erlang Loss formula to appropriately select μ so as to have a target packet drop rate α when using buffering to enhance privacy. This observation is powerful as it allows us adjust the buffer delay parameter μ at different locations in the sensor network, while maintaining a desired buffer performance. In particular, the expression for $E(\rho, k)$ implies that, as we approach the sink and the traffic rate λ increases, we must decrease the average delay time $1/\mu$ in order to maintain $E(\rho, k)$ at a target packet drop rate α.

14.5 RCAD: Rate-Controlled Adaptive Delaying

A consequence of the results of the previous section is that nodes close to the sink will have high buffer demands and their buffers may be full when new packets arrive. In practice, we need to adjust the delay distribution as a function of the incoming traffic rate and the available buffer space.

In order to accomplish this adjustment, we propose *RCAD*, a Rate-Controlled Adaptive Delaying mechanism, to achieve privacy and desirable buffer performance simultaneously. The main idea behind RCAD is buffer preemption– if the buffer is full, a node should select an appropriate buffered packet, called the *victim packet*, and transmit it immediately rather than drop packets. Consequently, preemption automatically adjusts the effective μ based on buffer state. In this chapter, we have proposed the following buffer preemption policies:

- *Longest Delayed First (LDF).* In this policy, the victim packet is the packet that has stayed in the buffer the longest. By doing so, we can ensure that each packet is buffered for at least a short duration. The implementation of this policy requires that each node record the arrival time of every packet.

- *Longest Remaining Delay First (LRDF).* In this policy, the victim packet is the packet that has the longest remaining delay time. Preempting such packets can lessen the buffer load more than any other policy because such packets would have resided in the buffer the longest. The implementation of this scheme is straightforward because each node already keeps track of the remaining buffer time for every packet.

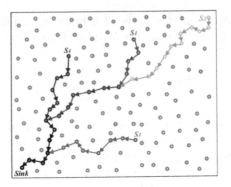

FIGURE 14.2. Simulation topology

- *Shortest Delay Time First (SDTF).* In this policy, the victim packet is the one with the shortest delay time. By lessening an already short delay time, we expect that the overall performance will remain roughly the same. The implementation of this policy requires that each node record the delay of every packet.

- *Shortest Remaining Delay First (SRDF).* In this policy, the victim packet is the packet that has the shortest remaining delay time. In this way, the resulting delay times for that node are the closest to the original distribution. As in the case of the LRDF policy, the implementation is straightforward.

14.6   Evaluating RCAD Using Simulations

In this study, we have developed a detailed event-driven simulator to study the performance of RCAD. We have set the simulation parameters such as buffer size and traffic pattern following measurements from an actual sensor platform, i.e. Berkeley motes, to model realistic network/traffic settings. Finally, we have measured important performance and privacy metrics.

14.6.1   Privacy and Performance Metrics

In our simulation studies, we have measured both temporal privacy and network performance of RCAD. As discussed in Section 14.2.1, we use the mean square error to quantify the error the adversary has in estimating the creation times of each packet. After introducing delays at intermediate nodes, we need to modify our adversary estimation model to accommodate this additional delay process. In addition to the knowledge the adversary has in Section 14.2.1, i.e. hop-count to the source and the average transmission delay at each node, we also assume the adversary knows the delay

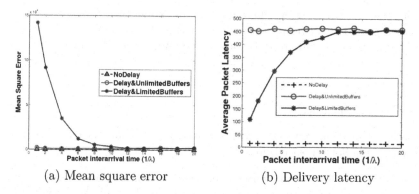

(a) Mean square error   (b) Delivery latency

FIGURE 14.3. Comparing no privacy with basic exponential delay in intermediate nodes, with and without unlimited buffers

process for each flow, i.e the delay distribution. Consequently, for an observed packet arrival time z, our adversary now estimates the creation time of this packet as $x' = z - y$, where y includes both the transmission delay and the additional delay. Here, our baseline adversary model uses the original delay distribution to calculate his estimations, neglecting the fact that some packets may have shorter delays than specified by the original delay distributions due to packet preemptions. For example, our adversary estimates the delay for flow i as h_i/μ, where h_i is the hop count of flow i and $1/\mu$ is the average per hop delay. For a sequence of packets coming from one source, we use the mean square error to measure the temporal privacy of the underlying network.

Additionally, we note that it is desirable to achieve privacy while maintaining tolerable end-to-end delivery latency for each packet. Hence, in our studies, for a network performance metric we use the average end-to-end delivery latency for packets coming from a particular flow versus the underlying traffic rate and the RCAD strategies employed.

14.6.2 Simulation Setup

The topology that we considered in our simulations is illustrated in Figure 14.2. Here, nodes S_1, S_2, S_3, and S_4 are source nodes and create packets that are destined for the sink. Thus, we had four flows, and these flows had hop counts 15, 22, 9 and 11 respectively. Each source generated a total of 1000 packets (at periodic intervals) with an interarrival time of $1/\lambda$ time units. In our experiments we varied $1/\lambda$ from 2 (i.e. the highest traffic rate) time units to 20 (the slowest traffic rate) to generate different cases of traffic loads for the network. The main focus of our simulator is the scale of the network, so we simplified the PHY- and MAC-level protocols by adopting a constant transmission delay (i.e. 1 time unit) from any node to its neighbors. When a packet arrives at an intermediate node, the intermediate node introduces a random delay following an exponential distribution with mean $1/\mu$. Unless mentioned otherwise we took $1/\mu = 30$ time units in the simulations. The results reported are for the flow S_1 to the sink.

14.6.3 Simulation Results

Effectiveness of RCAD

To study the effectiveness of RCAD strategies, we compare temporal privacy in the following situations:

- Nodes in the network forward packets as soon as they receive them. This scenario is the baseline case with no effort made to explicitly provide any temporal privacy.

- Each node in the path of a network packet introduces an exponential delay with mean $1/\mu = 30$, before forwarding the packet. This scenario adds uncertainty to the adversary's inference of the time of origin of a packet. In this case, we assume that the nodes have unlimited buffers.

- Same as previous case except each node now has limited buffers. Specifically, we assume each node can buffer 10 packets, which approximates the buffers available on the mica-2 motes. This models the real-world scenario with resource constrained sensor nodes. Here we studied the effect of the simplest RCAD preemption strategy, Shortest Remaining Delay First (SRDF) (see section 14.5), which preempts (and forwards ahead of schedule) the packet with the shortest remaining time before its scheduled departure.

We used the topology in figure 14.2 with 4 different traffic flows. Figure 14.3(a) shows the mean square error in the adversary estimate for the 3 situations above, with regards to flow from S1. We can see that the error is very small in both cases 1 and 2 (it may appear to be zero, but that's only because of the relative scale as compared to case 3). In case 3 however, the adversary, attempts to use his knowledge of the intermediate delay distributions (specifically the knowledge of $1/\mu$) to estimate the time of origin of packets. But the preemptions at higher traffic rates (small inter-arrival times), mean that the effective latencies of the packets are much lower than the expected latencies and this results in very high error in the adversary's estimate. Figure 14.3(b) shows the average latency for packets to reach from the source to the sink. As expected, case 1 shows the lowest latency, because no artificial delays have been introduced. Note that case 2 shows the highest latency which is the average of the combined delay distribution of all the nodes in the path of flow from S1. Further, in case 3, we find that the preemptions due to limited buffers actually help reduce the average delivery latency, especially in case of high source traffic rates (smaller inter-arrival

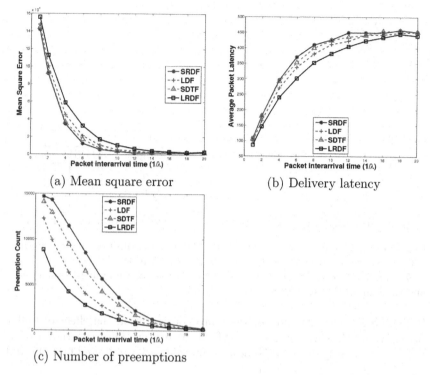

(a) Mean square error   (b) Delivery latency

(c) Number of preemptions

FIGURE 14.4. Comparison of the four RCAD strategies and the scenario with unlimited buffers. We note the x-axis is in terms of average source interarrival time $1/\lambda$.

times). For example at $1/\lambda = 2$, case 3 reduces the end-to-end latency by a factor of 2.5. These results clearly demonstrate the efficacy of delaying algorithms in providing temporal privacy and the added effectiveness of the simplest of our preemption strategies to do so with a controlled overhead in terms of end-to-end packet latency.

Comparison of RCAD Strategies

After demonstrating the effectiveness of RCAD, we now study the four RCAD strategies in depth and compare their performance relative to each other. Figures 14.4(a) and (b) present the mean square error and the delivery latency of the four RCAD strategies where we assume each sensor node has a buffer of 10 slots (which is typical for a Mica2 mote). In this set of experiments, we used the baseline adversary model that estimated the delay for flow i as $30h_i$ because the average delay on each node is 30 time units. At low traffic rates ($1/\lambda = 16, 18, 20$), these four strategies perform the same because the average buffer requirement per node is less than 10. As the traffic increases, the four preemption strategies lead to much higher mean square error, thus providing better temporal privacy. This is because

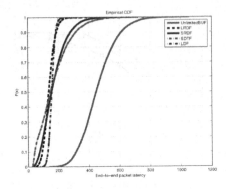

Y_i	$Y_{unbuf} - Y_i$
Y_{lrdf}	317.61
Y_{srdf}	287.84
Y_{sdtf}	289.95
Y_{ldf}	299.44

(a) CDFs of latency distributions   (b) Deviation of average latency

FIGURE 14.5. Comparison of the effective latency distributions caused by pre-emption strategies with the expected latency distributions. Y_{unbuf}, Y_{lrdf}, Y_{srdf}, Y_{sdtf} and Y_{ldf} are the average latencies for the baseline unlimited buffers case and the four RCAD strategies respectively.

the baseline adversary did not take into consideration the effect of buffer preemptions. Among the four RCAD strategies, we observed that LRDF policy consistently delivers the best temporal privacy, followed by LDF and SDTF, while SRDF is the worst.

In order to understand the difference between these four strategies, let us look at more detailed statistics. Figure 14.4(c) presents the number of preemptions that occurred during the experiments. We observe the opposite order here: the strategy that provides the most privacy incurred the least number of preemptions. This may appear counter-intuitive at first glance, but can be simply explained: the strategy that leads to more preemptions tends to alter the original delay distribution less, and thus confuses the adversary less. For example, LRDF selects the packet that has the longest remaining delay time as the victim packet. Preempting these packets will have two effects:

- it will alter the original delay distribution more, and

- it will reduce the number of preemptions.

Moving our attention to delivery latency in figure 14.4(b), among the four preemption strategies, LRDF has the shortest latency because it tends to reduce the delay times in the buffer the most.

Figure 14.5(a) shows the distribution of effective end-to-end packet latencies with different RCAD strategies at high traffic rates ($1/\lambda = 2$). We know that the expected distribution of latencies is the one with unlimited buffers case and all other distributions are the results of various preemption strategies. We observe from figure 14.5(b) that all the RCAD strategies ef-

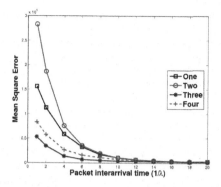

FIGURE 14.6. Performance of LRDF strategy for different source-sink separations

fectively reduce the packet latencies with LRDF leading the pack with the lowest average latency. As noted above, this fact coupled with the result that LRDF having the least overhead in terms of the preemptions, make it the ideal RCAD strategy to use.

Reducing preemption overhead using an improved model for delay distributions

Buffer preemption is necessary to avoid dropping packets due to buffer saturation, and we have just seen that it can help provide better temporal privacy. Buffer preemption, however, also has disadvantages, especially as it introduces additional protocol overhead at each sensor node associated with the selection of victim packets. Hence, in order to reduce protocol overhead we must reduce the frequency of preemption, but at the same time strive to maintain the same level of temporal privacy.

One strategy for reducing preemption is to let each node employ a different delay distribution. Since sensor networks usually have many more sources than sinks, as illustrated in Figure 14.1 (c), nodes closer to the sink experience higher traffic volumes than nodes closer to the source. As a result, the nodes closer to the sink should delay packets for a much shorter interval in order to relieve the buffer requirements at these nodes. Our objective with this approach is to keep the buffer usage the same across all the nodes, irrespective of the number of packets they serve. As discussed in Section 14.4, the number of buffered packets at node i can be estimated as $\overline{N_i} = \rho_i = \frac{\lambda_i}{\mu_i}$.

Suppose we consider a flow with h hops (i.e. h nodes before the sink), and use node 1 to denote the last node before the sink and node h to denote the source node. To keep $\overline{N_i}$ constant across all nodes while having a target overall average delay of D, we choose the average delay time $1/\mu_i$ for node i as β/h_i, where β is the coefficient and h_i is the hop count between node i and the sink. Thus, we have $\sum_{i=1}^{h} \frac{\beta}{h_i} = \sum_{i=0}^{h-1} \frac{\beta}{i+1} = \beta(\gamma + \psi(h+1))$,

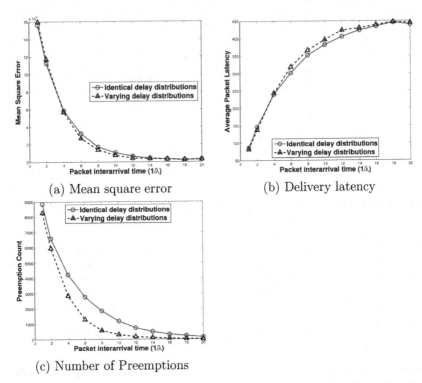

(a) Mean square error   (b) Delivery latency

(c) Number of Preemptions

FIGURE 14.7. Comparison of the performance of the LRDF policy when all nodes have identical delay distributions and when nodes have varying delay distributions.

where γ is the Euler-Mascheroni constant and $\psi(x)$ is the digamma function. Hence, the average delay time $1/\mu_i$ for node i is calculated as $1/\mu_i = \frac{D}{(i+1)(\gamma+\psi(h+1))}$.

We conducted a set of experiments to study the performance of RCAD strategies when using varying delay distributions chosen as above. The results with LRDF are presented in Figure 14.7. We observe that employing variable delays can **significantly reduce the number of preemptions**, especially for mid-range traffic rates. At the same time, having variable delays will not degrade either the mean square error or latency much. Although the preemptions were reduced by an amount up to 70%, the largest estimation error reduction we observed was 16%, while the largest latency increase was only 4%.

14.6.4   The Adaptive Adversary Model

Since RCAD schemes dynamically adapt the delay processes by adopting buffer preemption strategies, it is inadequate for the adversary to estimate the actual delay times using the original delay distributions before preemp-

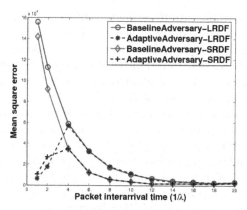

FIGURE 14.8. The estimation error for the two adversary models, when LRDF and SRDF RCAD are employed.

tion. Hence, we also enhance the baseline adversary to let the adversary adapt his estimation of the delays depending on the observed rate of incoming traffic at the sink. We call such an adversary as an *adaptive* adversary.

In order to understand our adaptive adversary model, let us first look at a simple example. Let us assume there is only one node with one buffer slot between the source and sink. Further, assume that the packet arrival follows a Poisson process with rate λ, and the buffer generates a random delay time that follows an exponential distribution with mean $1/\mu$. If the buffer at the intermediate node is full when a new packet arrives, the currently buffered packet will be transmitted. In this example, if the traffic rate is low, say $\lambda < \mu$, then the packet delay time will be $1/\mu$. However, as the traffic increases, the average delay time will become $1/\lambda$ due to buffer preemptions. Following this example, our adaptive adversary should adopt a similar estimation strategy: at low traffic rates, he estimates the overall average delay y by h/μ, while at higher traffic rates, he estimates the overall average delay y as a function of the buffer space and the incoming rate, i.e. hk/λ, where h is the flow hop count, k is the number of buffer slots at each node, and λ is the traffic rate of that flow. Given an aggregated traffic rate λ_{tot} from n sources converging at least one-hop prior to the sink, the adversary can compute the probability of buffer overflow via the Erlang Loss formula in equation (14.9). He then can compare this against a chosen threshold and if the probability is less than the threshold, he will assume the average delay introduced by each hop is $1/\mu$. However, if the probability is higher than the threshold, the average delay at each node is calculated to be nk/λ_{tot}.

We studied the ability of an adaptive adversary to estimate the time of creation when using RCAD with identical delay distributions across the network. The resulting estimation mean square errors are presented in Figure 14.8. The adaptive adversary adopts the same estimation strategy as the baseline adversary at lower traffic rates, i.e. h_i/μ for flow i, but it uses the incoming traffic rate to estimate the delay at higher traffic rates, i.e. $h_i k/\lambda_i$ for the average delay of flow i. The switch between estimation strategies used the Erlang Loss formula for a threshold preemption rate of 0.1. Figure 14.8 shows that the adaptive adversary can significantly reduce the estimation errors, especially at higher traffic rates (lower interarrhival times) where preemption is more likely.

An interesting observation is that at high traffic rates, the adaptive adversary can more accurately estimate the delay times generated by the LRDF policy when compared to other RCAD policies. Recall that earlier, LRDF had the highest estimation error against a baseline adversary while the SRDF policy had the least error (and hence the least privacy). Now, at high traffic rates against an adaptive adversary, the trend is reversed-- SRDF has the best privacy while LRDF has the worst.

14.7 Conclusion

Preventing an adversary from learning the time at which a sensor reading was measured cannot be accomplished by merely using cryptographic security mechanisms. In this chapter, we have proposed a technique complimentary to conventional security techniques that involves the introduction of additional delay in the store-and-forward buffers within the sensor network. We formulated the objective of temporal privacy using an information-theoretic framework, and then examined the effect that additional delay has on buffer occupancy within the sensor network. Temporal privacy and buffer utilization were shown to be objectives that conflict, and to effectively manage the tradeoffs between these design objectives, we proposed an adaptive buffering algorithm, RCAD (Rate-Controlled Adaptive Delaying) that preemptively releases packets under buffer saturation. We then evaluated RCAD using an event-driven simulation study for a large-scale sensor network. We observed that, when compared with a baseline network consisting of unlimited buffers, RCAD was able to provide enhanced temporal privacy (the adversary had higher error in estimating packet creation times), while reducing the end-to-end delivery latency. Further, by adopting variable delays among the nodes along a routing path, we can reduce the number of buffer preemptions in RCAD without noticeably affecting privacy and network performance. Finally, we also devised an improved adversary model that can better estimate the delays produced by RCAD when compared to a naive adversary. In spite of the improved adversary model, RCAD is still able to protect the temporal privacy of sensor flows.

15

Securing Wireless Systems via Lower Layer Enforcements

15.1 Introduction

In this chapter we present the viewpoint that there are new modalities for securing wireless systems that can turn the nature of the wireless medium from a disadvantage into an advantage. In essence, rather than rely solely upon generic, higher-layer cryptographic mechanisms, as has been the norm, we will show that it is possible to achieve a lower-layer approach that supports important security objectives, such as authentication and confidentiality. The enabling factor in our approach is that, in the rich multipath environment typical of wireless scenarios, the response of the medium along any transmit-receive path is *frequency-selective* (or in the time domain, *dispersive*) in a way that is *location-specific*. In particular, channel characterizations (e.g. a set of complex gains at different frequencies, or the impulse response at different time delays) decorrelate from one transmit-receive path to another if the paths are separated by the order of an RF wavelength or more.

These unique space, time, and frequency characteristics of the wireless physical layer can be used to augment traditional higher-layer authentication and confidentiality methods. Two wireless entities can identify or authenticate each other's transmitter by tracking each other's ability to produce an appropriate received signal at the recipient. Similarly, the fact that pairwise radio propagation laws between two entities are unique and decorrelate quickly with distance can serve as the basis for establishing shared secrets. These shared secrets may be used as encryption keys for higher-layer applications or wireless system services that need confidential-

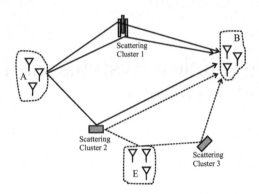

FIGURE 15.1. The adversarial multipath environment involving multiple scattering surfaces. The transmission from Alice (A) to Bob (B) experiences different multipath effects than the transmission by the adversary, Eve (E).

ity. In short, these two strategies suggest that merely using cryptographic methods does not capture the full spectrum of possible solutions that are available to the wireless engineer.

We begin this chapter in Section 15.2 by providing an overview of our proposed PHY-layer security services. We then turn to focus on PHY-layer authentication/identification services in Section 15.3, and then examine PHY-layer protocols that support confidentiality services in Section 15.4. After describing the protocols and theoretical underpinnings, we proceed to describe our initial validation efforts in Section 15.5, where we have conducted experiments using the USRP/GNURadio platform. We wrap up the chapter by providing conclusion remarks in Section 15.6.

15.2 Alice, Bob and Eve Get Physical

We now set the stage for the discussion in the remainder of this chapter by describing the basic communication scenario we are considering. Further, in order to make the chapter more accessible, we provide a brief survey of the relevant principles of radio propagation. A more detailed and precise exposition on propagation can be found in [82].

Both the authentication and confidentiality services we describe make use of the complexity associated with multipath propagation. Throughout our discussion, we shall employ the popular security convention by introducing three different parties: Alice, Bob and Eve. For our purposes, these three entities may be be thought of as wireless transmitters/receivers that are potentially located in spatially diverse positions, as depicted in Figure 15.1. Our two "legal" protagonists are the usual Alice and Bob, and for the sake of discussion throughout this chapter, Alice will serve as the transmitter that initiates communication, while Bob will serve as the intended receiver.

Their nefarious adversary, Eve, may either be a passive eavesdropper or an active adversary that injects undesirable communications into the medium. We note that, in this communication scenario, all entities are considered to be within radio range of one another, and hence the techniques presented in this chapter are meant to serve as local "one-hop" security mechanisms.

Across the wireless channel, RF signals transmitted from Alice to Bob are affected by a variety of factors, ranging from attenuation due to conservation of energy as the wavefront expands, to large and small-scale fading. Fading arises from a transmitted signal traversing different paths (multipaths) that combine constructively or destructively at the receiver. The effect of multipath for a specific transmitter-to-receiver scenario can be represented as a system where the input $u(t)$ is the transmitted signal and the received signal is

$$r(t) = \int_{-\infty}^{\infty} h(t,\tau)u(t-\tau)d\tau.$$

In this system, the *channel response* is the time-varying function $h(t,\tau)$. A direct formulation of $h(t,\tau)$ from underlying physics is generally unwieldy, and the practice is to apply reasonable stochastic assumptions to simplify the model. The wide-sense stationary with uncorrelated scatterers (WSSUS) assumption implies that the channel response can be modeled (in the time-domain sense) as a tapped-delay line

$$h(t,\tau) = \sum_{i=1}^{N} h_i(t)\delta(t-\tau_i),$$

and with the additional assumption of a Rayleigh fading model, the $h_i(t)$ become a zero-mean complex Gaussian stochastic process. Thus, the channel response can be interpreted as the sum of N delayed, attenuated and phase-shifted versions of the original signal. Since $h(t,\tau)$ is itself stochastic, there is a temporal and spectral variability of the channel response, i.e. the multipath profile will change with time and affect different frequency components of $u(t)$ differently. The fading effects experienced at two different times or at two different frequencies is closely related to the separation between these times/frequencies. The level of temporal and spectral variability is reflected by two notions, the *coherence time* and *coherence bandwidth* of the channel. Coherence time is a measure of the difference in time that is needed in order for the fading correlation to drop below a threshold amount, and coherence bandwidth is similarly defined. Finally, we note that, at a specific instance and frequency, we may examine the fading correlation between a source and two different receiver locations. In this case, the common rule of thumb (c.f. the well-known Jakes uniform scattering model [175]), is that the received signal rapidly decorrelates over a distance of roughly half a wavelength, and that spatial separation of one to two wavelengths is sufficient for assuming independent fading paths.

Turning back to physical layer security, our security objective is to provide authentication and confidentiality to Alice and Bob, in spite of the presence of Eve. Authentication is associated with the assurance that a communication comes from a specific entity, while confidentiality is concerned with the assurance that communication between entities is illegible to eavesdroppers [170]. In our communication scenario, these two objectives may be interpreted as follows. Since the adversary, Eve, is within range of Alice and Bob, and can inject her own signals into the environment (perhaps for the purpose of impersonating Alice), it is desirable for Bob to have the ability to differentiate between legitimate signals from Alice and illegitimate signals from Eve. He therefore needs some form of evidence that the signal he receives did, in fact, come from Alice. On the other hand, for the purpose of confidentiality we wish to ensure that the adversary Eve cannot decipher the communication between Alice and Bob.

The focus of this chapter is to further develop these two security objectives at the PHY-layer. Towards this end, we discuss the following:

- *Channel-based Authentication:* Rather than employ a shared "cryptographic authentication key" between Alice and Bob, we instead exploit the uniqueness of the Alice-Bob channel relative to the Eve-Bob channel. We will outline techniques to distinguish between legitimate transmissions from Alice and anomalous traffic from an adversary Eve.

- *Secret Key Establishment via Multipath Channels:* Confidentiality is traditionally achieved through encryption using a shared key between Alice and Bob that is unknown to Eve. In multipath environments, the unique characteristics of the channel between Alice and Bob can provide parameters that create a unique private key between them– a key that cannot be created from any other location.

These topics are related– each is based based upon the ability of the multipath environment to provide a waveform whose structure an adversary cannot measure or model accurately. Our assumption throughout this chapter is that the radio environment is both quasi-static and richly scattered. These conditions are highly favorable to the effectiveness of the techniques we propose, and correspond to a wide range of practical scenarios: The Rayleigh scattering nature of cellular channels, for example, is well-known [82, 175–177] and the slow temporal variations of channel responses on fixed outdoor links have been reported by many researchers [178–180].

15.3   PHY-Enhanced Authentication

Authentication and identification services deal with verifying the identity of an entity involved in a transaction. Such a notion of authentication can be addressed through traditional techniques. Wireless authentication, however, can be expanded to include new functionalities, such as recognizing a particular device based upon its unique channel characteristics. It is the authentication of the actual *transmitter* that we now discuss.

15.3.1   Channel-based Authentication

We seek to exploit the uniqueness of the Alice-Bob channel as an authenticator to distinguish between a legitimate transmitter and an illegitimate transmitter. The ability to distinguish between different transmitters would be particularly valuable for preventing spoofing attacks, in which one wireless device claims to be another wireless device. Currently, spoofing attacks are very easy to launch in many wireless networks. For example, in commodity networks, such as 802.11 networks, it is easy for a device to alter its MAC address by simply issuing an `ifconfig` command. This weakness is a serious threat, and there are numerous attacks, ranging from session hijacking [88] to attacks on access control lists [89], that are facilitated by the fact that an adversarial device may masquerade as another device.

To illustrate how the property of rapid spatial decorrelation can be used to authenticate a transmitter, let us return to Figure 15.1 and consider a simple transmitter identification protocol in which Bob seeks to verify that Alice is the transmitter. Suppose that Alice probes the channel sufficiently frequently to assure temporal coherence between channel estimates and that, prior to Eve's arrival, Bob has estimated the Alice-Bob channel [181]. Now, Eve wishes to convince Bob that she is Alice. Bob will require that each information-carrying transmission be accompanied by an authenticator signal. The channel and its effect on a transmitted signal between Alice and Bob is a result of the multipath environment. Bob may use the received version of the authenticator signal to estimate the channel response and compare this with a previous record for the Alice-Bob channel. If the two channel estimates are close to each other, then Bob will conclude that the source of the message is the same as the source of the previously sent message. If the channel estimates are not similar, then Bob should conclude that the source is likely not Alice. Here we have achieved unilateral authentication. Mutual authentication can be achieved by having Bob subsequently send Alice an authenticator signal.

Realizing channel-based authentication in a time-varying radio environment involves two aspects. One is the authenticator signaling technique for a fixed instantiation of the channel, and the other is the necessary measures for ensuring the continuity of such an authentication procedure when the channel changes in subsequent epochs. We first discuss approaches for authenticator signaling and then techniques for maintenance of such authentication.

We now describe two strategies for authenticator signaling, but note that other forms of channel sounding, such as used for multiple-input multiple-output (MIMO) channels, are also appropriate.

Temporal (Pulse-type) Probing: Ideally, Bob's received signal $r(t)$ will be the convolution of Alice's signal $u(t)$ with the channel response plus the addition of receiver-side noise. In order for Bob to measure the channel response, Alice will send a probing pulse $u(t)$. The pulse bandwidth is critical to the ability to resolve the multipath environment. If $u(t)$ has sufficiently wide bandwidth W, i.e. $1/W$ is small compared to the temporal width of the impulse response, then the multipath profile can be resolved [82,176]. Consequently, wideband channel probing strategies, such as direct RF pulsing [176,182] or spread spectrum methods [183,184], can be used to construct channel estimates for authentication.

Suppose that the channel impulse response between Alice and Bob is time-invariant over the time period of interest τ, i.e., $h_{AB}(t,\tau) = h_{AB}(\tau)$. Once Bob has an estimate of $h_{AB}(\tau)$, there are two approaches to performing authentication at a later time. The first method involves the claimant (the entity wishing to be authenticated) transmitting another probe, allowing Bob to build a candidate response $\tilde{h}(\tau)$. Bob would compare $\tilde{h}(\tau)$ with $h_{AB}(\tau)$ and decide the claimant is Alice if they are sufficiently similar. The second method involves the claimant sending a known authenticator signal $g(t)$, which would ideally lead to Bob receiving $r_g(t) = (h_{AB} \star g)(t) + n(t)$, where $\star$ denotes convolution. However, if the transmitter is not Alice, then what is observed is $\tilde{r}(t) = (h_E \star g)(t) + n(t)$. To authenticate, Bob compares $\tilde{r}(t)$ with $(h_{AB} \star g)(t)$. The main difference between these two variations is whether the discrimination is performed directly on the received signal, or on parameters derived from the received signal.

Multiple Tone Probing: In this approach, the authenticator signal consists of multiple, simultaneous carrier waves. To ensure independent fading across these different carriers, we require that the carrier frequencies f_i are separated by an amount greater than the channel coherence bandwidth [82,176]. Let us suppose that Alice has initially sent Bob N carrier waves. Disregarding noise, on the ith carrier, Bob has received

$$\Re\left\{e^{j(2\pi f_i t - \phi_0)}\left(\sum_k \alpha_k e^{-j\phi_k}\right)\right\} = \Re\left\{\tilde{H}_i e^{j(2\pi f_i t - \phi_0)}\right\},$$

where ϕ_0 is the phase of Alice's carrier wave at transmission, the summation is over the amount of multipaths for that carrier, and the complex factor $\tilde{H}_i$ is the gain of the Alice-Bob multipath channel at the ith carrier. Bob measures all $H_i = \tilde{H}_i + z_i$, where z_i is noise and interference that is modeled as a complex Gaussian $\mathcal{N}(0, \sigma^2)$. At a later time, the claimant will send Bob N carrier waves with the same carrier frequencies, and Bob will measure the corresponding set of complex gains $\{G_i\}$. The verification process involves testing $\{G_i\}$ against $\{H_i\}$. Under

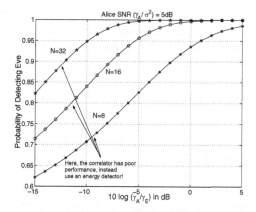

FIGURE 15.2. Probability of detecting Eve as a function of different power ratios γ_A/γ_E for varying number of carriers, N, used in a multiple tone authentication scheme. The probability of false alarm is 0.01.

the null hypothesis $\mathcal{H}_0$, the claimant is Alice, and $G_i = \tilde{H}_i + n_i$, for measurement noise $n_i \sim \mathcal{N}(0, \sigma^2)$, while under $\mathcal{H}_1$ the claimant is Eve and $G_i = \tilde{G}_i + n_i$. Here, over i, $\tilde{H}_i$ has average power γ_A, while $\tilde{G}_i$ has average power γ_E. We may choose, for example, a normalized correlation statistic, $T = (\sum_i H_i G_i^*)/(N \gamma_A)$ for discrimination. If we assume that we have a uniform scattering environment [175], and that Eve is several wavelengths away from Alice, then we can assume independence between $\tilde{H}_i$ and $\tilde{G}_i$. In this case $E[T|\mathcal{H}_0] = 1$, $Var(T|\mathcal{H}_0) = (2\gamma_A\sigma^2 + \sigma^4)/(N\gamma_A^2)$, while $E[T|\mathcal{H}_1] = 0$ and $Var(T|\mathcal{H}_1) = (\sigma^2 + \gamma_A)(\sigma^2 + \gamma_E)/(N\gamma_A^2)$. The variance of the test statistic decreases as we increase the number of carriers N. In Figure 15.2, we present the probability of detecting Eve versus the (adversarial) power ratio γ_A/γ_E for a 1% false alarm rate with the number of carriers N as a parameter. If we make assumptions on Eve's largest likely channel gain power γ_E, then these results serve as guidelines for choosing the number of carriers needed for reliable discrimination and authentication. Similarly, if we have limits on N, such as might arise from regulatory or hardware constraints, then we may use these results to assert Eve's ability to successfully forge a single authentication challenge, thereby quantifying the additional security gain provided by physical layer authentication. It should be noted that, when Eve has a large power γ_E, the correlator alone performs poorly. However, Eve can then be detected through energy detection techniques.

15.3.2 Maintenance of the channel authenticator

In PHY-layer authentication, we can assume that Bob uses traditional higher-layer authentication procedures to associate the initial link between Alice and Bob with Alice's identity. The PHY-layer authentication described compares a new measurement with a prior channel estimate, thereby verifying whether the new measurement likely came from the source of the prior measurement. The newly verified channel estimate then replaces the prior measurement, allowing for the verification of the next channel estimate. Using $y \leadsto x$ to denote "y is verified from x", we thus have the verification chain $h(t_M, \tau) \leadsto h(t_{M-1}, \tau) \leadsto \cdots \leadsto h(t_0, \tau)$, where $t_M > t_{M-1} > \cdots > t_0$ and t_0 is the time of the initial channel estimate, and thus we only authenticate the transmitter of the initial communication. For this reason, it is necessary to employ traditional higher layer methods to do the initial link association.

However, even in the absence of an initial "cryptographically" verifiable association between $h(t_0, \tau)$ and Alice's identity, we may still use the verification chain to detect whether there has been a change in the transmitter. For example, by maintaining the transmit-receive channel history between a transmitter and receiver, we can detect scenarios where Eve tries to act as Alice. Such an approach could find application in detecting device spoofing.

In either case, the utility of the PHY-layer verification chain is related to the time-varying nature of the channel. In particular, in implementing these techniques, it is necessary to probe the channel at time intervals less than the channel's coherence time in order to support valid comparisons. We note that, for unilateral identification, this process only needs to be one-directional, that is Alice transmits to Bob and Bob maintains the verification chain. On the other hand, for full mutual identification (as opposed to two separate unilateral identifications), the exchanges must be both bidirectional and it is necessary that they have the same channel with which to base their verification chains upon. This implies that channel reciprocity should apply[1], and hence Alice and Bob must use the same carriers in a time-division-duplexing (TDD) manner. In this case, Bob transmits a channel-probing set of tones over some interval $(t_0, t_0 + T]$. Alice transmits the same tones over $(t_0 + T, t_0 + 2T]$. Data is transmitted in one or both directions before the process repeats. The temporal width of the exchange, $2T$ plus the data transmission interval, must be small compared to the correlation time of the channel response. This condition can be met using realistic channel probing/data transmission times (e.g., see [178, 179]).

Finally, we note that a single verification failure $h(t_M, \tau) \not\leadsto h(t_{M-1}, \tau)$ does not definitively indicate the presence of an adversary, but rather should serve as a warning flag. It may be that $h(t_M, \tau) \not\leadsto h(t_{M-1}, \tau)$ is merely a false alarm, and the warning flag should trigger more careful analysis by the software. For example, as Alice and Eve are both communicating, there will be a repetition of failures, and Bob may record a history of channel estimates in order to enhance the detection of Eve's intrusion. Thus, just

---

[1]We note the difference between channel reciprocity and the notion of asymmetric links, e.g. [185]. Channel reciprocity is the equivalence of the channel transfer function, and not a statement about noise conditions relative to each entity.

as in any intrusion detection system, it is not a single event that should trigger a response, but rather it is the persistence of anomalous events that should serve to indicate verification failure and set off warning messages.

15.4 PHY-Enhanced Confidentiality

Confidentiality services, like encryption and key management, are the work horses for many security protocols. A fundamental belief held by the security community is that, when designing confidentiality services, one should not replace traditional ciphers, such as AES, with new ciphers as existing ciphers are very thoroughly cryptanalyzed and designed for bulk-data processing. Hence, our approach to achieving confidentiality focuses on the issue of establishing keys between wireless entities. In one sense, the methods we describe are analogous to Diffie-Hellman key establishment, and can be considered as building blocks rather than complete security solutions.

Referring back to Figure 15.1, our communication scenario for our confidentiality methods involves Alice transmitting an encoded message to Bob, the intended recipient. Bob receives a signal that is a result of the Alice-Bob channel, while Eve receives a signal that follows from the Alice-Eve channel. Alice's objective is to maximize the rate at which she communicates with Bob (i.e. the key establishment rate), while simultaneously minimizing the information that Eve learns. There are two different extremes to using the wireless channel to establish keys: *extraction* and *dissemination*. In extraction, Alice's signal may be a probing signal that Bob uses to estimate channel state information h_{AB}, from which keys are extracted. In dissemination, however, Alice transmits a signal that is an appropriately coded version of the information Alice wishes to give to Bob. We will present several constructions that represent a variety of methods ranging between these two extremes.

In all of the methods we describe, we assume as a starting point that Alice and Bob each have estimates of their shared channel, e.g. by probing in a TDD fashion. We will denote h_{AB} to be Bob's estimate of the Alice-Bob channel, and h_{BA} to be Alice's estimate of the Bob-Alice channel. Similarly, we will denote h_{AE} to be Eve's estimate of the Alice-Eve channel. The channel estimates may correspond to scalar or vector channel estimates.

15.4.1 Key Extraction from Channel Estimates

Once channel state information has been estimated, the process of key extraction is rather straight-forward. One simple approach for extracting shared keys would employ cryptographic one-way functions [186]. For example, once Alice and Bob have converted h_{BA} and h_{AB} to a binary representation (requiring quantization of the channel state information), Alice can calculate $K_A = f(h_{BA})$, while Bob can calculate $K_B = f(h_{AB})$, where

f is a one-way function. If $h_{BA} = h_{AB}$, then they will have arrived at the same result. To prevent scenarios where a key from a previous time period was captured yet the channel state had not changed enough to make the subsequent key extraction yield a different result, we can employ nonces and achieve added security. In this case, Alice will send a random bit sequence r, and Alice then calculates $K_A = f(h_{BA}\|r)$ and Bob calculates $K_B = f(h_{AB}\|r)$, where $\|$ is concatenation.

Ideally, $K_A = K_B$, and hence they have a shared key. A challenge may arise because Alice and Bob could have slightly different channel estimates. Since we use a pseudo-random one-way function, if $h_{AB} \neq h_{BA}$ then K_A and K_B will be dramatically different, i.e. a single bit difference will produce dramatically different outputs. In order to fix this, one strategy that can be used is a subsequent higher-layer challenge-response verification protocol: Alice sends Bob $E_{K_A}(r_1)$ and Bob responds with $E_{K_B}(r_1 + 1)$. If Alice decrypts and verifies $r_1 + 1$, then she accepts that Bob has obtained the correct key. Similarly, Bob can do likewise.

15.4.2   Key Dissemination via Channel State Masking

The previous scheme extracted keys from common information shared between Alice and Bob, much like is done in Diffie-Hellman key agreement, and neither Alice nor Bob have control over what the key will be. At the other extreme, are key dissemination techniques where it is possible for one entity to choose the key and distribute it to the other party. We now look at a simple approach to key dissemination that uses the channel state to mask the key being distributed.

In channel state masking, Alice and Bob convert h_{AB} and h_{BA} into binary representations, and then using them as the key sequence in a one-time pad to mask the key being distributed. For example, suppose Alice creates a vector of bits x that she wishes to use as a shared key with Bob. She forms a message $y = x \oplus h_{BA}$, where $\oplus$ is bitwise XOR. Now, if Alice sends this to Bob, say over a public channel, then ignoring errors in the transmission (through the use of an error correcting code on y), Bob calculates $z = y \oplus h_{AB}$, and can recover x if $h_{AB} = h_{BA}$. For the adversary, Eve, her channel estimate h_{AE} will be quite different from h_{AB}, and thus her attempted decoding will be corrupted with errors.

In practice, Alice's and Bob's channel estimates are merely correlated and $h_{AB} \approx h_{BA}$, causing $z \neq x$. To cope with this, an error correcting code must be applied to x, producing a codeword $\tilde{x}$, and now $y = \tilde{x} \oplus h_{BA}$. Bob will calculate $\tilde{z} = \tilde{x} \oplus h_{AB} \oplus h_{BA} = \tilde{x} \oplus e_{AB}$, where e_{AB} captures the mismatch between h_{AB} and h_{BA}. Now, Bob will decode $\tilde{z}$ as x as long as the Hamming weight of e_{AB} is sufficiently small. On the other hand, the sequence $h_{BA} \oplus h_{AE}$ should have a larger Hamming weight than e_{AB}. If the Hamming weight of $h_{BA} \oplus h_{AE}$ is beyond the error correction capability, then Eve will fail to recover the key.

Several interesting issues arise regarding the choice of the error correcting code. In particular, it is desirable to choose an error correcting code that guarantees the formation of the Alice-Bob key while minimizing the likelihood of Eve being able to form the key. On the other hand, error correcting codes work by mapping observed data to a "best guess" codevector. Even should an error correcting code fail to decode, it can still provide some information about the original message (such as parity checks). This also raises an interesting question in the formal context of cryptography about the semantic security of such a scheme. A semantically secure protocol is one in which the communications do not reveal *any* advantageous information to the adversary [187]. In particular, there may be some correlation between h_{AB} and h_{AE}, especially if Eve is physically close to Bob, and thus it might be possible for Eve to estimate certain bits (such as sign bits) of h_{AB} and this, combined with partial decoding, can allow her to infer some bit values of x, or to narrow down the possible decodings of $y \oplus h_{BA} \oplus h_{AE}$ to codewords close to $\tilde{x}$.

15.4.3 Key Dissemination via Probabilistic Encoding

The next approach that we describe takes a different approach to disseminating keys, and achieves semantic security by using techniques from the theory of probabilistic encryption, and hence inherits the advantages of distinguishability and robustness to the leakage of partial information [188] associated with probabilistic encryption. Further, we note that the scheme we describe is motivated by constructions for Wyner's classical wiretap channel [189], and insight gained from Csiszar and Korner [190] extensions of Wyner's work.

Our approach makes use of trapdoor functions and hard core predicates. Loosely speaking, a trapdoor function $f(x)$ is one for which calculating $c = f(x)$ is easy, but determining x from observing c is difficult without additional, *private* knowledge. A hard core predicate for the trapdoor function $f(x)$ is a Boolean function $G(x)$ such that calculating $G(x)$ is easy, but calculating $G(x)$ from just $f(x)$ is computationally hard. The basic probabilistic encryption of a single plaintext bit $m \in \{0, 1\}$ via hard core predicates is presented in Figure 15.3 (a). Here, Alice chooses a *random* $x \in \{0, 1\}^N$ such that $G(x) = m$. Now, Alice sends Bob the ciphertext $c = f(x)$. Since Bob possesses the trapdoor, he may calculate $f^{-1}(c) = f^{-1}(f(x)) = x$, and from x he calculates the plaintext via $G(x) = m$.

We seek a similar construction that uses wireless-specific components. Our basic strategy is portrayed in Figure 15.3 (b). Here, suppose Alice and Bob have estimated their channel conditions, and have produced a channel with error rate p_{AB}, and that the Alice-Eve channel has an error rate p_E, with $p_E > p_{AB}$. Alice will send a plaintext bit m to Bob by mapping it to an appropriate random code word. Let us look at an extreme case where $p_{AB} = 0$. Suppose Alice wishes to send Bob a 0 or a 1. She will encode this

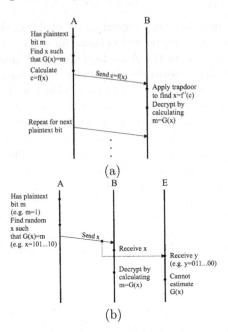

(a)

(b)

FIGURE 15.3. (a) Probabilistic encryption via trapdoor functions, (b) Key dissemination motivated by probabilistic encryption.

with a *random* N bit sequence x of the same parity, i.e. $G(x) = m$ is the parity function. Since Bob's channel is error-free, he calculates the parity of x, and recovers m. The situation is different for Eve. For sufficiently large N, the probability that there is an even amount of bit errors is arbitrarily close to the probability of an odd amount of bit errors (the probability of an even amount of bit errors is $0.5[1 + (1 - 2p_E)^N] \to 0.5$). Hence, Eve does not learn anything about the actual value of the bit, and since x was random it is very unlikely that she will witness the same ciphertext again (thus providing semantic security). This basic scenario can be modified for the general $p_{AB} > 0$ case.

The heart of our probabilistic construction boils down to the validity of the $p_{AB} < p_E$ assumption. If we can make $p_{AB} < p_E$, then sending multiple codebits through the communication medium naturally makes it harder for Eve to decode than Bob.

Thus, we must utilize the channel in a manner that forces $p_{AB} < p_E$, even if Eve is closer to Alice than Bob. To this end, our strategy uses a multicarrier system (similar methods can be applied in a MIMO system). Here, information is encoded across k out of K carriers with a separation larger than the channel coherence bandwidth in order to guarantee independent fading. The information received by Bob on subchannel i may be modeled via $y_i = \alpha_i x_i + n_i$, where $|\alpha_i|$ is a Rayleigh random variable describing the subchannel gain, and n_i is noise that is assumed to have equal power σ^2 across all subchannels. The quantity $|\alpha_i|^2/\sigma^2$ reflects the quality of subchannel i, and will vary across subchannels, as depicted in Figure 15.4 (a). Since Alice and Bob share an estimate of the same channel, they know which channel is the best, and may send their information on that subcarrier.

Assuming that Eve is far enough from Bob so that the Alice-Eve and Alice-Bob subchannel gains are independent (i.e. on the order of one wavelength, $\lambda \approx 0.1-1.0\text{m}$ for 300MHz - 3GHz.), then there is a good chance that one of the better Alice-Bob subchannels will be bad for Eve and hence, if we code only on this channel, then $p_{AB} < p_E$. More specifically, we may code only on Alice-Bob's best channel, in which case the question of whether $p_{AB} < p_E$ boils down to whether Alice-Bob's best subchannel is better than a random subchannel for Alice-Eve. For Rayleigh fading, Bob's gains $|\alpha_i|$ are Rayleigh (and hence $|\alpha_i|^2$ are exponential) with average power γ_B, while Eve's gains $|\alpha_E|$ are Rayleigh with average power γ_E and Eve's noise has power σ^2, we may calculate the probability of $p_{AB} < p_E$ by finding $Pr\left(\max\{|\alpha_i|^2\} > |\alpha_E|^2\right)$:

$$\beta_1 = Pr\left(\max\{|\alpha_i|^2\} > |\alpha_E|^2\right) \tag{15.1}$$

$$= 1 - \int_0^\infty (1 - e^{-x/\gamma_B})^K \frac{1}{\gamma_E} e^{-x/\gamma_E} dx \tag{15.2}$$

$$= 1 - \sum_{n=0}^K \binom{K}{n} (-1)^n \frac{\gamma_B}{n\gamma_E + \gamma_B}. \tag{15.3}$$

We present a plot of β_1 versus $10\log(\gamma_B/\gamma_E)$ in Figure 15.4 (b), for the case of $K = 32$ subcarriers. From this, we see that in many cases, even when Eve has a better average gain than Bob (i.e. $\gamma_E > \gamma_B$), it is likely that $Pr(p_{AB} < p_E)$. Still, for enhanced secrecy, it is desirable to improve this likelihood and instead of using one subcarrier, we may choose the k best Alice-Bob subcarriers, and code in such a way that Eve must have better gains on *all* of these k subcarriers in order for $p_E < p_{AB}$, for example by employing a hash function. The corresponding probabilities $\beta_k = Pr(p_{AB} < p_E)$ are presented for $k = 2$ and $k = K$ in Figure 15.4 (b).

There are many possible variations for how we select carriers and allocate resources across these channels. For example, rather than choose the best k channels, we may instead apply a waterfilling strategy [171]. Even with such constructions, though, we cannot absolutely guarantee that Eve has a worse decoding capability than Bob. However, by tying in a propagation and shadowing model, we may relate γ_B/γ_E to the relative distances that Bob and Eve are from Alice. This allows us to define a *threat region* about Alice where, with a prescribed likelihood, Eve may be able to successfully decipher the key bits that Alice is sending Bob. To assure complete security, Alice would then need to physically guarantee that Eve is not within this threat region.

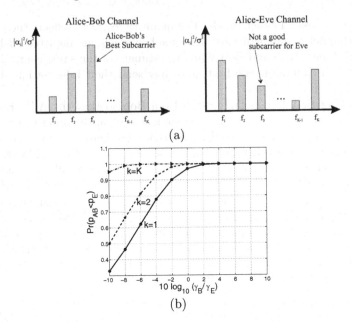

(a)

(b)

FIGURE 15.4. (a) Alice-Bob and Alice-Eve channel gains across different subcarriers. (b) Probability that $p_{AB} < p_E$ versus $10\log(\gamma_B/\gamma_E)$ for $K = 32$ subcarriers, and for different choices of k (the number of subcarriers used).

15.5 Experimental Validation

The proposed PHY-layer security methods introduce new functionalities to the radio that are not possible to validate using conventional off the shelf equipment, and it was therefore necessary to conduct our experimental efforts using a software defined radio (SDR) platform that allows for access to waveform-level details. In our experiments, we used the Universal Software Radio Peripheral (USRP) board in conjunction with GNURadio running on a personal computer. GNURadio is an open source, free software toolkit that provides a library of signal processing blocks for developing communications systems and experiments. Our experiments were implemented using GNURadio under Debian GNU/Linux (kernel 2.6).

The USRP supports the simultaneous transmit and receive of four real or two complex channels in real-time. For reception it utilizes four 12-bit analog-to-digital converters (ADCs) operating at 64MHz, and four digital-downconverters with programmable decimation rates. The transmit side of the USRP incorporates four 14-bit DACs that operate at 128MHz, and two digital-upconverters with programmable interpolation rates. Data is transferred between the computer and the USRP via a USB2.0 interface. Given a sustainable data rate of 32 MBps and complex 16-bit samples, the effective total spectral bandwidth is limited to 8MHz. The USRP itself is

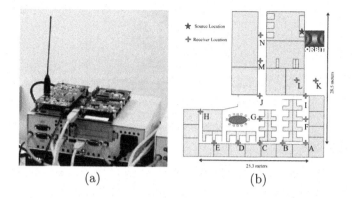

FIGURE 15.5. (a) USRP/GNU Radio platform and the RX400 RF Frontend used in experiments. (b) Relative locations of transmitter and receiver during experiments.

not directly capable of RF input/output, and it is necessary to interface the USRP with RF daughterboard modules for actual RF transmit/receive. In our experiments, we used the the RFX400 transceiver, which operates in the 400-500MHz band, and is capable of outputting 100 mW (+20 dBm) with a Noise Figure of 3-5 dB. The platform used as the transmitter is depicted in Figure 15.5 (a).

15.5.1 Fundamental Measurements

We first conducted several experiments to evaluate the spatio-temporal coherency properties of the indoor wireless channel. We conducted all experiments in our building, as depicted in Figure 15.5 (b). For all experiments, we fixed the location of the transmitter in a laboratory (upper right hand corner), and then varied the positions of the receiver platform throughout the building. For each position, we measured data at a net sampling rate of 2MHz (following decimation by 32). The following experiments were performed:

Experiment 1: In this experiment, we were interested in the temporal coherency of the channel for a fixed location. We used three carrier frequencies at 420MHz, 450MHz and 480MHz. The 30MHz separation between these carriers was chosen to ensure fading independence across the carriers. Data was measured at location G. For this experiment, we sampled the carriers for a duration of 1 second every 15 minutes, collecting roughly 16MB for each carrier frequency and each sampling interval. This sampling was carried out over an experimental period of 1 hour. We calculated the magnitude of the channel gain for each frequency versus time across our 1sec interval, using a sliding window with an integration time of 1msec.

We present the magnitude of the channel gains for 0.1 seconds for each of four sampling intervals (a period of 1 hour) in Figure 15.6 (a), for each

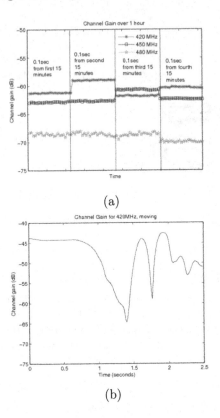

(a)

(b)

FIGURE 15.6. (a)Magnitude of the channel gains for location G over 1hour. (b) Channel gain for 420MHz as receiver moves from N to J.

of the three carrier frequencies. From this figure, we can see that there is some variability in the channel across time, as illustrated by the 420 and 450MHz gains crossing each other in between the second and third interval. We also note that the 480MHz gains were much lower. Overall, although there is some temporal variability, this variability is still within a constrained dynamic range, implying that the variation occurs around a mean response profile for this location.

Experiment 2: In this experiment, we examined the spatial nature of the channel by collecting data from each of the three carriers at different locations in the building. For this experiment, we again collected roughly 16MB for each carrier frequency at each location. One challenge that we encountered was the fact that we could not measure in different locations simultaneously. To minimize the temporal variations, we conducted our experiments during a time of no activity in the building. Even so, we note that we encountered difficulty discriminating between channel phase and oscillator drift, and thus restricted our attention to comparing the magnitudes of the gain across locations. We observed several instances where different pairs of locations would exhibit rather diverse properties across different carrier frequencies. In general, the fact that two locations might exhibit correlation for one carrier but not for another is an important observation that is critical to the success of using the channel to identify the transmitter. It implies that, with the application of more carriers, we should be able to improve our ability to discriminate between the channels associated with different locations.

Experiment 3: In this experiment, we examined the effect of mobility on the channel coherence by gathering data on just the 420MHz carrier as we moved the receiver from location N to location J (as depicted in Figure 15.5 (b)). For this experiment, a roughly 100MByte trace was collected. We present the channel gain versus time for 2.5 seconds of data in Figure 15.6(e). In this trace we see that there is variability of the channel as we move. We note that there are two periods of "deep" fades, which we conjecture are due to shadowing as we passed by two metal doors. However, even so, a visual inspection suggests that for the most part the channel is coherent over time intervals of roughly 20-100msec, which would imply that in a mobile environment we should conduct channel probing at intervals of tens of milliseconds.

15.5.2 Evaluation of PHY Authentication

The spatial and temporal correlation properties across the three carriers we examined are encouraging and suggest that it should be possible to discriminate between two transmitters at two different locations.

Applying the principle of channel reciprocity, we used our traces from Experiment 2 to construct a synthetic trace to explore the utility of PHY-layer authentication techniques to discriminate between transmitters in a spoofing scenario. In our trace, we placed Bob (the receiver) at the location of the transmitter in Experiment 2, and we placed both Alice and Eve (the transmitters) at two receiver locations from Experiment 2. We chose Alice to be located at position A, while we placed Eve at location H. In our synthetic trace, we assume that Alice has started communication for a short period before introducing Eve. Following the Alice-only period, we employed a randomized schedule corresponding to Alice and Eve alternately communicating. This randomized schedule is meant to reflect the fact that Alice would most likely continue transmitting while Eve is conducting a spoofing attack. In the trace, we assume that Eve knows that verification requires transmission on 420MHz, 450MHz, and 480MHz, and hence Eve transmits on all these carriers.

To detect Eve, we estimated the magnitude of the channel gains versus time $|\alpha_j(t)|$ across the three carriers, and constructed a feature vector $\mathbf{v}(t) = [|\alpha_1(t)|, |\alpha_2(t)|, |\alpha_3(t)|]^T$. Since our objective is to detect a change in

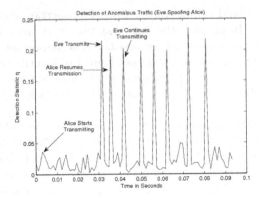

FIGURE 15.7. Value of the change point statistic $\eta(t)$. After a short time, Eve commences spoofing Alice, and then Eve and Alice alternately transmit for random durations.

the channel state versus time, we employed a simple change-point detector

$$\eta(t) = \frac{\|\mathbf{v}(t) - \mathbf{v}(t-1)\|}{\|\mathbf{v}(t-1)\|}.$$

We present $\eta(t)$ for our synthetic trace in Figure 15.7. We note that during periods where just Alice transmits, the value of $\eta(t)$ is small, but that $\eta(t)$ exhibits large spikes in its value at times where there is a change in the transmitter, thereby facilitating the detection of device spoofing. Similar change point detection was observed for all other Alice-Eve location pairs.

15.5.3 Evaluation of PHY Confidentiality

We also used the traces from Experiment 2 to explore the feasibility of using the PHY layer to establish cryptographic keys. We calculated the dynamic range for the magnitude of the channel gains over the entire data set from Experiment 2 (across frequency, time and location). Using this, we built an 8-bit uniform quantizer to quantize channel gains for each of the three carriers at each location, producing a total of 24 bits of channel symbol information. In order to remove any correlation from channel symbol information across locations, we used the SHA-1 hash function in conjunction with a random seed r that was transmitted via a public channel (i.e. we assume each location receives r perfectly). Our keys were 24bits, as calculated by

$$K = M_{24}\left[f_{SHA-1}([Q(\mathbf{v}(t))\|r])\right],$$

where $Q()$ is the 8-bit quantizer, M_{24} is extraction of the middle 24 bits, and $\|$ denotes concatenation. As an example of the output, we present results of the 24-bit key sequence generated at each of the locations A through N

using the first channel state estimates from Experiment 2. Using a random seed $r = 0x374573$, the resulting keys were

Location	Key Sequence
A	1001 1001 1101 1000 0101 1011
B	0110 0101 0010 1000 1110 1110
C	1101 0111 0000 0011 0011 1010
D	1111 0001 1101 1001 1111 0111
E	0100 0001 1101 1101 1000 0001
F	1111 1000 1101 1101 0001 1011
G	1100 0001 1110 1010 0110 1101
H	0000 1101 0000 1000 0000 0101
I	1111 0010 1011 0010 0111 1111
J	0010 1110 0011 0101 1111 0001
K	1010 1010 0101 0111 0100 0000
L	1011 1000 0110 0001 1001 0101
M	0101 1010 1010 1101 1010 1000
N	1100 1011 0101 1100 0111 1000

In particular, in spite of the correlation that exists between certain pairs of locations on individual carriers, the fact that we are utilizing more than one carrier with different correlation properties causes the quantized feature vector, and hence the resulting bit sequences, to differ.

15.6 Conclusion

Wireless networks represent a dramatically different type of communication system than traditional "wired" networks. Typically, these differences have been seen as a challenge making it harder to secure wireless systems. In this chapter, however, we have proposed that it is possible to exploit the underlying properties of the wireless medium in order to support security objectives. Channel probing techniques, such as wideband pulsing and multitone probing, may be used to estimate the channel state, which can be compared against a history of transmitter-receiver channel states in order to detect anomalous behavior. Further, the rapid decorrelation properties of the multipath channel allows for either the extraction of key material from channel state information, or the secret dissemination of keying information across the channel through suitable encoding.

We have presented the results of an initial validation effort using the USRP/GNURadio SDR platform. The objective behind this effort was to support the feasibility of physical layer identification and confidentiality techniques. Our initial results indicate the merit of physical layer security, and in particular illustrate that it is possible to detect change points associated with instances where entity spoofing is performed. Additionally, we presented a simple example of how key information could be shared across a public channel once the transmit-receiver channel has been estimated. However, our experience has also shown the importance of employing probing techniques with sufficient transmit bandwidth (e.g. more carriers separated by the coherence bandwidth in our multicarrier techniques). This, however, necessitates a more powerful SDR platform capable of processing hundreds of MHz of bandwidth, or even a UWB platform. Our current efforts are focused on developing these methods for more powerful SDRs. One further direction that we are exploring is the degree to which temporal variability affects our proposed schemes. We have observed more temporal variability at 5GHz, which would thus require more frequent channel probing at these frequencies. We believe there is a point where too little temporal coherency can make our techniques impractical, and we are currently working on exploring this conjecture.

Finally, we note that the two security objectives that we have focused on are a fraction of what can be accomplished at the physical layer of the protocol stack. For example, a non-repudiation service can exploit the broadcast nature of the wireless medium by introducing witnesses, making it harder for wireless entities to deny carriage of information. An availability service can use spreading and power control to maintain network connectivity in the presence of RF interference attacks. Overall, we envision that it will be possible to develop a suite of lower-layer enforcement strategies that can complement traditional methods, and ultimately lead to more secure wireless systems.

16
Concluding Remarks

As wireless technologies rapidly advance, new forms of wireless systems will continue to emerge and have an increasing impact on our life. Already, new applications, such as reverse-911 or the ubiquitous media delivery services point to a future where wireless systems will be an inseparable part of our social fabric. As users become less tethered to a specific location for accessing information, they will be free to roam and interact with each other. The possibilities of this freedom is great, but so is the potential for damage. Just as consumers will increasingly move to mobile platforms for accessing the Internet, so too will adversaries and the result will be a broad range of new threats facing consumers. Conventional approaches to security, which are built upon cryptographic primitives and the notions of distinct layers of a network protocol stack, must serve as a first-line of defense against these emerging threats. However, there are many unique aspects related to wireless and mobile communications that can also be leveraged to enhance the security of wireless systems.

The purpose behind this book has been to provide an overview of security issues that are facing emerging wireless technologies, and to survey a broad array of "lower-layer" security mechanisms that can augment traditional approaches to network security. The methods described in this book involve many aspects that are unique to wireless systems, such as issues related to the propagation of communication signals or the notion of a wireless device's location. The strategies discussed in this book involve aspects of geographical location, physical layer communication, link-link layer and medium access, and routing within wireless systems. As such, the methods discussed in this book either exist outside of the network stack or operate at the lowest three layers of the OSI stack. In particular, we have not examined any issues related to transport-layer phenomena in wireless systems, though certainly many transport layer issues can be well-addressed through conventional security protocols, such as SSL/TLS.

At a high-level, we have examined localization, identity, availability, privacy, and physical layer security. Location is a property that is tightly tied to wireless systems. Wireless systems will allow users to access content and services from anywhere at anytime, and this will allow for new services to be developed that specifically target consumers based on where they are located, not just merely who they are or whether or not they can connect. For example, already there are many advertisement services being developed for cellular systems that target customers if they are within a certain distance of a store. Additionally, it is easy to envision applications where users are only allowed to access content if they are at the right place at the right time. Together, these scenarios point towards a need to ensure the trustworthiness of location information. The first part of this book has focused on methods for ensuring the accuracy of localization in spite of malicious attempts to subvert localization. Additionally, we have examined how applications can use location information as the basis for controlling access to content.

The identities used in wireless networks can be the source of a broad array of attacks on wireless networks. Although it is true that identities can be easily spoofed in wired networks, the fact that wireless devices can be easily acquired, easily reconfigured, and easily relocated provides an easy avenue for adversaries to launch attacks where they monitor communications, learn the identities of other entities, and then imitate these entities. Cryptographic tools can be used to defend against spoofing, but also require the establishment of a priori cryptographic material, which is often not possible in wireless systems. The lack of an always-present entity, such as a certificate authority, to supply and update authentication keys suggests that it is desirable to use light-weight mechanisms that use non-cryptographic properties to detect these identity attacks. We have examined the use of properties inherent in wireless packet transmission and location information as a means to defend against spoofing threats.

Wireless networks are considered valuable because they will provide the pervasive opportunity for connectivity to other devices or to the Internet. Without guarantees associated to the availability of wireless network connectivity, the very utility of wireless communications can be undermined. In particular, wireless communication is very susceptible to interference, whether intentional or accidental. We have examined issues related to interference and provided an overview of different jamming strategies that an adversary may use to undermine link-connectivity. We then examined different methods for coping with this interference, starting with detecting the presence of jamming using an appropriate collection of statistics, and then moving to methods that can be used to re-establish link connectivity. Notably, we presented a collection of channel adaptation strategies, known as channel surfing, that can restore network link functionality in a distributed manner, in spite of the presence of interference.

Privacy is a severe concern for wireless networks, especially if the network is being used to monitor precious assets. Compared to traditional networks, wireless networks witness a new set of privacy concerns, which we refer to as contextual privacy. Contextual privacy does not focus on protecting the privacy of message content, but on protecting the context information around exchanged messages. The two types contextual information that are of particular concern are the location of the data source (i.e. the monitored object), as well as the time when a certain event is observed. Violation of each of these two types contextual privacy has its own problem, but when they are both violated, the aggregated effect is much more adverse as the adversary now can keep track of the movement of the monitored object. The fourth part of this book discusses solutions to address these two types of privacy concerns. In particular, we show that the discussed methods are effective in protecting both source-location privacy and temporal privacy in large-scale wireless sensor networks. These methods can apply to wireless technologies where there is a need to obfuscate the origin of a communication or the time of creation of a communication.

Lastly, we examined an exciting, new area for securing wireless networks that uses the unique properties associated with fading and propagation. In a multipath rich environment, as is typical in mobile outdoor or indoor scenarios, the fading phenomena between a transmitter and a receiver is a powerful source of shared information that can be used to devise new forms of authentication and confidentiality services. In terms of authentication, another method for detecting spoofing was presented, whereby a receiver tracks a transmitter's channel response versus time and compares newly arriving signals with prior signals in order to decide whether the transmission is from the claimed entity or not. Further, we may use the fact that the channel response between a transmitter and receiver rapidly decorrelates in distance from either the transmitter or receiver as a source of nonpredictable common-randomness. This common-randomness may be used as the basis for extracting a cryptographic key, and thus support confidentiality services. Lastly, by exploiting the spectral and temporal variability properties of the wireless channel, we showed that one can use appropriate variations of Wyner coding to secretly disseminate messages (rather than extract keys) between two participants.

Looking forward, wireless systems will continue to evolve and there will be an ever-increasing need to develop new security solutions that address emerging threats. Although we have mapped out a broad variety of defense mechanisms that operate at the lower-layers of the communication stack, there are many other avenues available for utilizing domain-specific information associated with wireless systems to enhance security. Notably, one direction for enhancing the security of wireless systems is to exploit the inherent radiative properties of wireless communication. By building upon this simple property, it may be possible to develop new forms of non-repudiation or forensic services for wireless systems. In particular, as RF energy radiates, and wireless entities within the radio coverage pattern may serve as witnesses for the actions of the transmitter. This makes it harder for radio entities to deny receiving a message or having performed an action. We may introduce communication auditors into the wireless infrastructure to assist in quantifying the trust of wireless entities.

We conclude our discussion, by providing an example of such a future wireless security technology and hope that the idea will be examined by the community as part of an ongoing effort to develop new wireless security solutions. One concern facing the eventual deployment of new wireless technologies is being able to support billing and guarantee that services were provided to a user. New forms of service non-repudiation might be possible in various wireless networks by adding caching functionality to the network layer software located at intermediate nodes in the wireless network. In the case of a cellular network, these nodes would simply correspond to base stations, while in wireless LANs these nodes might be access points. The choice of nodes in ad hoc wireless networks might be more challenging, though, as the network topology is generally dynamic. However, it is known that flat network topologies introduce network inefficiencies, especially when mobility and network capacity are considered. Recent research comparing the network capacity of flat and hierarchical networks suggests that wireless networks, and particularly ad hoc networks, ultimately will adopt a hierarchical strategy. As a result, there will likely be a class of network nodes with less mobility and that might serve to cache packets or hashes of these packets, and thus we may assume there will be a collection of well-resourced nodes capable of supporting non-repudiation. Regardless of what type of network, at these packet caching nodes, interesting challenges related to selecting the appropriate sampling resolution needed to reconstruct network characteristics, such as link packet loss, would have to be addressed. Overall, the objective would be to use caching and collaboration between different entities within the wireless network to provide proof that content was delivered to the end-user, even though the user might falsely claim to never have received the content. Such a service would be a valuable security functionality for wireless systems, supporting the implementation of better Authentication Authorization Accounting (AAA) entities within the wireless network.

References

[1] "IEEE 802.11 Standards," http://standards.ieee.org/getieee802/802.11.html.

[2] "IEEE 802.15.1 Standards," http://standards.ieee.org/getieee802/download/802.15.1-2003.pdf.

[3] "IEEE 802.15.4 Standards," http://standards.ieee.org/getieee802/download/802.15.4-2003.pdf.

[4] "Moteiv Corporation," White paper available at http://www.moteiv.com.

[5] "Crossbow Technology Inc.," White paper available at http://www.xbow.com.

[6] P. Enge and P. Misra, *Global Positioning System: Signals, Measurements and Performance*, Ganga-Jamuna Pr, 2001.

[7] Dragos Niculescu and Badri Nath, "Ad hoc positioning system (APS)," in *Proceedings of the IEEE Global Telecommunications Conference (GLOBECOM)*, 2001, pp. 2926–2931.

[8] K. Langendoen and N. Reijers, "Distributed localization in wireless sensor networks: a quantitative comparison," *Comput. Networks*, vol. 43, no. 4, pp. 499–518, 2003.

[9] Z. Li, W. Trappe, Y. Zhang, and B. Nath, "Robust statistical methods for securing wireless localization in sensor networks," in *Proceedings of the Fourth International Symposium on Information Processing in Sensor Networks (IPSN 2005)*, 2005, pp. 91–98.

[10] K.K. Chintalapudi, A. Dhariwal, R. Govindan, and G. Sukhatme, "Ad hoc localiztion using ranging and sectoring," in *Proceedings of the IEEE International Conference on Computer Communications (INFOCOM)*, March 2004.

[11] D. Niculescu and B. Nath, "Vor base stations for indoor 802.11 positioning," in *in Proceedings of the Annual ACM International Conference on Mobile Computing and Networking (MOBICOM)*, 2004, pp. 2926–2931.

[12] Moustafa Youssef, Ashok Agrawal, and A. Udaya Shankar, "Wlan location determination via clustering and probability distributions," in *Proceedings of the First IEEE International Conference on Pervasive Computing and Communications (PerCom)*, Mar. 2003, pp. 143–150.

[13] T. Roos, P. Myllymaki, and H.Tirri, "A Statistical Modeling Approach to Location Estimation," *IEEE Transactions on Mobile Computing*, vol. 1, no. 1, pp. 59–69, Jan-March 2002.

[14] D. Madigan, E. Elnahrawy, R.P. Martin, W. Ju, P. Krishnan, and A. S. Krishnakumar, "Bayesian indoor positioning systems," in *Proceedings of the IEEE International Conference on Computer Communications (INFOCOM)*, March 2005, pp. 324–331.

[15] E. Elnahrawy, X. Li, and R. P. Martin, "The limits of localization using signal strength: A comparative study," in *Proceedings of the First IEEE International Conference on Sensor and Ad hoc Communcations and Networks (SECON 2004)*, Oct. 2004, pp. 406–414.

[16] P. Bahl and V. N. Padmanabhan, "Radar: An in-building rf-based user location and tracking system," in *Proceedings of the IEEE International Conference on Computer Communications (INFOCOM)*, March 2000, pp. 775–784.

[17] N. Priyantha, A. Chakraborty, and H. Balakrishnan, "The cricket location-support system," in *Proceedings of the ACM International Conference on Mobile Computing and Networking (MobiCom)*, Aug 2000, pp. 32–43.

[18] A. Harter, A. Hopper, P. Steggles, A. Ward, and P.Webster, "The anatomy of a context-aware application," in *Proceedings of the MOBICOM 99*, 1999.

[19] Andreas Savvides, Chih-Chien Han, and Mani Srivastava, "Dynamic Fine-Grained Localization in Ad-Hoc Networks of Sensors," in *in Proceedings of the Seventh Annual ACM International Conference on Mobile Computing and Networking (MobiCom)*, Rome, Italy, July 2001.

[20] D. Nicelescu and B. Nath, "Ad hoc positioning (APS) using AOA," in *Proceedings of IEEE Infocom 2003*, 2003, pp. 1734 – 1743.

[21] J. Hightower, G. Borriello, and R. Want, "Spoton: An indoor 3d location sensing technology based on rf signal strength," Technical Report 00-02-02, University of Washington, Dept. of Computer Science and Engineering, February 2000.

[22] R. Volpe, T. Litwin, and L. Matthies, "Mobile robot localization by remote viewing of a colored cylinder," in *Proceedings of IEEE/RSJ International Conference on Robots and Systems (IROS)*, 1995.

[23] C. Savarese, J. Rabay, and K. Langendoen, "Robust positioning algorithms for distributed ad-hoc wireless sensor networks.," in *USENIX Technical Annual Conference*, 2002.

[24] D. Nicelescu and B. Nath, "DV based positioning in ad hoc networks," *Telecommunication Systems*, vol. 22, no. 1-4, pp. 267–280, 2003.

[25] A. Savvides, H. Park, and M. Srivastava, "The bits and flops of the n-hop multilateration primitive for node localization problems," in *Proceedings of First ACM International Workshop on Wireless Sensor Networks and Application (WSNA)*, 2002, pp. 112–121.

[26] N. Bulusu, J. Heidemann, and D. Estrin, "Gps-less low cost outdoor localization for very small devices," *IEEE Personal Communications Magazine*, vol. 7, no. 5, pp. 28–34, 2000.

[27] Tian He, Chengdu Huang, Brian Blum, John A Stankovic, and Tarek Abdelzaher, "Range-free localization schemes in large scale sensor networks," in *Proceedings of the Ninth Annual ACM International Conference on Mobile Computing and Networking (MobiCom'03)*, San Diego, CA, Sept. 2003.

[28] Roberto Battiti, Mauro Brunato, and Alessandro Villani, "Statistical Learning Theory for Location Fingerprinting in Wireless LANs," Technical Report DIT-02-086, University of Trento, Informatica e Telecomunicazioni, Oct. 2002.

[29] Y. Shang, W. Ruml, Y. Zhang, and M. P. J. Fromherz, "Localization from mere connectivity," in *Proceedings of the Fourth ACM International Symposium on Mobile Ad-Hoc Networking and Computing (MobiHoc)*, Jun 2003, pp. 201–212.

[30] L. Doherty1, K. S. J. Pister, and L. ElGhaoui, "Convex position estimation in wireless sensor networks," in *Proceedings of the IEEE International Conference on Computer Communications (INFOCOM)*, Apr. 2001, pp. 1655–1663.

[31] Mike Hazas and Andy Ward, "A high performance privacy-oriented location system," in *Proceedings of the First IEEE International Conference on Pervasive Computing and Communications (PerCom)*, Dallas, TX, Mar. 2003.

[32] Roy Want, Andy Hopper, Veronica Falcao, and Jonathon Gibbons, "The active badge location system," *ACM Transactions on Information Systems*, vol. 10, no. 1, pp. 91–102, Jan. 1992.

[33] J. Werb and C. Lanzl, "Designing a positioning system for finding things and people indoors," *IEEE Spectrum*, pp. 71–78, 1998.

[34] J. Newsome, E. Shi, D. Song, and A. Perrig, "The sybil attack in sensor networks: analysis and defenses," in *Third International Symposium on Information Processing in Sensor Networks*, 2004, pp. 259–268.

[35] S. Capkun and J.P. Hubaux, "Secure positioning in sensor networks," Technical report EPFL/IC/200444, May 2004.

[36] W. Xu, T. Wood, W. Trappe, and Y. Zhang, "Channel surfing and spatial retreats: defenses against wireless denial of service," in *Proceedings of the 2004 ACM workshop on Wireless security*, 2004, pp. 80–89.

[37] Y.C. Hu, A. Perrig, and D. Johnson, "Packet leashes: a defense against wormhole attacks in wireless networks," in *Proceedings of IEEE Infocom 2003*, 2003, pp. 1976–1986.

[38] S. Zhu, S. Xu, S. Setia, and S. Jajodia, "LHAP: A lightweight hop-by-hop authentication protocol for ad-hoc networks," in *Proceedings of the IEEE International Workshop on Mobile and Wireless Network (MWN)*, 2003, pp. 749–755.

[39] S. Garg, N. Singh, and T. Tsai, "Schemes for enhancing the denial of service tolerance of srtp," in *Proceedings of the International Conference on Security and Privacy for Emerging Areas in Communication Networks*, 2005.

[40] M. bohge and W. Trappe, "An authentication framework for hierarchical ad hoc sensor networks," in *Proceedings of the ACM Workshop on Wireless Security (WiSe)*, 2003, pp. 79–87.

[41] B. Wu, J. Wu, E. Fernandez, and S. Magliveras, "Secure and efficient key management in mobile ad hoc networks," in *Proceedings of the 19th IEEE International Parallel and Distributed Processing Symposium (IPDPS)*, 2005.

[42] L. Zhou and Z. Haas, "Securing ad hoc networks," *IEEE Network*, vol. 13, no. 6, pp. 24–30, 1999.

[43] A. Khalili and J. Katz, "Toward secure key distribution in truly ad-hoc networks," in *Proceedings of IEEE workshop on Security and Assurance in Ad-Hoc Networks*, 2003.

[44] A. Wool, "Lightweight key management for ieee 802.11 wireless lans with key refresh and host revocation," *ACM/Springer Wireless Networks*, vol. 11, no. 6, pp. 677–686, 2005.

[45] S. Brands and D. Chaum, "Distance-bounding protocols," in *Proceedings of the Workshop on the Theory and Application of Cryptographic Techniques on Advances in Cryptology*, 1994, pp. 344–359.

[46] N. Sastry, U. Shankar, and D. Wagner, "Secure verification of location claims," in *Proceedings of the ACM workshop on wireless security*, 2003, pp. 1–10.

[47] S. Capkun and J. P. Hubaux, "Secure positioning of wireless devices with application to sensor networks," in *Proceedings of the IEEE International Conference on Computer Communications (INFOCOM)*, 2005, pp. 1917–1928.

[48] S. Capkun and J.P. Hubaux, "Securing localization with hidden and mobile base stations," in *Proceedings of the IEEE International Conference on Computer Communications (INFOCOM)*, March 2006.

[49] L. Lazos, R. Poovendran, and S. Capkun, "Rope: robust position estimation in wireless sensor networks," in *Proceedings of the Fourth International Symposium on Information Processing in Sensor Networks (IPSN 2005)*, 2005, pp. 324–331.

[50] D. Liu, P. Ning, and W. Du, "Attack-resistant location estimation in sensor networks," in *Proceedings of the Fourth International Symposium on Information Processing in Sensor Networks (IPSN 2005)*, 2005, pp. 99–106.

[51] D. Liu and P. Ning and W. Du, "Detecting malicious beacon nodes for secure location discovery in wireless sensor networks," in *Proceedings of the 25th IEEE International Conference on Distributed Computing Systems (ICDCS 05)*, June 2005, pp. 609–619.

[52] M. Bishop, *Computer Security: Art and Practice*, Addison Wesley, 2003.

[53] S. Gavrila D. Ferraiolo, R. Sandhu, D. Richard Kuhn, and R. Chandramouli, "Proposed NIST standard for Role-Based Access Control," *ACM Transactions on Information and System Security*, vol. 4, no. 3, pp. 224–274, 2001.

[54] L. Lazos and R. Poovendran, "SeRLoc: Secure range-independent localization for wireless sensor networks," in *Proceedings of the 2004 ACM Workshop on Wireless Security*, 2004, pp. 21–30.

[55] D. B. Faria and D. R. Cheriton, "No Longterm Secrets: Location-based Security in Overprovisioned Wireless LANs," in *Proceedings of the Third ACM Workshop on Hot Topics in Networks*, 2004.

[56] N. Michalakis, "PAC: Location Aware Access Control for Pervasive Computing Environments," 16 September 2002.

[57] Nikolaos Michalakis, "Location-aware Access Control for Pervasive Computing Environments," .

[58] W. Han, J. Zhang, and X. Yao, "Context-sensitive access control model and implementation," in *The Fifth International Conference on Computer and Information Technology*, pp. 757–763.

[59] R. J. Hulsebosch, A. H. Salden, M. S. Bargh, P. W. G. Ebben, and J. Reitsma, "Context sensitive access control," in *SACMAT '05: Proceedings of the tenth ACM symposium on Access control models and technologies*, New York, NY, USA, 2005, pp. 111–119, ACM Press.

[60] E. Bertino, B. Catania, M. L. Damiani, and P. Perlasca, "GEO-RBAC: a spatially aware RBAC," in *SACMAT '05: Proceedings of the tenth ACM symposium on Access control models and technologies*, New York, NY, USA, 2005, pp. 29–37, ACM Press.

[61] J. Joshi, E. Bertino, U. Latif, and A. Ghafoor, "A generalized temporal role-based access control model," vol. 17, pp. 4–23, 2005.

[62] B. Schilit, J. Hong, and M. Gruteser, "Wireless Location Privacy Protection," *Computer*, vol. 36, no. 12, pp. 135–137, 2003.

[63] M. Gruteser, G. Schelle, A. Jain, R. Han, and D. Grunwald, "Privacy-aware location sensor networks," in *Workshop on Hot Topics in Operating Systems (HotOS)*, 2003.

[64] M. Gruteser and D. Grunwald, "Anonymous Usage of Location-based Services through Spatial and Temporal Cloaking," in *Proceedings of the International Conference on Mobile Systems, Applications, and Services (MobiSys)*, 2003.

[65] S. Duri, M. Gruteser, X. Liu, P. Moskowitz, R. Perez, M. Singh, and J. Tang, "Context and Location: Framework for security and privacy in automotive telematics," in *Proceedings of the 2nd international workshop on Mobile commerce*, 2002.

[66] J. Bellardo and S. Savage, "802.11 denial-of-service attacks: Real vulnerabilities and practical solutions," in *Proceedings of the USENIX Security Symposium*, 2003, pp. 15 – 28.

[67] F. Ferreri, M. Bernaschi, and L. Valcamonici, "Access points vulnerabilities to dos attacks in 802.11 networks," in *Proceedings of the IEEE Wireless Communications and Networking Conference*, 2004.

[68] T. Aura, "Cryptographically generated addresses (cga)," *RFC 3972, IETF*, 2005.

[69] E. Kempf, J. Sommerfeld, B. Zill, B. Arkko, and P. Nikander, "Secure neighbor discovery (send)," *RFC 3971, IETF*, 2005.

[70] W. Du, L. Fang, and P. Ning, "Lad: Localization anomaly detection for wireless sensor networks," in *Proceedings of the 19th IEEE International Parallel and Distributed Processing Symposium (IPDPS 05)*, April 2005.

[71] R. Wilson, "Propagation Loss through Common Building Materials, 2.4GHz vs. 5GHz," 2002, White paper available at http://www.magisnetworks.com.

[72] S. Lang, *Real and Functional Analysis*, Springer, 1993.

[73] Y. Chen, J. Francisco, W. Trappe, and R. P. Martin, "A practical approach to landmark deployment for indoor localization," in *Proceedings of the Third Annual IEEE Communications Society Conference on Sensor, Mesh and Ad Hoc Communications and Networks (SECON)*, September 2006.

[74] Y.C. Hu, A. Perrig, and D. Johnson, "Packet leashes: a defense against wormhole attacks in wireless networks," in *Proceedings of the IEEE International Conference on Computer Communications (INFOCOM)*, 2003, pp. 1976–1986.

[75] S. Weisberg, *Applied Linear Regression*, Wiley Series in Probability and Mathematical Statistics, 2005.

[76] A.S. Krishnakumar and P. Krishnan, "On the accuracy of signal strength-based location estimation techniques," in *Proceedings of the IEEE International Conference on Computer Communications (INFOCOM)*, March 2005.

[77] K. Langendoen and N. Reijers, "Distributed localization in wireless sensor networks: a quantitative comparison," *Comput. Networks*, vol. 43, no. 4, pp. 499–518, 2003.

[78] B. Przydatek, D. Song, and A. Perrig, "SIA: secure information aggregation in sensor networks," in *SenSys '03: Proceedings of the 1st International Conference on Embedded Networked Sensor Systems*, 2003, pp. 255–265.

[79] D. Wagner, "Resilient aggregation in sensor networks," in *SASN '04: Proceedings of the 2nd ACM workshop on Security of ad hoc and sensor networks*, 2004, pp. 78–87.

[80] P. Rousseeuw and A. Leroy, "Robust regression and outlier detection," Wiley-Interscience, September 2003.

[81] P. Bahl, V.N. Padmanabhan, and A. Balachandran, "Enhancements to the RADAR User Location and Tracking System," Tech. Rep. Technical Report MSR-TR-2000-12, Microsoft Research, February 2000.

[82] A. Goldsmith, *Wireless Communications*, Cambridge University Press, 2005.

[83] J. E. Hopcroft and J. D. Ullman, *Introduction to Automata theory, languages and computation*, Addison-Wesley Publishing Company, 1979.

[84] M. Gruteser and D. Grunwald, "Anonymous Usage of Location-Based Services through Spatial and Temporal Cloaking," in *Proceedings of First ACM/USENIX International Conference on Mobile Systems, Applications, and Services (MobiSys)*, 2003, pp. 31–42.

[85] Y. Gertner, S. Goldwasser, and T. Malkin, "A random server model for private information retrieval or how to achieve information theoretic PIR avoiding database replication," *Lecture Notes in Computer Science*, vol. 1518, 1998.

[86] G. D. Crescenzo, Y. Ishai, and R. Ostrovsky, "Universal service-providers for database private information retrieval (extended abstract)," in *Symposium on Principles of Distributed Computing*, 1998, pp. 91–100.

[87] C. Cachin, S. Micali, and M. Stadler, "Computationally private information retrieval with polylogarithmic communication," *Lecture Notes in Computer Science*, vol. 1592, 1999.

[88] A. Mishra and W. A. Arbaugh, "An initial security analysis of the ieee 802.1x standard," Tech. Rep. CS-TR-4328, University of Maryland, College Park, 2002.

[89] W. A. Arbaugh, N. Shankar, Y.C.J. Wan, and Kan Zhang, "Your 802.11 network has no clothes," *IEEE Wireless Communications*, vol. 9, no. 6, pp. 44–51, Dec. 2002.

[90] G. Montenegro and C. Castelluccia, "Statistically unique and cryptographically verifiable," 2002.

[91] H. V. Poor, *An Introduction to Signal Detection and Estimation*, Springer Verlag, 2nd edition, 1994.

[92] F. Guo and T. Chiueh, "Sequence number-based mac address spoof detection," in *Proceedings of the 8th International Symposium on Recent Advances in Intrusion Detection*, 2005.

[93] Joshua Wright, "Detecting wireless LAN MAC address spoofing," http://home.jwu.edu/jwright/papers/wlan-mac-spoof.pdf.

[94] J. Fifield T. Weingart D. C. Sicker C. Doerr, M. Neufeld and D. Grunwald, "Multimac - an adaptive mac framework for dynamic radio networking," in *DySpan 2005, 1st IEEE International Symposium on New Frontiers in Dynamic Spectrum Access Networks, November 8-11, 2005, Baltimore, USA*, Nov 2005.

[95] A. Rao and I. Stoica, "An overlay mac layer for 802.11 networks," in *MobiSys '05: Proceedings of the 3rd international conference on Mobile systems, applications, and services*, 2005, pp. 135–148.

[96] W. Trappe and L.C. Washington, *Introduction to Cryptography with Coding Theory*, Prentice Hall, 2002.

[97] A. Perrig, R. Canetti, D. Song, J. D. Tygar, and B. Briscoe, "TESLA: Multicast source authentication transform introduction," *IETF working draft, draft-ietf-msec-tesla-intro-01.txt*.

[98] A. Perrig, R. Canetti, J. D. Tygar, and D. Song, "The TESLA broadcast authentication protocol," *RSA Cryptobytes*, vol. 5, no. 2, pp. 2–13, 2002.

[99] Y.C. Hu, M. Jakobsson, and A. Perrig, "Efficient constructions for one-way hash chains," in *Applied Cryptography and Network Security (ACNS)*, 2005.

[100] A. Law and W. D. Kelton, *Simulation Modeling and Analysis*, McGraw Hill, 2nd edition, 1991.

[101] C. Ko, M. Ruschitzka, and K. Levitt, "Execution monitoring of security-critical programs in distributed systems: A specification-based approach," in *Proceedings of the IEEE Symposium on Security and Privacy*, 1997.

[102] C. Taylor and J. Alves-Foss, "Nate- network analysis of anomalous traffic events, a low-cost approach," in *Proceedings of the New Security Paradigms Workshop*, 2001.

[103] W. Lee and S. Stolfo, "Data mining approaches for intrusion detection," in *Proceedings of the USENIX Security Symposium*, 1998.

[104] K. Pawlikowski, H. D. J. Jeong, and J. S. R. Lee, "On credibility of simulation studies of telecommunication networks," *IEEE Communications Magazine*, vol. 40, no. 1, pp. 132–139, 2002.

[105] "NSF workshop on network research testbeds," http://www.net.cs.umass.edu/testbed_workshop, Chicago, Ill., Oct. 2002.

[106] "Orbit: Open access research testbed for next-generation wireless networks," http://www.orbit-lab.org.

[107] M. Singh, M. Ott, I. Seskar, and P. Kamat, "ORBIT measurements framework and library (OML): Motivations, design, implementation and features," in *First International Conference on Testbeds and Research Infrastructures for the Development of Networks and Communities (IEEE Tridentcom 2005)*, Feb. 2005.

[108] D.B. Faria and D.R. Cheriton, "Detecting identity-based attacks in wireless networks using signalprints," in *Proceedings of the ACM Workshop on Wireless Security (WiSe)*, September 2006.

[109] A. Haeberlen, E. Flannery, A. Ladd, A. Rudys, D. Wallach, and L. Kavraki, "Practical robust localization over large scale 802.11 wireless networks," in *Proceedings of the Annual ACM/IEEE International Conference on Mobile Computing and Networking (MobiCom)*, September 2004.

[110] T. Roos, P. Myllymaki, H.Tirri, P. Misikangas, and J. Sievanen, "A probabilistic approach to WLAN user location estimation," *International Journal of Wireless Information Networks*, vol. 9, no. 3, pp. 155–164, July 2002.

[111] T. Hastie, R. Tibshirani, and J. Friedman, *The Elements of Statistical Learning, Data Mining Inference, and Prediction*, Springer, 2001.

[112] Y. Chen, G. Chandrasekaran, E. Elnahrawy, J. Francisco, K. Kleisouris, X. Li, R. P. Martin, R. S. Moore, B. Turgut, "Grail: A general purpose localization system," *Sensor Review*, 2008.

[113] Q. Huang, H. Kobayashi, and B. Liu, "Modeling of distributed denial of service attacks in wireless networks," in *IEEE Pacific Rim Conference on Communications, Computers and Signal Processing*, 2003, vol. 1, pp. 41–44.

[114] L. Kleinrock and F. Tobagi, "Packet switching in radio channels: Part i–carrier sense multiple-access modes and their throughput-delay characteristics," *IEEE Trans. on Communications*, vol. 23, no. 12, pp. 1400 – 1416, 1975.

[115] L. Kleinrock, *Queueing Systems, Volume 2: Computer Applications*, John Wiley & Sons, 1976.

[116] F.H.P. Fitzek and M. Reisslein, "MPEG-4 and H.263 video traces for network performance evaluation," *IEEE Network*, vol. 15, no. 6, pp. 40–54, November/December 2002.

[117] K. Ma, Y. Zhang, and W. Trappe, "Mobile network management and robust spatial retreats via network dynamics," in *Proceedings of the The 1st International Workshop on Resource Provisioning and Management in Sensor Networks (RPMSN05)*, 2005.

[118] B. Karp and H. T. Kung, "GPSR: greedy perimeter stateless routing for wireless networks," in *Proceedings of the Sixth Annual ACM/IEEE International Conference on Mobile Computing and Networks (MobiCOM)*, August 2000.

[119] A. Goldsmith, "Stanford University EE 359 Wireless Communications Course Notes," http://www.stanford.edu/class/ee359/.

[120] S. Pack and Y. Choi, "Pre-authenticated fast handoff in a public wireless lan based on ieee 802.1x model," in *Proceedings of the IFIP TC6/WG6.8 Working Conference on Personal Wireless Communications*. 2002, pp. 175–182, Kluwer, B.V.

[121] X. Fu, T. Chen, A. Festag, H. Karl, G. Schäfer, and C. Fan, "Secure, QoS-enabled mobility support for IP-based networks," in *Proc. IP Based Cellular Network Conference (IPCN)*, Paris, France, 2003.

[122] A. Wood, J. Stankovic, and S. Son, "JAM: A jammed-area mapping service for sensor networks," in *24th IEEE Real-Time Systems Symposium*, 2003, pp. 286 – 297.

[123] J. G. Proakis, *Digital Communications*, McGraw-Hill, 4th edition, 2000.

[124] A. Wood and J. Stankovic, "Denial of service in sensor networks," *IEEE Computer*, vol. 35, no. 10, pp. 54–62, October 2002.

[125] G. Noubir and G. Lin, "Low-power DoS attacks in data wireless lans and countermeasures," *SIGMOBILE Mob. Comput. Commun. Rev.*, vol. 7, no. 3, pp. 29–30, 2003.

[126] W. Xu, T. Wood, W. Trappe, and Y. Zhang, "Channel surfing and spatial retreats: defenses against wireless denial of service," in *Proceedings of the 2004 ACM workshop on Wireless security*, 2004, pp. 80 – 89.

[127] J. Polastre, J. Hill, and D. Culler, "Versatile low power media access for wireless sensor networks," in *SenSys '04: Proceedings of the 2nd international conference on Embedded networked sensor systems*. 2004, pp. 95–107, ACM Press.

[128] J. L. Hill and D. E. Culler, "Mica: A wireless platform for deeply embedded networks," in *IEEE Micro*, 2002, pp. 12–24.

[129] S. Kay, *Fundamentals of Statistical Signal Processing: Detection Theory*, Prentice Hall, 1998.

[130] B. Kedem, *Time Series Analysis by Higher Order Crossings*, IEEE Press, 1994.

[131] "Chipcon CC1000 datasheet," http://www.chipcon.com/files/.

[132] C. Intanagonwiwat, R. Govindan, and D. Estrin, "Directed Diffusion: A Scalable and Robust Communication Paradigm for Sensor Networks," in *Proceedings of the Sixth Annual ACM/IEEE International Conference on Mobile Computing and Networks (MobiCOM)*, August 2000.

[133] A. Perrig, R. Szewczyk, D. Tygar, V. Wen, and D. Culler, "SPINS: security protocols for sensor networks," *Wireless Networks*, vol. 8, no. 5, pp. 521–534, 2002.

[134] P. Juang, H. Oki, Y. Wang, M. Martonosi, L. Peh, and D. Rubenstein, "Energy-Efficient Comiputing for Wildlife Tracking: Design and Tradeoffs and Early Experiences with Zebranet," in *Proceedings of the Tenth International Conference on Architectural Support for Programming Languages and Operating Systems*, 2002, pp. 96–107.

[135] S. Madden, M. Franklin, J. Hellerstein, and W. Hong, "TAG: a Tiny Aggregation Service for Ad-Hoc Sensor Networks," in *Proceedings of the Usenix Symposium on Operating Systems Design and Implementation*, 2002.

[136] C. Schleher, *Electronic Warfare in the Information Age*, MArtech House, 1999.

[137] W. Xu, W. Trappe, Y. Zhang, and T. Wood, "The feasibility of launching and detecting jamming attacks in wireless networks," in *MobiHoc '05: Proceedings of the 6th ACM international symposium on Mobile ad hoc networking and computing*, 2005, pp. 46–57.

[138] Y. Law, P. Hartel, J. den Hartog, and P. Havinga, "Link-layer jamming attacks on s-mac," in *Proceedings of the 2nd European Workshop on Wireless Sensor Networks (EWSN 2005)*, 2005, pp. 217 – 225.

[139] B. Potter, "Wireless security's future," *IEEE Security and Privacy Magazine*, vol. 1, no. 4, pp. 68–72, 2003.

[140] W. Trappe and L. Washington, *Introduction to Cryptography with Coding Theory*, Prentice Hall, 2002.

[141] J. Zhao and R. Govindan, "Understanding packet delivery performance in dense wireless sensor networks," in *SenSys '03: Proceedings of the 1st international conference on Embedded networked sensor systems*, 2003, pp. 1–13.

[142] A. Woo, T. Tong, and D. Culler, "Taming the underlying challenges of reliable multihop routing in sensor networks," in *SenSys '03: Proceedings of the 1st international conference on Embedded networked sensor systems*, 2003, pp. 14–27.

[143] M. Bellare, R. Canetti, and H. Krawczyk, "Keying hash functions for message authentication," *Advances in Cryptology - Crypto '96*, vol. 1109, pp. 1–15, 1996, Lecture Notes in Computer Science.

[144] M. Bohge and W. Trappe, "An authentication framework for hierarchical ad hoc sensor networks," in *Proc. of the 2003 ACM Workshop on Wireless Security*, 2003, pp. 79–87.

[145] A. Perrig, R. Canetti, D. Song, and J.D. Tygar, "Efficient and secure source authentication for multicast," in *Proceedings of Network and Distributed System Security Symposium*, February 2001.

[146] S. Zhu, S. Xu, S. Setia, and S. Jajodia, "LHAP: A lightweigth hop-by-hop authentication protocol for ad-hoc networks," in *International Workshop on Mobile and Wireless Network (MWN 2003)*, 2003.

[147] A. Weimerskirch and G. Thonet, "A distributed light-weight authentication model for ad-hoc networks," in *The 4th International Conference on Information Securtiy and Cryptology (ICISC 2001)*, December 2001.

[148] W. Ye, J. Heidemann, and D. Estrin, "An energy-efficient mac protocol for wireless sensor networks," in *Proceedings of the IEEE INFOCOM*, 2002, vol. 3, pp. 1567– 1576.

[149] S. Ganeriwal, R. Kumar, and M. Srivastava, "Timing-sync protocol for sensor networks," in *SenSys '03: Proceedings of the 1st international conference on Embedded networked sensor systems*, 2003, pp. 138–149.

[150] "Tinyos homepage," http://webs.cs.berkeley.edu/tos/.

[151] L. Eschenaur and V. Gligor, "A key-management scheme for distributed sensor networks," in *Proceedings of the 9th ACM conference on Computer and communications security*, 2002, pp. 41–47.

[152] D. Chaum, "Untraceable electronic mail, return addresses, and digital pseudonyms," *Communications of the ACM*, vol. 24, pp. 84–88, 1981.

[153] M.Reed, P. Syverson, and D. Goldschlag, "Anonymous connections and onion routing," *IEEE Journal on Selected Areas in Communications*, vol. 16, pp. 482–494, May 1998.

[154] "Mixmaster remailer," http://mixmaster.sourceforge.net/.

[155] "WWWF - the conservation organization," http://www.panda.org/.

[156] A. Cerpa and D. Estrin, "ASCENT: Adaptive Self-Configuring Sensor Networks Topologies," in *Proceedings of IEEE INFOCOM'02*, June 2002.

[157] W. Ye, J. Heidemann, and D. Estrin, "An Energy-Efficient MAC Protocol for Wireless Sensor Networks," in *Proceedings of IEEE INFOCOM'02*, June 2002.

[158] C. L. Barrett, S. J. Eidenbenz, L. Kroc, M. Marathe, and J. P. Smit, "Parametric probabilistic sensor network routing," in *Proceedings of the 2nd ACM international conference on Wireless sensor networks and applications*, 2003.

[159] Z. Cheng and W. Heinzelman, "Flooding Strategy for Target Discovery in Wireless Networks," in *proceedings of the Sixth ACM International Workshop on Modeling, Analysis and Simulation of Wireless and Mobile Systems (MSWiM 2003)*, 2003.

[160] H. Lim and C. Kim, "Flooding in Wireless Ad-hoc Networks," in *IEEE computer communications*, 2000.

[161] D. Braginsky and D. Estrin, "Rumor routing algorthim for sensor networks ," in *Proceedings of the 1st ACM international workshop on Wireless sensor networks and applications*, 2002.

[162] P. Th. Eugster, R. Guerraoui, S. B. Handurukande, P. Kouznetsov, and A.-M. Kermarrec, " Lightweight probabilistic broadcast," *ACM Transactions on Computer Systems (TOCS)*, vol. 21, no. 4, pp. 341 – 374, November 2003.

[163] D. Niculescu and B. Nath, "Trajectory Based Forwarding and its Applications," in *Proceedings of the Ninth Annual ACM/IEEE International Conference on Mobile Computing and Networks (MobiCOM)*, September 2003, pp. 260–272.

[164] Vincent Tseng and Kawuu Lin, "Mining temporal moving patterns in object tracking sensor networks," in *International Workshop on Ubiquitous Data Management, 2005.*, 2005.

[165] Pandurang Kamat, Yanyong Zhang, Wade Trappe, and Celal Ozturk, "Enhancing source-location privacy in sensor network routing," in *ICDCS '05: Proceedings of the 25th IEEE International Conference on Distributed Computing Systems (ICDCS'05)*, 2005.

[166] Celal Ozturk, Yanyong Zhang, and Wade Trappe, "Source-location privacy in energy-constrained sensor network routing," in *SASN '04: Proceedings of the 2nd ACM workshop on Security of ad hoc and sensor networks*, 2004.

[167] J. Deng, R. Han, and S. Mishra, "Intrusion tolerance and anti-traffic analysis strategies for wireless sensor networks," in *IEEE International Conference on Dependable Systems and Networks (DSN)*, 2004.

[168] J Deng, R Han, and S. Mishra, "Countermeasures against traffic analysis attacks in wireless sensor networks," in *First IEEE/CreateNet Conference on Security and Privacy for Emerging Areas in Communication Networks (SecureComm)*, 2005.

[169] R. Szewczyk, A. Mainwaring, J. Polastre, J. Anderson, and D. Culler, "An analysis of a large scale habitat monitoring application," in *SenSys '04: Proceedings of the 2nd international conference on Embedded networked sensor systems.* 2004, pp. 214–226, ACM Press.

[170] W. Trappe and L.C. Washington, *Introduction to Cryptography with Coding Theory*, Prentice Hall, 2002.

[171] T. Cover and J. Thomas, *Elements of Information Theory*, John Wiley and Sons, 1991.

[172] D. Guo, S. Shamai, and S. Verdu, "Mutual information and minimum mean-square error in gaussian channels," *IEEE Trans. on Information Theory*, vol. 51, no. 4, pp. 1261–1282, 2005.

[173] V. Anantharam and S. Verdu, "Bits through queues," *IEEE Trans. on Information Theory*, vol. 42, pp. 4–18, 1996.

[174] D. Bertsekas and R. Gallager, *Data Networks*, Prentice Hall, 1992.

[175] W.C. Jakes Jr., *Microwave Mobile Communications*, Wiley, 1974.

[176] T.S. Rappaport, *Wireless Communications- Principles and Practice*, Prentice Hall, 2001.

[177] A. F. Molisch, Ed., *Wireless Communications*, John Wiley and Sons, 2005.

[178] A. Domazetovic, L. J. Greenstein, I. Seskar, and N.B. Mandayam, "Propagation models for short range wireless channels with predictable path geometries," *IEEE Trans. on COM*, vol. 53, no. 7, pp. 1123–1126, July 2005.

[179] A. Domazetovic, L. J. Greenstein, I. Seskar, and N.B. Mandayam, "Estimating the doppler spectrum of a short range fixed wireless channel," *IEEE COM Letters*, vol. 7, no. 5, pp. 227–229, May 2003.

[180] V. Erceg et. al., "Channel Models for Fizxed Wireless Applications," IEEE 802.16 Broadband Wireless Access Working Group, July 27, 2003.

[181] J. Tugnait, L. Tong, and Z. Ding, "Single-user channel estimation and equalization," *IEEE Signal Processing Magazine*, pp. 17–28, 2000.

[182] T. S. Rappaport, "Characterization of UHF multipath radio channels in factory buildings," *IEEE Trans. on Antennas and Propagation*, vol. 37, pp. 1058–1069, 1989.

[183] D. C. Cox, "Delay doppler characteristics of multipath delay spread and average excess delay for 910 MHz urban mobile radio paths," *IEEE Trans. Antennas and Propagation.*, vol. 20, pp. 625–635, 1972.

[184] R. J. C. Bultitude and G.K Bedal, "Propagation characteristics of microcellular mobile radio channels at 910 Mhz," *IEEE J. Sel. Areas Commun.*, vol. 7, pp. 31–39, 1989.

[185] G. Zhou, T. He, S. Krishnamurthy, and J. Stankovic, "Impact of radio irregularity on wireless sensor networks," in *MobiSYS '04: Proceedings of the 2nd international conference on Mobile systems, applications, and services*, 2004, pp. 125–138.

[186] A. Menezes, P. vanOorschot, and S. Vanstone, *Handbook of Applied Cryptography*, CRC Press, 1997.

[187] "Lecture notes on cryptography," MIT Summer Course, available at http://www.cs.ucsd.edu/users/mihir/papers/gb.html, 2001.

[188] S. Goldwasser and S. Micali, "Probabilistic encryption," *Journal of Computer and System Sciences*, vol. 28, pp. 270–299, 1984.

[189] A. D. Wyner, "The wire-tap channel," *Bell Syst. Tech. Journal*, vol. 54, pp. 1355–1387, 1975.

[190] I. Csiszar and J. Korner, "Broadcast channels with confidential messages," *IEEE Trans. Inform. Theory*, vol. 24, pp. 339–348, 1978.

Author Biography

Yingying Chen received her Ph.D. degree in Computer Science from Rutgers University. She is currently an assistant professor in the Department of Electrical and Computer Engineering at Stevens Institute of Technology. Her research interests include wireless and system security, wireless networking, and distributed systems. Prior to joining Stevens Institute of Technology, Dr. Chen was with Bell Laboratories and the Optical Networking Group, Lucent Technologies. She also worked as a system researcher for embedded networks at Alcatel Network Systems. Dr. Chen received the IEEE Outstanding Contribution Award from IEEE New Jersey Coast Section each year from 2000 to 2005. She is the recipient of the Best Technological Innovation Award from the 3rd International TinyOS Technology Exchange in 2006. She is a member of the IEEE and ACM. She has been an executive officer for IEEE Computer and Communication Joint Chapter at New Jersey Coast Section since 2000, and has served on the technical programs for several IEEE/ACM conferences on wireless networking and security.

Wenyuan Xu received her B.S. degree with the highest honor from Zhejiang University, Hangzhou, China, in 1998, M.S. degree in computer science and engineering from Zhejiang, China in 2001, and the Ph.D. degree in electrical and computer engineering from Rutgers University in 2007. She is currently an assistant professor in the Department of Computer Science and Engineering, University of South Carolina. Her research interests include wireless networking and network security. Dr. Xu received the outstanding undergraduate scholarship at Zhejiang University from 1994 to 1998. She is a member of the ACM and IEEE, and N^2 Women Community, and has served on the technical programs for IEEE/ACM conferences on wireless networking and security.

Wade Trappe received his B.A. degree in Mathematics from The University of Texas at Austin in 1994, and the Ph.D. in Applied Mathematics and Scientific Computing from the University of Maryland in 2002. He is currently an assistant professor at the Wireless Information Network Laboratory (WINLAB) and the Electrical and Computer Engineering Department at Rutgers University. His research interests include wireless security, wireless networking, multimedia security, and network security. While at the University of Maryland, Dr. Trappe received the George Harhalakis Outstanding Systems Engineering Graduate Student award. Dr. Trappe is a co-author of the textbook Introduction to Cryptography with Coding Theory, Prentice Hall, 2001, which is currently in its 2nd edition and has been translated into Mandarin Chinese. He is also co-author of the book Multimedia Fingerprinting Forensics for Traitor Tracing. He is the recipient of the 2005 Best Paper Award from the IEEE Signal Processing Society, as well as the 2005 Best Paper award from the European Association for Signal, Speech and Image Processing (EURASIP). He is a member of the

IEEE Signal Processing and Communications societies, a member of the ACM, and has served on the technical programs for several IEEE/ACM conferences on wireless networking and security.

Yanyong Zhang received her Ph.D. degree in computer science and engineering from Pennsylvania State University in 2002. She is an assistant professor in the Department of Electrical and Computer Engineering, Rutgers University, and she is also a faculty member at Wireless Information Network Laboratory (WINLAB). Her current research interests include sensor networks, sensor network security and privacy, and fault-tolerant sensor networks. She has received several US National Science Foundation (NSF) grants on these topics, including an NSF Faculty Early Career Development (CAREER) Award. She is a member of the ACM, the IEEE, and the IEEE Computer Society.

Index